Textbook on

Energy Resources and Management

Textbook on
Energy Resources and Management

Renu Dhupper MSc, PhD

Assistant Professor
Department of Environmental Science
Amity University
Noida, UP

CBS Publishers & Distributors Pvt Ltd

New Delhi • Bengaluru • Chennai • Kochi • Mumbai • Pune
Hyderabad • Kolkata • Nagpur • Patna • Vijayawada

Textbook on Energy Resources and Management

ISBN: 978-81-239-2575-2

Copyright © Author and Publisher

First Edition: 2015

Published by Satish Kumar Jain and produced by Varun Jain for
CBS Publishers & Distributors Pvt Ltd
4819/XI Prahlad Street, 24 Ansari Road, Daryaganj, New Delhi 110 002, India.
Ph: 23289259, 23266861, 23266867   Fax: 011-23243014   Website: www.cbspd.com
e-mail: delhi@cbspd.com; cbspubs@airtelmail.in.
Corporate Office: 204 FIE, Industrial Area, Patparganj, Delhi 110 092
Ph: 4934 4934   Fax: 4934 4935   e-mail: publishing@cbspd.com; publicity@cbspd.com

Branches

• **Bengaluru:** Seema House 2975, 17th Cross, K.R. Road,
  Banasankari 2nd Stage, Bengaluru 560 070, Karnataka
  Ph: +91-80-26771678/79   Fax: +91-80-26771680   e-mail: bangalore@cbspd.com
• **Chennai:** No. 7, Subbaraya Street, Shenoy Nagar, Chennai 600 030, Tamil Nadu
  Ph: +91-44-26260666, 26208620   Fax: +91-44-42032115   e-mail: chennai@cbspd.com
• **Kochi:** 36/14 Kalluvilakam, Lissie Hospital Road, Kochi 682 018, Kerala
  Ph: +91-484-4059061-65   Fax: +91-484-4059065   e-mail: kochi@cbspd.com
• **Mumbai:** 83-C, Dr E Moses Road, Worli, Mumbai-400018, Maharashtra
  Ph: +91-22-24902340/41   Fax: +91-22-24902342   e-mail: mumbai@cbspd.com
• **Pune:** Bhuruk Prestige, Sr. No. 52/12/2+1+3/2 Narhe, Haveli
  (Near Katraj-Dehu Road Bypass), Pune 411 041, Maharashtra
  Ph: +91-20-64704058/59, 32392277   Fax: +91-20-24300160   e-mail: pune@cbspd.com

Representatives

• **Hyderabad** 0-9885175004   • **Kolkata** 0-9831437309, 0-9051152362
• **Nagpur** 0-9021734563   • **Patna** 0-9334159340   • **Vijayawada** 0-9000660880

Printed at: Repro India Ltd., Navi Mumbai

Preface

Energy is the primary and universal measure of all kinds of work by humans and nature. Everything that happens in the world is the expression of flow of energy in one of its forms. Most people use the word 'energy' for input to their bodies or to the machines and thus think about fuels and electric power. Energy consumption of a nation is usually considered as an index of its development as almost all developmental activities, directly or indirectly, are dependent on energy.

It is a well-known fact that the known sources of fossil fuels in the world are depleting very fast and by the turn of the century, mankind will have to increasingly depend upon renewable resources of energy. For developing countries like India, large scale demand of heat energy for meeting day-to-day domestic, institutional and industrial requirements can be met by utilizing solar, thermal systems, biogas, photovoltaic (PV) cells, wind energy, geothermal, magnetohydrodynamic (MHD) generation, etc., that is, from non-conventional sources of energy. Non-conventional sources of energy are the areas of emerging technologies which have higher priority with reference to national needs.

We are aware of several activities taking place in nature. These activities have some form of motion of particles or objects. Energy is the cause behind the motion of particles or objects. Energy is the capability to produce motion, force, work, change in shape and form, etc. Energy technology is an applied science dealing with various alternative energy routes comprising exploration and extraction of primary raw energy, conversion to intermediate or secondary forms of energy and by-products, transportation alternatives, storage, distribution and supply of secondary forms of energy. The studies enable judicious and economic choice of energy routes and mastery over energy management and conversion of energy.

Keeping the above consideration in view, a number of universities, colleges and polytechnics are now offering this course to students. It appeared to me that there is a need for a concise and practical book on conventional and non-conventional energy sources. The treatment of this book begins with fundamental and develops in a way that allows the practical man as well as the student to expand their knowledge progressively. Throughout the text, attempt has been made to present the subject matter in a simple, lucid and precise manner. Chapters in the book deal with the different non-conventional and conventional energy systems, giving latest data on some of the relevant parameters. In addition to serving as a textbook, the contents will also be useful for anyone planning to take examination on environmental sciences.

The salient features of this book are: Simple language, the concept of energy-chains and links, precise terms and definition, illustrations, flow diagrams, figures, schematics, layouts, graphs, description of modern plants and reference data, tables and practical useful information, SI units, theory, equations and practical aspects.

Suggestions for further improvement of the book will be gratefully acknowledged.

Renu Dhupper

Contents

Section 1

Introduction

Introduction

1.1 INTRODUCTION

Energy is the primary and universal measure of all kinds of work by human being and nature. Everything that happens in the world is the expression of flow of energy in one of its forms. Most people use the word *energy* for input to their bodies or to machines and thus think about crude fuels and electric power. Electricity is the only form of energy which is easy to produce, transport, use, and control. So, it is mostly the terminal form of energy for transmission and distribution. Electricity consumption per capita is the index of the living standard of people of a place or country.

Energy consumption of a nation is usually considered as an index of its development. This is because almost all the developmental activities are directly or in directly dependent upon energy. We find wide disparities in per capita energy use between the developed and the developing nations.

The energy sources available, can be divided into 3 types:

1. **Primary energy source:** Primary energy sources can be defined as sources which provide a net supply of energy. Coal, oil, uranium are examples of this type. The energy required to obtain these fuels is much less than what they can produce by combustion or nuclear reaction. The primary fuels only can accelerate growth but their supply is limited. It becomes very essential to use these fuels sparingly.

2. **Secondary fuel:** Produce no net energy. Though it may be necessary for the economy. Intensive agriculture is an example, where *in terms of energy* the yield is less than the input.

3. **Supplementary source:** These sources are defined as those whose net energy yield is zero and requiring highest investment in terms of energy of insulation, thermal energy is an example of this source.

Coal, natural gas, oil and nuclear energy using breeder reactor are net energy yielders and are primary sources of energy. Secondary sources are solar energy, wind energy, water energy etc. Geothermal and ocean thermal are the other sources which may well prove worthwhile. It may be necessary in future to develop these secondary sources like solar, wind etc. The standard of living of the people of any country is considered to be proportional to the energy consumption by the people of that country. In one sense, the disparity one feels from country to country arises from the extent of accessible energy for the citizens of each country. Unfortunately the world energy demands are mainly met by the fossil fuels today. The geographical non-equidistribution of this source and the ability to acquire and control the production and supply of this energy source have given rise to man of issues and also the disparity in the standard of living.

The world energy consumption pattern is also increasing as shown in Fig. 1.1. The energy consumption has been increasing and

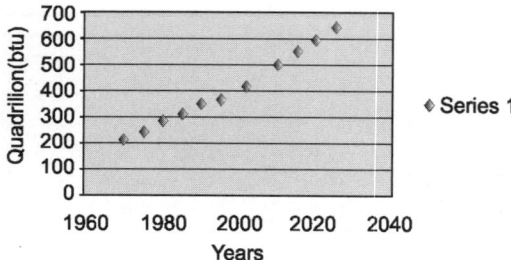

Fig. 1.1: Energy consumption (1970–2025)

it will triple in a period of 50 years by 2025 as shown in Fig. 1.1. The fossil fuel used as energy source has many limitations. There are a number of pollutants that have been identified as coming out of the use of fossil fuels and are casue of serious health hazards.

1.2 ENERGY CONSUMPTION AS A MEASURE OF ECONOMIC PROSPERITY

Energy is an important input in all sectors of any country's economy. Energy demand in developing countries will rise enormously as per capita income and population grow. The standard of living of a given country can be directly related to per capita energy consumption. Energy crisis is due to two reasons; firstly that the population of the world has increased rapidly and secondly the standard of living of human beings has increased. If we take the annual per capita income of various countries and plot them against per head energy consumption, it will appear that the per capita energy consumption is a measure of the per capita income or the per capita energy consumption is an measure of the prosperity of a nation. With reference to the situations of people without access to modern energy forms, the chapter shows why energy is an economic 'good' and thus why energy supplies will need to be expanded to meet emerging demands, if living standards are to be improved and developing countries are to achieve prosperity. Energy demand in industrialized countries is also likely to remain strong, notwithstanding and to some extent because of enormous gain in the efficiency

with which energy is produced and used. Both energy and financial resources are amply available to meet market needs.

Energy is an important measure of prosperity of a nation. Energy has been the life blood for continual progress of human civilization. Since the beginning, industrial consumption has increased by leaps and bounds to accelerate the human living standard, particularly in the industrialized nations of the world. Infact, per-capita energy consumption has been a barometer of a nation's economy, and prosperity. The higher the per capita consumption of energy of a country the more prosperous is the nation. Therefore, one might conclude that to be materially prosperous, one needs to consume more energy than the per capita requirement.

Although, modern energy forms contribute appreciably to economic welfare, they are not affordable until incomes rise above a certain threshold. Technical progress and falling costs are lowering this threshold, but ultimately income growth is what matters. Countries that have been able to raise productivity and incomes on a broad basis—through good macroeconomic management, trade, and investment in human and physical resources—have been able to extend service most rapidly.

Developed nations have much higher per capita energy consumption than the developing world. The available resources of energy are limited. There is a need to locate and harness new sources of energy or to use the available ones judiciously so as to make them last longer. In order to have a energy secure future for our coming generations, we need to take a look into our usage of energy. Such is the thrust on promoting sustainable energy that the United Nation has constituted special task force to find means of sustainable energy.

The sustainable energy includes finding new sources of energy as well as conserving the available ones. Renewable sources of energy, viz. wind, solar, tidal, biomass etc. are being given preference as energy producers. Another way out is conservation. Energy *conservation* is a tool to save the scarce available resources of energy. Broadly stated, *energy conservation* means the practice of decreasing the quantity of energy used. It may be achieved through efficient energy use, in which energy use is decreased while achieving a similar outcome, or by reduced consumption of energy.

The energy consumption of a nation can be broadly divided into the following areas or sectors depending on energy related activities. These can further be subdivided into sub-sectors as given:

1. Domestic sector
2. Transportation sector
3. Agriculture sector
4. Industrial sector

Consumption of a large amount of energy in a country indicates increased activities in these sectors. This may apply better comforts at home due to use of various appliances, better transport facilities and more agricultural and industrial production. All this amounts to a better quality of life. Therefore, the per capita energy consumption of a country is an index of the standard of living or prosperity (i.e. income) of the people of that country.

1.3 WORLD ENERGY RESERVES

If the trend continues, the world in the coming year will be more crowded than that of today, and increasing day by day at an alarming frequency. The conventional sources of energy are depleting and may be exhausted by the end of the century or beginning of the next century. Nuclear energy requires skilled technicians and poses the safety with regard to radioactive waste disposal. Solar energy and other non-conventional energy sources are the sources, those are to be utilized in future.

World needs more and more energy. Increase in population also increases demand for energy and is always looking for new solutions that would ensure adequate global energy supply. There are also times when global energy demand is experiencing a decline (global financial crisis, global recession) but these are only temporary happenings, however, the hunger for more energy is even more than it was before these temporary situations.

World satisfies its energy needs mainly from non-renewable energy sources—fossil fuels, mostly coal, oil and natural gas. Not only that, these energy sources are non-renewable which means that they cannot last for eternity, they are also not ecologically acceptable because they are not only responsible for different forms of pollution but also for one of the biggest challenges in human history; climate change and global warming. Majority agrees that global warming phenomenon is mostly result of man-made activities due to excessive fossil fuels burning, and the only thing that can slow down the impact of global warming is drastic cut in CO_2 (carbon dioxide), and other greenhouse gas emissions. In order to do so we need to stop being so dependent on fossil fuels, because as long as coal, oil, and natural gas hold "top energy sources" spot, world won't make that step forward that we so desperately need. This is where renewable energy sources should step in and make the difference. *Can we really expect this in near future*?

Renewable energy sources have a great potential in years to come, but mimetically only they have very limited possibilities and have much more expensive gained energy. Because of this fact, years will pass before any significant use of these energy sources. Till that day comes we have to rely only on non-renewable energy sources. These are:

1. Nuclear energy
2. Coal
3. Oil
4. Natural gas

Coal, oil and natural gas are together called "fossil fuels". The name "fossil fuel" reflects itself about its origin. Main energy source of fossil fuels is carbon, and its combustion emits a lot of CO_2 into atmosphere. Thus is the biggest ecological problem of using fossil fuels. Nuclear power plants aren't releasing CO_2, but after its use, nuclear fuel is still extremely radioactive and is necessary to keep it in store for more decades (the most radioactive element, even more than 100 years) in safe concrete pools of stainless steel. In the normal condition, nuclear energy presents a very clean energy source but potential hazard of possible nuclear disaster has caused smaller number of newly installed power plants in the last couple of years. Conclusions of the study on alternate energy strategies are:

1. Demand for energy will continue to grow, so, governments should adopt vigorous policies to conserve energy. This growth must increasingly satisfied by energy resources other than oil, which will be progressively reserved for further uses.

2. The continued growth of energy demand requires energy resources to be developed with the utmost vigour. The alternative requires 5 to 15 years to develop and so, the need for replacement fuels will increase rapidly as the last decade of the century is approached.

3. Electricity from nuclear power is capable of making an important contribution to the global energy supply, although, worldwide acceptance of it is still awaited. Natural gas reserves are large enough to meet projected demand provided the incentives are sufficient to encourage the development of extensive and costly intercontinental gas transportation systems.

4. Other than hydroelectric power, renewable resources of energy, e.g. solar, wind, wave are unlikely to contribute significant quantities of additional energy during the century at the global level.

There are a variety of energy resources and energy forms. These include hydro power, wind, solar, biomass and geothermal resources and in the energy forms—light, heat, electricity, hydrogen and fuel. How this transition has to occur depends on many factors but surely the transition has to take place sooner or later. What kind of mix will be required also depends on the location and the availability of the resources. Photovoltaic devises have been advocated as a powerful energy source, but the technology still needs high investment and also the reliability and sustainability questions have to be addressed. It was concluded that world oil production, if likely to level off very shortly and that alternative fuels will have to be meet growing energy demand. The task for the world will be to manage a transition from dependence on oil and natural gases to a greater reliance on other sources like fossil fuels, nuclear energy and later, renewable energy system

1.4 ENERGY, MAN AND ENVIRONMENT

Man extracts energy from the nature in the form of raw energy (primary energy resources). The primary energy resources are processed and transformed to intermediate and finally useable energy forms. Energy conversion processes are accompanied with pollution problems. A major portion of energy is transformed to electrical form by power plants. Coal fired power plants emit solid particles, SOx, NOx, CO, CO_2 and heat/chemicals etc. into the environment. Pollution of environment disturbs the ecological balance.

The world's annual energy consumption rate is increasing at a rate of 2 to 4 percent. Nuclear power plants, thermal power plants, chemical conversion plants etc. are emitting solid, liquid and gaseous pollutants to the environment. Gaseous pollution is causing greenhouse effect and global warming.

Environmental restrictions are being made more stringent throughout the world to overcome this issue.

1.5 ENERGY MANAGEMENT

Energy management is defined as "the art and science of optimum use of energy to maximize profits (minimize costs) and thereby improving the economic competitiveness." The energy should be used efficiently, economically and optimally.

The following steps are involved in energy management of an enterprise:

1. Commitment to 'energy management' as a corporate/government/business policy.
2. To have a well compiled energy policy.
3. For selection of energy managers and to assign responsibilities alternatively, and also to appoint energy consultants.
4. Making efforts for conservation of energy.
5. Carry out energy audits.
6. Improve communication on energy matters.

1.6 ENERGY CONSERVATION OPPORTUNITIES

The energy conservation opportunities can be broadly divided into the following categories:

- Opportunities of reducing/eliminating wastage. The investment required may be very modest
- To eliminate leakages in compressed air, oils, lubrication, water, and thermal systems etc.
- Use recycled material
- Stop wastage by switching off electric circuit, water tap, oil tap etc., when not in use
- Use of thermal insulation
- Use of alternate energy resources
- Use of biodiesel

Section 2

Non-renewable Energy Resources

Coal as an Energy Source

2.1 INTRODUCTION

Renewable energy sources have a great potential in years to come, momentarily they have very limited possibilities and have much more expensive gained energy. Because of this fact years will pass before any significant use of these energy sources. Till that day comes we have to rely on non-renewable energy sources. These are:

1. Nuclear energy
2. Coal
3. Oil
4. Natural gas

Coal, oil and natural gas are together called *"fossil fuels."* The name "fossil fuel" reflets itself about its origin. Many million years ago, animal and plant remains began to sediment on the ocean's bottom or on the ground. As the years went by, remains were covered with layers of mud, sludge and sand. In these conditions of extreme temperature and pressure, ideal conditions were developed to turned remains of animals and plants into fossil fuels. The primary fuels which are burned to release heat and generate steam in boilers are the fossil fuels in the form of coal, fuel oil and natural gas, which represent the remains of plant and animal life that are preserved in the sedimentary rocks.

2.2 HISTORICAL BACKGROUND

Is it not surprising to convert a rock into liquid! Coal is quite often termed as a heterogeneous rock of organic origin containing significant inorganic mineral matter. The thought of transforming coal into liquid began in 1869 when **Bertholet** demonstrated that coal could be hydrogenated. However, it is only with **Bergius** in 1913, high pressure hydrogenation and hydrodesulphurization of coal began on practical scale. Bergius employed the high-pressure technology of the Haber's ammonia process to hydrogenate coal at elevated temperature and pressure. The major commercial developers of hydrogenation were IG Farben industrie AG (IG Farben) and BASF. Initial goals of research at IG Farben were to design an effective slurry-phase hydrogenation catalyst and a sulphur-resistant vapour-phase catalyst that would survive the inherent sulphur in the coals. The development of a supported catalyst composed on pelleted Mo and Zn oxides was a breakthrough by Pier in 1926. In 1934, IG Farben switched from molybdenum to iron oxide as a slurry-phase catalyst. In slurry-phase mode disposable catalysts, such as inexpensive ores, coal minerals and metallic wastes will be employed. Catalyst self-sufficiency and no concern for catalyst deactivation or recovery are the advantages in this mode. Slurry-phase catalysts are coal minerals and iron sulfides. Even though the use of 'a disposable' or a slurry-phase catalyst in direct coal liquefaction was initiated in the Bergius process, currently SRC-II technology is being employed, where the presence of recycled coal mineral matter considerably enhanced coal liquefaction.

2.3 TYPES OF NON-RENEWABLE ENERGY SOURCES

Worldwide, there is a range of energy resources available to us. These energy resources fall into two main categories, called renewable and non-renewable energy sources. Each of these sources can be used to generate electricity, which is a very useful way of transferring energy from one place to another such as to the home or to industry.

Non-renewable sources of energy can be divided into two types, i.e. *Fossil fuels* and *Nuclear fuel*. Fossil fuels like coal, oil and gas generate a considerable amount of energy when they are bound (the process of combustion). Non-renewable resources have high carbon content because their origin lies in the photosynthetic activity of plants millions of years ago. The fuels release this carbon back into the atmosphere as CO_2. The rate at which such fuels are being burnt is thus resulting in eleviating the concentration of CO_2 in the atmosphere, a cause of the *greenhouse effect*.

2.3.1 Coal

Clearly, energy security and energy independence are the two challenges ahead of any nation in this new millennium. The global appetite for energy is simply too large and recurring as well. There is an abrupt need to look something beyond incremental changes because the additional energy needed is greater than the total of all the energy currently produced. Coal is the principal energy source, especially in India in view of its extensive reserves and accessibility. Coal was made out of the old plants, 300 million years prior, even before dinosaurs, hence gigantic plants began sedimentation in swamps. It started from dead organic matter which developed a huge number of years back. Trees and plants falling into water rotted and later generated peat lowlands. Immense topographical changes buried these swamps under layers of silt. Underground hotness, soil pressure and movement of Earth's crust distilled off a portion of the marsh dampness and solidified it to structures like tan coal or lignite. Today coal is found basically beneath the layers of rocks and mud and to get to it, mines are excavating. Coal is imperative especially as a result of the two explanations; steel processing and electrical energy, coal is giving 23% of the planet's aggregate essential energy, 38% of the planet's created electrical energy is gained up from coal. For 70% of the planet's steel production, coal is required as the key element. As per geographical request of structuring, coal may be of the following types:

1. Peat
2. Lignite
3. Sub-bituminous
4. Bituminous
5. Subanthracite
6. Anthracite, with increasing % of carbon.

After anthracite, graphite is formed. Anthracite burns slowly without smoke. It has a high energy value. It contains more than 86% fixed carbon and less volatile matter. Volatile matter helps in the ignition of coal.

Bituminous has 50%–80% carbon and 20–40% of volatile matter with a good energy value. Bituminous coal can be subdivided into (a) cooking coal, which is used in blast furnaces (b) gas coal, which is used for making gas (c) steam coal, which is nearly smokeless and is used as household coal, and in steam powered vessels. Lignite is brown in colour. It has 60% carbon and gives off a low heat. It is used in the production of thermal electricity. Peat is not regarded as a rank of coal, as it gives less heat, and leaves a lot of ash after burning. *Rank carrier* defines the meaning of degree of maturation (carbonization) and is a measure of carbon content in coal. Lignite is considered as a low rank coal and anthracite as high rank.

2.3.2 Molecular Structure of Coal

Coal is thought to constitute a substantial polymeric framework of aromatic structures, usually called the coal macromolecule. This macromolecular system comprises dusters of aromatic carbon that are linked at the chance to other aromatic structures by extension. Bridges between the aromatic dusters are framed from a wide mixture of structures. Most bridges are thought to be aliphatic in nature, yet might likewise incorporate different molecules, for example, oxygen and sulphur. Those extensions that hold oxygen as others are thought to have relatively weak bond strength. Different bridges are made up of aliphatic practical groups only. Because of the huge variety of functional groups that make up the extension structures of coal, bridges have an extensive distribution of bond qualities. This conveyance of bond strengths gets paramount throughout the pyrolysis transform as the weakest bonds are broken first. There are different connections to the aromatic dusters that don't form bridges. These connections are referred to as side ties and are thought to comprise principally of aliphatic and carbonyl functional groups gathering. A portable stage exists in coal. This portable stage is trespassed with the coal macromolecule. This versatile stage is thought to be composed of little sub-atomic structures that are not firmly bonded to the macro-molecule. The portable stage is acknowledged to be either (a) trapped in the sub-atomic struc-ture of the coal or (b) feebly, weakly bonded to the coal macromolecule with hydrogen bonds or van der Waal's force of interaction.

Lignite: The macromolecular structure of lignite would consist of small aromatic units (mainly single rings) joined by crosslinks of aliphatic (methylene) chains or aliphatic ethers. If the polymerization were to be random with cross links heading off in all directions.

Bituminous coal: Compared to lignites, bituminous coals have higher carbon content and lower O_2 content. The progression of changes that occur in the structure leads to increase on coal rank. When bituminous coals are examined by X-ray diffraction, it is possible to detect weak graphite like signals emerging from the amorphous background. This information indicates that in bituminous coals the aromatic ring systems are beginning to grow and to become aligned. The structure of bituminous coals with carbon content is in the range of 85% to 91%.

Anthracites: Anthracites have carbon content over 91%. The structure of anthracite is approaching that of graphite. X-ray diffraction data shows increased alignment of the aromatic rings with little contribution from aliphatic carbon.

2.4 COAL ANALYSIS

There are two types of coal analysis: *proximate* and *ultimate*, both done on a mass percent basis. Both these types may be based on: (a) received basis, useful for combustion calculations, (b) dry or moisture free basis, and (c) dry mineral matter free or combustible basis.

2.4.1 Proximate Analysis

A proximate analysis is not an estimated analysis! It is unfortunate that the name proximate sounds much like the expression approximate. The methods for the proximate analysis are thoroughly secured by the American Society for testing and materials (ASTM) as well as norms for the adequate levels of error inside a laboratory and between distinctive research centers. The proximate examination shows the conduct of coal when it is warmed. The point when 1 gm specimen of coal is subjected to a temperature of 105°C for 1 hour, the loss in weight gives the moisture content of the coal, and when 1 gm sample of coal is placed in a covered platinum crucible and heated to 950°C and maintained the temperature for about

7 minutes, there is a loss in weight due to the elimination of moisture. The various parts of a proximate analysis can be undertaken in a lab, and involve a variety of tests and measurements. Coming from underground, coal is wet when it is mined. Groundwater and other liquids add to the moisture level within coal, which is known as *inherent moisture*. Coal analysis attempts to gauge how much inherent moisture is in a particular sample. As logic would dictate, the less moisture in a piece of coal, the better. The volatility of coal is measured by the proportion of volatile matter, which includes various types of hydrocarbons and sulphur in a sample. This measure basically indicates how completely a piece of coal burns when air is not present. It is tested by heating a sample to 1740°F (9500°C). Subtracting what remains after a volatility test from the original mass of a sample also provides measures of what is known as *fixed carbon content*. This generally makes up about half the overall mass of a given sample of coal. The measure of ash in a coal analysis simply determines how much material remains after burning. Since virtually all the carbon, sulphur, and moisture is burned off when ignited, the ash that remains is only a small percentage of the original amount of coal.

2.4.2 Ultimate Analysis

The determination of the foremost components of coal, in particular, carbon, hydrogen, oxygen, nitrogen and sulphur is known as *ultimate analysis*. It is not 'extreme', in the sense of figuring out totally the elemental arrangement of coal. The ultimate analysis gives the chemical components that contain the coal substances, together with ash and moisture. Coal composed of organic compounds of carbon, hydrogen and oxygen, determined from the definitive vegetable matter. The analysis demonstrates the accompanying parts on mass basis. Carbon (C), hydrogen (H), oxygen (O), nitrogen (N), sulphur (S), moisture (*M*) and ash (*A*) along these lines,

$C+H+O+N+S+M+A = 100\%$ by mass. The dry and ash free analysis on combustible basis is obtained on dividing C, H, O, N and S by the fraction

$$\frac{(1 - M + A)}{100}$$

2.4.3 Mineral Matter of Coal

All elements in coal except C, H, N, O and S will be termed *mineral matter* even if they are present as organometallic chelates or absorbed species. Inorganic S and inorganic O_2 (sulphur and oxygen not present in heteroaromatic structure) will also be considered as mineral matter. Iron silica and alumina constitute the major portion of mineral matter in coal. The major groups of mineral matter include aluminum silicates (clay minerals), carbonates, sulfides and silicates (mainly quartz). There are certain properties of coal which are important in power plant applications. They are swelling index, grind ability, weather ability, sulphur content, heating value and ash softening temperature.

Swelling index: A few types of coal throughout and after release of volatile matter get delicate and pale and the structure agglomerates. These are called *solidifying coal*. The caking conduct of coal is measured in USA by the free swelling index (FSI), it indicates the solidifying limit of coal. In a fixed bed, for example, a travelling grate stokes, the coal must not cake as it blazes. The subsequent agglomeration disturbs significantly the accessibility of air burn, yielding low combustion efficiency. Coal that does not cake is called *free blazing coal*. It breaks throughout burning uncovering expansive surface region to the air, accordingly upgrading the ignition process. Caking up of coals are utilized to generate coke by heating the coke in the absence of air, with the volatile matter determined off. Coal without volatile matter is called *coke*, which is generally required in steel plants. A qualitative assessment strategy, called the swelling index, has been implemented to figure out the degree with respect to consuming of a coal. A free smoldering coal has a high value of swelling index, which demonstrates that it stretches in volume throughout ignition. The point when modern pulverized coal burners are utilized, the swelling property of coal is, nonetheless, of less significant.

Grindability: Grindability is a paramount criterion for selecting a coal. This property of coal is measured by the standard grindability index, which is contrarily corresponding to the force needed to crush the coal to a specified molecule size for blazing. The grindability of a coal is a measure of its resistance to pulverizing. Two elements influencing grindability are moisture and ash of a coal. The lignites and anthracites are more resistant to grinding than bituminous coals. Grindability of a standard coal is characterized as 100. Assuming that the coal selected for utilization at a force plant has a grindability list of 50, it might oblige double the grinding force of the standard coal to generate a specified molecule size.

Weatherability: It is measured when stockpiled for long period of time without disintegrating to pieces. Advanced force plants regularly stockpile coals to 60 to 90 days in a huge heap close to the force plant. The coal emptied from wagons is packed in a long trapezoidal heap. Over the top, disintegrating or weathering of the coal because of climatic conditions may bring about little particles of coal which could be apportioned by wind as rain.

Sulphur content: Sulphur content in coal is ignitable and produces some energy by its oxidation to SO_2. SO_2 is a significant source of climatic contamination. There is an ecological regulation on SO_2 discharge. The working expense of SO_2 evacuation needs to be considered while selecting a coal with high sulphur content.

Heating value: The heating value or calorific quality of coal is a property of principal imperativeness. The energy worth of coal, or the fuel substance, is the measure of potential energy in coal that might be changed over into real heating capacity. The quality could be ascertained and contrasted different grades of coal or significantly different materials but will generate contrasting measures of hotness for a given mass. It may be determined on appropriated dry and ash free basis. It is the hotness exchanged, when the results of complete ignition of specimen of coal (or other fuel) are cooled to the starting temperature of air and fuel. It is regularly determined through a standard test in a bomb calorimeter, where a coal sample of known mass is blazed with pure O_2 supply totally in a stainless steel bomb or vessel encompassed by a known mass of water, and the ascent in water temperature is noted. Two distinctive heating valves are cited. The higher heating value (HHV) states that the water vapour in the products consolidates and hence, incorporates the inert high temperature of vaporization of the water vapour shaped by ignition. The lower heating value (LHV) states that the water vapour structured by ignition leaves as vapour itself. Although coal doesn't absorb moisture, it is best to store it undercover. This can be in bins or a bunker covered with a plastic hoop house. Outdoor storage may result in frozen lumps that don't feed well. Combustion of coal is different than fossil fuels and wood. There are very little volatiles in coal, about 30% in bituminous and 5% in anthracite as compared to 50% in wood. Due to this, most of the combustion air needed to burn coal has to be supplied from underneath the fire. A coal fire has to be started with wood, oil or gas to bring it upto the ignition temperature. Once started, it will burn as long as fuel is supplied.

Spontaneous combustion: Combustion (oxidation) of coal can take place rapidly as in a furnace or slowly on a stockpile. If it takes place slowly, there is a degradation or loss of energy content and hence in the value of fuel. The factors which influence spontaneous combustion and which can lead to a big fire are as follows:

a. Rank of coal, low rank coals are more susceptible because of their higher porosity.

b. Amount of surface area exposed to air.

c. Ambient temperature, with high solar insolutions aiding it.

d. O_2 content of coal.

e. Free moisture in coal.

To prevent spontaneous combustion, it is important to maintain a dry pile and compaction at regular intervals.

2.5 PROPERTIES OF COAL

For energy evaluation purposes, the most important properties of coal are:

1. Calorific value: CV (heating value HV). It is the heat evolved from unit quantity of fuel by complete combustion. It is expressed in kilocalorie per kg or joule per kg or BTU/Ib.

$$1 \text{ BTU} = 252 \text{ calories}$$
$$= 0.0002931 \text{ kWh} = 1055 \text{ J}$$

Indian classification of coal is based on comparison with *gross heating value* of pure coal.

Dry Mineral – Free Coal (d.m.f.)

$$\text{CV (d.m.f.)} = \frac{\text{CV (measured)} \times 100}{100 - (\text{moisture \%} - 1.1 \text{ ash \%})}$$

2. Volatile matter (d.m.f.)
$$= 100 - \text{fixed carbon (d.m.f.)}$$

3. Fixed carbon (d.m.f.)

$$= \frac{\text{Fixed carbon determined \%} \times 100}{100 - (\text{moisture \%} \times 1.1 \text{ Ash \%})}$$

4. Moisture (d.m.f.)
$$= \text{Mu}$$
$$= \frac{\text{moisture determined} \times 100}{100 - (\text{moisture determined} - \text{minerals \%})}$$

5. Minerals % determined as per IS:4311

2.6 COAL CONVERSION PRODUCTS AND APPLICATIONS

Coal is a general term for a number of solid blackish organic fossil minerals with different compositions and properties. Coal is essentially rich in amorphous carbon (carbon without regular structures). Coal contains several liquid and gaseous hydrocarbons (Fig. 2.1).

Liquefaction of coal gives several liquid, gaseous and solid fuels and chemicals. Thermal energy, work, electrical energy, chemical energy etc. are obtained from the coal and coal products. The solid, liquid, gaseous products of coal conversion processes shown are listed in Fig. 2.2. Coal is cleared, processed and then used in solid form as a fuel. Alternatively the coal may be gasified and then used as a gaseous fuel. Several processes of coal gasification have been developed for obtaining various synthetic gaseous fuels and chemical byproducts.

2.6.1 Coal Production and Processing

Coal production and processing technologies involve:
- Exploration
- Mining
- Preparation, sorting and cleaning
- Storage
- Transportation
- Supply – as solid fuel for consumption.
  - supply for conversion to coal products.

Coal conversion technologies are:
- Coal gasification processes
- Coal liquefaction processes
- Coal slurry for pipeline transport
- Coal carbonization for coke and coal gas production.

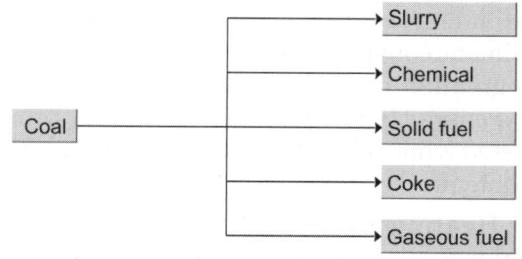

Fig. 2.1: Coal conversion products

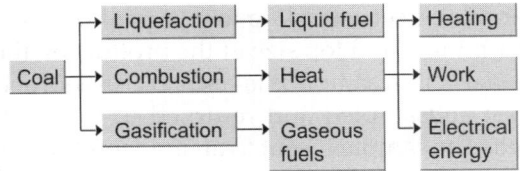

Fig. 2.2: Coal conversion processes

2.6.2 Exploration of Coals

Coal reserves are located in many place at varying depths and thickness of layers interspaced with sandstone and shale. Exploration of coal involves site surveys, data collection, drilling holes and collecting samples. Identifying areas of deposits, collecting more samples from varying depths of identified areas. Diamond core drill and drilling rigs are used for sample collection, several thousand samples are collected and data obtained is analysed.

2.6.3 Geo Logging or Electro Logging

The system consists of logging (recording) the geographical information from different penetrated openings. Transducers are suspended from the versatile truck into drilles opening. The information on geological properties at different depths is sent up by the transducer through control links in manifestation of electrical and computerized signs. The information is gathered, recorded, examined in the data processing instruments in the truck. Three imperative geographies are gamma rays, resistivity and density. Vicinity of coal is affirmed by low natural radiation of gamma beams with high resistivity and low density.

2.7 COAL PREPARATION

Coal preparation is an aggregate term for physical and mechanical methods connected to coal to make it suitable for specific utilization. Coal preparation plant (CPP) is a plant for cleaning and sorting (estimating) of coal before stacking on autos or trucks. Runoff mine coal is the crude coal concentrated from

mines before cleaning and arrangement. A CPP (Fig. 2.3) might additionally be known as a coal handling and arrangement plant (CHAP). The coal conveyed from the mine that shows up for the coal preparation plant is called *run-of-mine, or ROM coal*. This is the crude material for the CPP, and consists of coal, rocks, minerals and contamination. Contamination is normally presented by the mining procedure and may incorporate machine parts, utilized consumables and parts of ground engaging tools. ROM coal can have an expansive variability of moisture and most extreme molecular size. ROM coal is subjected to coal cleaning and preparation process methodologies before dispatch to the customer's premises. The fundamental destination of coal arrangement is to separate the coal and impurities , uproot its pollutants, earth, mud , sulphur and isolate the coal into high grade, mid-range review and low quality coals.

The coal cleaning and preparation process involves the following:

1. **Sorting-out:** Dull and hard lumps are sorted out from dark and softer coal, dirt and other non-coal lumps are removed.

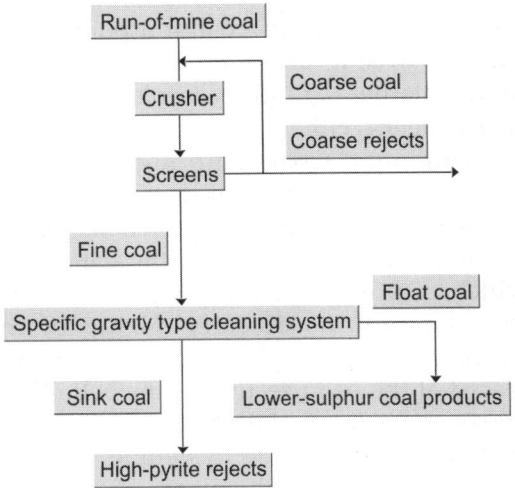

Fig. 2.3: Flow chart of coal preparation and cleaning

2. **Crushing:** Crushing reduces the overall topsize of the ROM coal, so that it can be more easily handled and processed within the CPP. Crushing requirements are an important part of CPP design and there are a number of different types. The equipment used for crushing includes—rotary crushers, double roller crushers, hammers, mills etc.

3. **Screening:** Screens in screening plant are used to group process particles into ranges by size. These size ranges are also called grades. Dewatering screens are used to remove water from the product. Screens can be static, or mechanically vibrated. Screen decks can be made from different materials such as high tensile steel, stainless steel, or polyethelene. The various sizes of coal pieces ranges from 200 mm to 13 mm are separated by means of screens.

4. **Cleaning:** This process is optional and is used when sulphur, sand, and washable impurities like dirt must be removed from the coal and the specified purity to be obtained.

2.8 COAL SAMPLING

Sampling of coal is an important part of the process control in the CPP. A grab sample is a one-off sample of the coal at a point in the process stream, and tends not to be very representative. A routine sample is taken at a set frequency, either over a period of time or per shipment. Coal sampling consists of several types of sampling devices. A "cross cut" sampler to mimic the "stop belt" sample according to ASTM. *ASTM* is the standard in which coal must be sampled. A cross cut sampler mounts directly on top of the conveyor belt. The falling stream sampler is placed at the head section of the belt. There are several points in the wash plant that many coal operations choose to sample. The raw coal, before it enters the plant, refuse, to see what the plant missed and then the clean coal, to see exactly what is being shipped. The

sampler is set according to tons per hour, feet per minute and top size of the product on the actual belt. A sample is taken and crushed, and then subsampled and returned to the main belt. The sample is sent to an independent lab for testing where the results will be shared with the buyer as well as the supplier. The buyer in many cases will also sample the coal again once it is received to "double check" the results. Continuous measurement of ash, moisture, kcal (BTU), S, Fe, Ca, Na, and other elements of the coal are reported by cross belt elemental analyzers. This information can be calibrated periodically to the lab data per ASTM methods. Some impurities which cannot be removed by physical cleaning require chemical cleaning processes.

2.8.1 Removal of Sulphur

The most hazardous impurities in coal is sulphur. When coal is burned, poisonous oxides of sulphur are released into the environment. High sulphur coals are washed and chemically treated for removal of sulphur. About 70% of sulphur impurities are found in coals. Several chemical coal cleaning processes have been developed for removal of sulphur and other impurities.

2.9. TRANSPORTATION OF COAL

Modes of coal transport incorporate continous land transport, river transport, ocean transport, pipeline transport etc. Since the separation between the energy supply point and the energy demand, focus for coal is frequently long, and numerous coal utilization projects require that transportation engineering and expenses to get acknowledged. For long distances the expense of transportation can incredibly exceed the mining expense. As the country uses and fares more stupendous measures of coal from additional remote areas, coal transportation will turn into a more important concern from energy, investment, and ecological standpoint. Coal could be moved directly by railroad, truck, pipeline and energy acquired from coal might be transported as a fluid or vaporous fuel or as power. The natural effects of coal transportation incorporate air contamination, water contamination, noise pollution, security and movement dangers. Immediate natural effects can happen at the mine, where the coal is continuously exchanged, transported or stacked. Indirect ecological effects from coal transportation all in effect from the burning of fuel for the transportation itself. As of late, the coal slurry transportation and gasified coal transportation advances are, no doubt considered as budgetary options, to transportation of solid coal. The main modes of transportation of coal are:

Transportation by road trucks:

- Transportation by rail wagons.
- Transportation by continuous velt/bucket

Sea/river transport:

- Transportation by barges
- Transportation by cargo ships

Land/sea pipelines continuous transport in coal slurry form.

The alternative to coal transport is conversion of coal to electrical energy in mine-mouth power plants and transport of energy in electrical form. This method is economical for large blocks of power and continuous transfer of energy.

2.10 COAL CONVERSION TECHNOLOGIES

Coal is converted from solid form to liquid or gaseous form into various solid, liquid and gaseous fuels. The various solid, liquid and gaseous fuels have their specific applications. Pollution is caused by burning of gaseous or liquid fuels. Direct burning of coal results in emission of particulates, smoke, SOx, NOx, CO and CO_2. The gaseous and liquefied fuels obtained from coal can be burned with lesser emission of toxic air pollutants. Environmental pollution controls are prohibiting the direct burning of coal in power plants. Coal conversion technologies are listed below:

1. **Coal gasification:** It is a conversion of coal to gaseous form for use as fuels. The gases released from this process are city gas, water gas, producer gas and methane etc.
2. **Coal liquefaction:** In this process coal is converted into liquid hydrocarbons and related liquid compounds by hydrogenation process.
3. **Carbonization of coal:** Coking, destructive distillation of coal—is heating of coal in absence of air, gives coal gas and coke.
4. **Coal gas:** An artificial gaseous fuel produced by heating coal in absence of oxygen. This fuel is also called *town gas*.
5. **Coal slurry pipeline:** Pipeline carries coal powder suspended in water. At sending end, slurry is prepared and pumped. At receiving end, the slurry is filtered and the coal is reclaimed.

2.11 COAL GASIFICATION

In the nineteenth century and the first part of the twentieth century there was widespread utilization of coal gas prepared by destructive distillation of coal for lighting up and cooking proposes. Liquid fuel from coal could be a future alternative to conventional petroleum. Despite the vast technology base for coal liquefaction, the efforts to construct commercial plants have financed. Coal gasification includes chemical reaction of coal, steam and air or oxygen at high temperatures to handle a mixture of hydrocarbon gases, ordinarily: Carbon monoxide, carbon dioxide, hydrogen and methane, additionally hydrogen sulfide. On heating coal without air, coal was carbonized to coke by remaining its assets and utilizing them as a byproduct gas. This gas was circulated in urban zones as town gas. The coke therefore generated was burned in big beds with less than the stoichiometric amount of air to yield the producer gas.

$$2C + (O_2 + 3.76\,N_2) \rightarrow 2CO + 3.76\,N_2$$
Coke $\quad$ Air $\quad\quad$ Producer gas

For each mole of oxygen in air there are 3.76 moles of nitrogen, when the bed was heated to a high temperature, the flow of air was replaced by the flow of steam and water gas produced.

$$C + H_2O \longrightarrow CO + H_2$$
Coke Steam   Water Gas

Water gas reaction is endothermic and the bed gets cooled. The steam flow is then replaced by air flow and exothermic partial combustion, e.g. reheated the bed, air flow is again replaced by steam flow. The alternate production of water gas and producer gas continued till the coke bed was exhausted. The gases obtained from coal are further processed and upgraded in several steps. The basic process involves several steps. The coal is first ground into powder and then preheated and dried to reduce caking during conversion. For a caking coal, a high temperature pretreatment is used to give the coal particles a thin coating of oxygen to prevent sticking. Figure 2.4. illustrates the basic steps in the gasification process.

Liquid fuel from coal could be a future alternative to conventional petroleum. From each ton of coal one to four barrels of oil can be produced. Inspite of the vast technology base for coal liquefaction, the efforts to build commercial plants have subsidized. High capital costs of synthetic oil plants and declining oil prices were the major obstacles. If costs would become competitive, coal could be the future source of liquid fuels. Coal gasification brings about three gas mixtures, grouped as per their heating value. They are called low heating value, medium heating quality and high heating value gas. The methodology, shown in Fig. 2.5 yields a low-heating value gas (which is a mixture of water gas and producer gas), called *synthesis gas*. The procedure illustrated by the upper flow sheet yields a high heating value gas, called *pipeline gas*, lands near that of natural gas. The essential procedure includes a few steps. The coal is first ground into powder and after that it is preheated and dried to diminish hardening throughout transformation. For a hardening coal, a high temperature pretreatment is used to give the coal particles a slim covering of oxygen to anticipate staying.

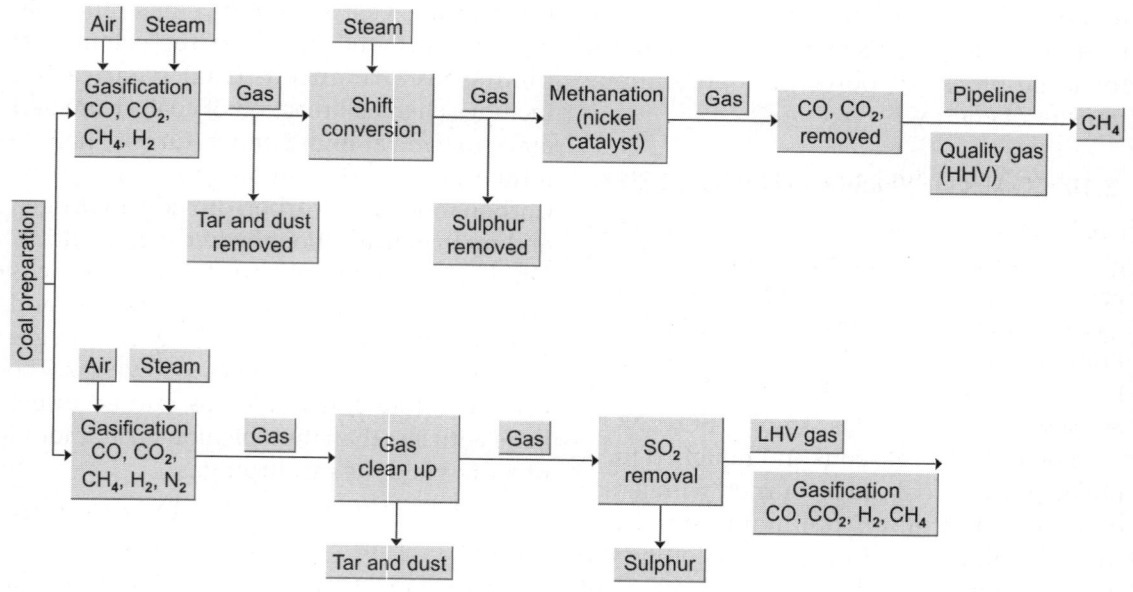

Fig. 2.4: Gasification process

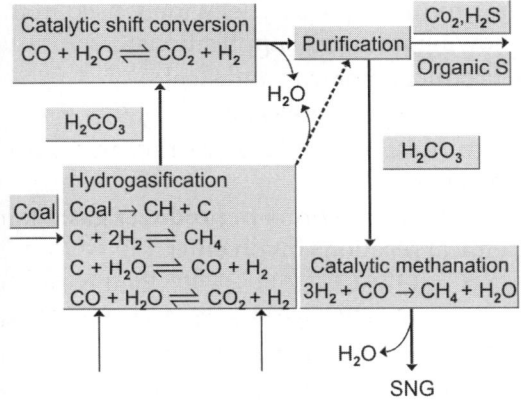

Fig. 2.5: Flow diagram showing methodology of obtaining synthesis gas

1. The feedstock is reacted with air and steam. The air quantity is less then stanchion metric.

$$C + O_2 + 3.76\,N_2 \longrightarrow CO_2 + 3.76\,N_2$$

   CO_2 from this reaction reacts further with additional carbon in the rich mixture to give nitrogen

$$CO_2 + C + 3.76\,N_2 \longrightarrow 2CO_2 + 3.76\,N_2$$

   In steam, $C + N_2O \longrightarrow CO + N_2$

   The result is low heating value syngas containing CO, N_2 and CO_2. It may also contain some CH_4.

2. To obtain medium-heating value gas, a shift reaction is used to produce additional hydrogen, and nitrogen is removed.

$$CO + H_2O \longrightarrow CO_2 + H_2$$

3. To produce pipeline gas, a catalytic methanation is carried out in which the products of water gas reaction are reacted over a nickel catalyst at 1100°C and a pressure of 6.8 bar.

$$CO + 3H_2 \longrightarrow CH_4$$

   The product gas is a high a quality gas having heating value of about 38 MJ/m³ and can be a direct substitute of natural gas. There is another method of classification, called *hydrogasification*, in which fluidized coal is gasified directly with hydrogen-rich steam to a methane-rich gas that requires very little additional shifting. The overall reaction is of the following form

$$CH_{0.8} + 0.55\,H_2O + 1.15\,H_2$$
$$\longrightarrow 0.575\,CH_4 + 0.425\,CO_2$$

The overall efficiency of conversion by this method is higher than the earlier method.

The gasification of coal involves chemical reaction between coal, steam and air (or oxygen) at high temperature.

2.12 UNDERGROUND COAL GASIFICATION

Deposits of coal that could not be mined competitively by other processes might be economically utilized by underground gasification. Coal is gasified *in situ* and the gas produced is conveyed to the surface and utilized to meet various needs. *In situ* coal gasification became attractive due to the following advantages:

1. It can extract energy from inaccessible reserves of coal that cannot be mined by conventional techniques.

2. It needs less coal-handling and transportation facilities.

3. It is safer and reduces the chance of occupational hazards.

4. From the environmental standpoint, H_2S rather than SO_2 is the predominant form of sulphur produced in underground coal gasification. It can be economically treated by hot carbonate scrubbing.

Two deep drilled holes are made, one for air and the other for product gas (Fig. 2.6). After initial ignition, combustion takes place at the bottom of the air hole and the combustion zone proceeds towards the product gas hole. In the combustion zone, CO_2 is formed in the reaction.

$$C + O_2 \longrightarrow CO_2 - \Delta H$$

Ahead of the combustion zone is a reduction zone, where CO is formed.

$$CO_2 + C \longrightarrow 2CO + \Delta H$$

From moisture in the coal, water gas reaction can take place

$$C + H_2O \longrightarrow CO + H_2 + \Delta H$$

Volatile matter is also released from the coal as the process proceeds.

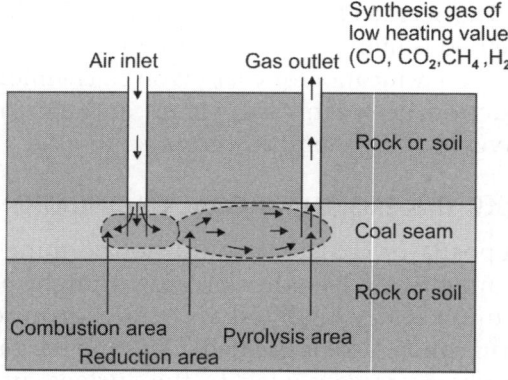

Fig. 2.6: *In situ* coal gasification

2.13 PRODUCER GAS

Producer gas is a mixture of nitrogen, carbon monoxide and hydrogen. It is a vaporous mechanical fuel from poor quality coals, peat, wood waste, coke and etc. The producer gas is obtained from low grade coals, peat, wood work, coke and etc. It is obtained from chemical reaction between strong coal, air or air and steam at temperature ~5000 °C. Just air may be utilized rather than mixture of air and steam. The plant for making producer gas is called *gas maker or producer*. The producer gas is a mixture of gases with high extent of carbon monoxide, nitrogen and hydrogen. The proportion varies with arrangement of coal. Rate of air and steam, supply temperature and pressure. Table 2.1 shows the composition of producer gas.

The producer gas is a fuel widely used in the following industrial applications:

- Production of iron and steel
- Production of coke and coal gas
- Production of glass, refractories, and ceramics.

Chemical reactions in producer gas: Main chemical reactions involved are:

$$C + O_2 = CO_2, \quad \wedge H = -405 \text{ kJ at base}$$
$$C + CO_2 = 2CO, \quad \wedge H = +162 \text{ kJ above base}$$

Producer gas: The plant which produces the producer gas is called *gas producer*. Figure 2.7 illustrates a cross-section of gas producer. The cylindrical vessel (1) is made up of fabricated, welded, mild steel plates. Refractory brick lining (2) is provided internally in (1). The size of the vessel (1) is about 4 m diameter and 6 m height. Air or mixture of air and steam is admitted through inlet pipe (3) from bottom. Producer gas is collected from outlet pipe (4). Ash slag is removed from bottom-pit (5) several poking holes (6) (H_1, H_2) are provided at intermediate levels and they are normally closed by iron balls and coal is admitted from top through hopper (7). The outlet producer gas has several impurities such as steam, tar, dust, H_2S, ammonia and are condensed on cooling. Producer gas is cooled by passing through heat recovery steam generator and impurities are separated from the commercial gas delivered to customers.

2.14 LURGI GASIFIERS

Lurgi procedure is a commercial methodology of coal gasification in which oxygen and steam

Composition	Producer gas form		
% volume	Mixed coal	Coke	Anthracite coal
Carbon monoxide (CO)	29	29	26
Hydrogen (H_2)	12	11	17
Nitrogen (N_2)	52	55	50
Methane (CH_4)	2.5	0.5	1.5
Other	4.5	4.5	5.5
Total	100	100	100

Table 2.1 Composition of producer gas

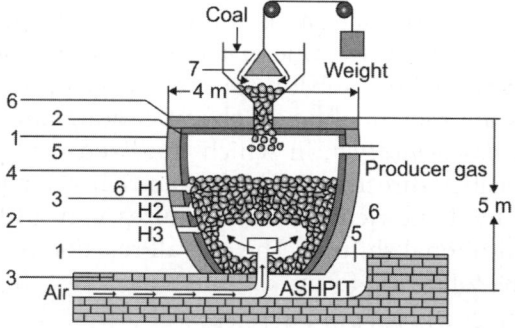

Fig. 2.7: Cross-section of gas producer

areas blown through hot coal rather than blowing air and steam. The gasifiers utilize high temperature (1090°C bottom and 250°C top). Lurgi gas has heating values of about 15 MJ/mq. This methodology began in Germany and is utilized as a part of Europe. It is currently being utilized as a part of modified structure for coal gasification (ICGCC power plants). This unit comprises of 12 parallel lurgi gasifiers, 9 working and 3 extra, to gasify measured coal (2–1/2″ × 1 × 4″). Several techniques of coal gasification utilizing steam and oxygen have been created in Europe, USA and are named by their inventer. Lurgi gas plants have been introduced in some nations to supply:

- Full gas for town pipeline fuel supply.
- Industrial fuel supply to chemical industry.

Lurgi process is further modified to obtain synthetic natural gas. Lurgi process is the most widely used coal gasification process in the world. The Lurgi gasifier is a moving bed reactor in which coal moves downward under gravity counter current to the upward flowing gasification agent, a mixture of superheated steam and 98.5% pure O_2. The coal bed is supported by a revolving grate through which superheated steam and oxygen are added and ash is removed. The gasifier is operated at a pressure of about 450 Psia. The gasifier is jacketed with about 10% of the feedsteam for the gasification agent being generated in the jacket. Composition of lurgi gas is given in Table 2.2.

Lurgi gas plant: The Lurgi gas plants have been installed in several countries.

The total plant has the following parts:

1. Lurgi gasification unit (generator) (Fig. 2.8)
2. Methanator
3. Cooler
4. Hot potash or water wash for CO removal
5. Scrubber for SOx removal etc

Chemical reactions that occur in the gasifier are described below:

1. The gasification of the carbon takes place by the reaction:

$$C + H_2O \text{ (steam)} + Heat \longrightarrow CO + H_2$$

2. Hydrogen is produced by the following reaction,

$$CO + H_2O \text{ (steam)} \longrightarrow CO_2 + H_2 + Heat \uparrow$$

3. Some carbon also reacts with CO_2 to form CO

$$C + CO_2 \longrightarrow 2 C(O)$$

4. Methane is produced by the hydro-gasification reaction

$$C + 2H_2 \longrightarrow CH_4 + Heat \uparrow$$

Table 2.2 Composition of lurgi gas		
Contents	**Lurgi-gas from raw bituminous coal% volume**	**From cleaned bituminous coal**
Carbon dioxide (CO_2)	28	1
Carbon monoxide (CO)	22	32
Hydrogen (H_2)	38	53
Methane (CH_4)	10	13
Other	2	1
Total	100% volume	100% volume

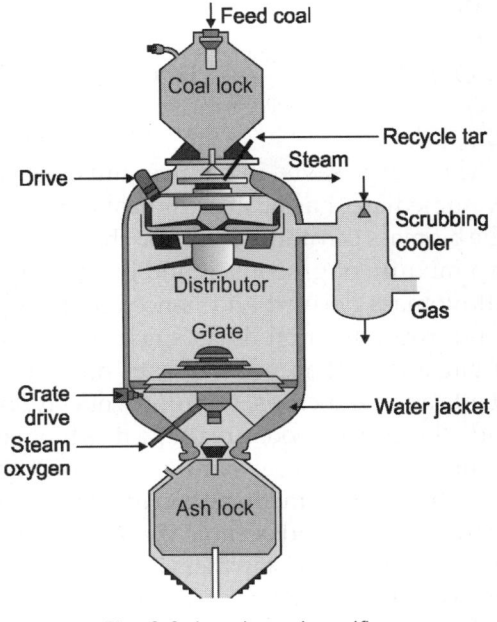

Fig. 2.8: Lurgi coal gasifier

5. The net heat input for the above reactions is supplied by the combustion of carbon with oxygen in the combustion zone.

$$C + O_2 \longrightarrow CO_2 + Heat$$

A Lurgi gasifier operates at a pressure of about 28 bar. Lump coal is admitted through the upper coal lock and feeds downward over the distributes into the gasifier. The gasifier is a water–jacketted vessel of about 4 m diameter and 8 m height. The coal is held on a rotating grate with the steam and oxygen coming up through the grate. Ash falls through the grate and is removed from the ash lock at the bottom. Caking coal cannot be normally handled in Lurgi gasifier.

2.15 BIOMASS

Biomass is the organic matter produced by plants, both in land and water. It includes forest crops, the crops which are grown in energy farms, and animal manure. While fossil fuels take millions of years to form, the biomass is an alternate fuel, the source of which may be considered renewable, since plant life renews and adds to itself every year. It is the solar energy stored by way of photosynthesis,

$$6\,CO_2 + 12H_2O \xrightarrow[\text{Chlorophyll}]{\text{Sunlight}} C_6H_{12}O_6 + 6H_2O + 6H_2$$

Air   Soil   (Glucose yielded by air or transparent starch)

The energy saved in biomass must be used by immediate sharing in areas near the source. The biomass can additionally be changed over to a mixture of gaseous, liquid or solid fuels. All biomass derived and hence all bio-fuels. Land crops, sugar products, sugarcane decline or bagasse, herbaceous harvests, non-woody plants and field crops, populous, eucalyptus, and other hard woods are utilized as biomass, animal and human wastes are indirect area crops from which methane and ethylene might be transformed and oceanic yields developed in ocean and brackish waters incorporate sea-weeds, marine green growth and perspectives make a difference.

There are 3 forms of bioconversion routes:
1. Direct combustion, such as wood waste.
2. Thermochemical conversion can take two routes, viz., gasification and liquefaction. Gasification is done by heating the biomass with limited oxygen to produce low heating value gas or by making the biomass react with steam and oxygen at a high pressure and temperature to produce medium heating value gas. The later may also be subjected to liquefaction by converting it to methanol and ethanol.
3. Biochemical conversion can take two ways: anaerobic digestion and fermentation. Anaerobic digestion is the bacterial decay of natural matter (biomass) without air or oxygen to transform a vaporous mixture (biogas) of methane and carbon dioxide in an around 2:1 volume ratio. Fermentation is the breakdown of complex atoms in natural mixes with the assistance of microed organisms, proteins, etc. grains and sugar yields are changed over by fermentation into ethanol. Ethanol in this way prepared could be blended with fuel to generate *gasohol* (99% gasoline, 10% ethanol), which might be utilized as a car fuel.

2.16 UNDERGROUND—*IN SITU* PRODUCTION OF COAL GAS

Underground—*In situ* coal gasification is an alternative to deep underground mining of coal, where mining is uneconomical or unsafe or practically difficult. Deposits of coal that could not be mined competitively by the processes might be economically utilized by underground gasification. The air or mixture of steam CO_2 are injected into underground coal after rejoining the coal. Coal is gasified *in situ* and the gas produced is conveyed to the surface and utilized to meet various needs. *In situ* coal gasification became attractive due to the following advantages:
1. No separate gasification plant necessary
2. Large quantity of coal gas can be produced

3. Simpler than deep coal mining
4. Cheaper than deep coal mining
5. Energy in poor quality coal at greater depth can be extracted
6. Safer than deep mines
7. Less pollution
8. Energy can be extracted from old, leftover mines.

Deep bore holes are drilled in the ground. The coal in the coal bed is ignited by electric means. Air or mixture of steam and oxygen are injected into the well through the injection well. The coal gas is pumped up through production wells. After initial ignition, combustion takes place at the bottom of the air hole and the combustion zone proceeds toward the product gas hole. In the combustion zone, carbon-dioxide is formed in the reaction:

$$C + O_2 \longrightarrow CO_2$$

Ahead of the combustion zone is a reduction zone, where carbon monoxide is formed:

$$CO_2 + C \longrightarrow 2CO$$

From moisture in the coal, watergas reaction can take place

$$C + H_2O \longrightarrow CO + H_2$$

Volatile matter is also released from the coal as the process proceeds. The product gases are very similar to the coal gas or water gas.

Composition: The composition of *in situ* coal gas depends on injected fluid (air or air and steam). Typical composition is:

Hydrogen	H_2	10%
Nitrogen	N_2	65%
Carbon dioxide	CO_2	10%
Carbon monoxide	CO	9.2%
Methane	CH_4	1.8%
Other		3%

2.17 LIQUEFACTION OF COAL

The procedures of change of coal into fluid hydrocarbons and identified mixes by hydrogenation is called *coal liquefaction*. Liquid fuel from coal could be a future alternative to conventional petroleum. From every ton of coal, 1 to 4 barrels of oil might have transformed. The lack of developing world petroleum supplies and the quickly expanding expense of the oil has revived investment in generating a fluid fuel from coal. Coal liquefactions engineering was stimulated in both Germany and Japan by the World War II. Japan generated flight fuel in a substantial plant in North Korea which changed over coke produced out of coal into calcium carbide in electric heaters, then to acetylene, acetaldehyde, butyraldehyde, octanol, and octane. The most important German procedure was the Fischer-Tropsch process, which is still utilized economically by the SASOL plant in South Africa. Liquefaction of coal gives different refinery items as petrol, diesel, fuel oil, paraffin waxes, LPG and chemicals like ethanol, propanol, butanol, acetone, mixed solvents, benzene, xylene and naphtha etc. Despite the boundless innovation base for coal liquefaction, the deliberations to fabricate business plants have sponsored. High capital expenses of manufactured oil plants and declining oil costs were the real deterrents. Assuming that expenses might get aggressive, coal could be what's to come source of fluid energies.

The conversion of coal into a liquid fuel requires the addition of hydrogen to the coal. Coal has a ratio of hydrogen to carbon only 0.8 to 1, while in petroleum this ratio is 1.75 to 1. There are three basic modes that have been used to liquefy coal. These are hydrogenation, catalytic conversion and hydropyrolysis.

1. **Hydrogenation:** In this process coal and catalyst are suspended as a slurry, which is reacted with hydrogen at high pressure and moderate temperature to form liquid hydrocarbons.

2. **Catalytic conversion process:** A synthesis gas is produced from the coal. The hydrogen and carbon monoxide in the gas are then combined in the presence of a catalyst to form a liquid hydrocarbon fuel.

3. **Hydropyrolysis:** Coal is heated beyond 450 °C, the fraction of coal volatilized greatly exceeds the volatile matter in coal. The hydrogen entrained pulverized coal is flash pyrolyzed. Up to 50% of the coal thus be liquefied.

2.17.1 Catalysis in the Liquefaction of Coal

One of the important strategies in the process of liquefaction of coal is hydrogen economy. That is the effective utilization of hydrogen. Now the question is which is the most efficient method of hydrogen transfer? This is where the role of catalyst becomes vital in the liquefaction of coal.

A catalyst need to efficiently transfer and distribute hydrogen.

An ideal catalyst should possess the following properties:

1. High activities for hydrogenation, cracking, and heteroatom removal.
2. Selectively for maximum liquid yield with minimum gas production.
3. Adequate physical strength to overcome mechanical degradation with aging and regeneration.
4. Resistance to deactivation caused by coke and metal deposition, poisoning, sintering and pose mouth blockage.

An effective catalyst should help liquid coal at lower severity conditions and improve yield and quality. Catalysts in direct coal liquefaction lowers the severity of the operation essentially by activating both the molecular hydrogen and also the coal molecules that to at a lower reaction temperature. They improve the product quality by cracking and reforming heavy molecules to desired products, and remaining heteroatom's (atoms other than C) namely N, O, S from coal and coal liquids. The advantages offered by catalysts in this regard are many. But all is not easy and hardly arise because of the complex nature of coals. Coal, frequently referred to as "naturals dump", tends to deactivate the catalyst. Most vital reasons for catalyst deactivation are coke formation as well as metal deposition apart from poisoning as sintering which are not as pernicious as these of the farmer.

2.17.2 The process of Liquefaction of Coal

Fluid fuel from coal could be a future alternative to conventional petroleum. From every ton of coal one to four barrels of oil might be prepared. Regardless of the limitless engineering base for coal liquefaction, the endeavors to raise commercial plants have sponsored. High capital expenses of engineered oil plants and declining oil costs were the major obstacles.

Major issues of concern in the liquefaction of coal are as follows:

1. Process thermal efficiency
2. Hydrogen utilization
3. Materials and components reliability
4. Solid-liquid separation
5. Product quality and flexibility
6. Feed coal flexibility and
7. Process severity

Among the major areas of concern listed above, the efficient utilization of hydrogen is one of the critical aspects in direct liquefaction. Effective utilization of hydrogen is important from the stand point of reaction chemistry, reaction mechanism and economics of the process. The cost of hydrogen production itself is 1/5 to 1/3 of the direct capital expenditure of coal liquefaction plants. Hydrogen stabilizes the reactive and unstable free radicals formed by thermal rupture of chemical bonds of the coal macromolecules. This prevents the occurrence of retrogressive reactions that produce the undesirable high molecular weight products that reduce the yield of desired liquid products. The overall conversion of coal can be regarded as a series of thermal decomposition/hydrogenation reactions. Preasphaltene, asphaltene and oil are defined as tetrahydrofuran soluble – toluene insoluble, toluene soluble – pentane insoluble and pentane soluble coal liquids respectively.

2.18 CARBONIZATION OF COAL

Carbonization of coal is the procedure of heating the coal without air. The significant matter in coal is discharged and the results of the gasification are coal gas, coke, liquid, ammonia, tar etc. Coal gas (town gas) and coke are obtained by heating coal without air, coal gas was utilized initially as town gas and as a fuel for domestic use. With accessibility of characteristic gas, the utilization of town gas has diminished. Coke is a porous solid residue resulting from heating bituminous coal in absence of air or with less air and incomplete combustion. Coke is mostly carbon. It is used as a fuel in certain metallurgical processes.

Carbonization is of the following categories:

- Low temperature carbonization (500° to 700 °C).
- Medium temperature carbonization (750° to 900 °C).
- High temperature carbonization (950° to 1300 °C).

The quality and quantity of the products of carbonization of coal (coke and coal gas, taro, liquid ammonia, etc.) depend upon:

- Type of coal—Bulk density kg/m.
- Temperature—Heating rate.

Fuel oils are added to reduce particulate emission and accelerate the process.

2.19 COAL SLURRY PIPELINE

The pulverised coal suspended in water is called *coal slurry*. The pipeline which transports coal slurry is called coal slurry pipeline. For the purpose of transport as slurry, coal is pulverised to the size of table sugar granules and mixed with water in equal volume. The slurry is pumped through pipeline. Pumping stations are located at sending end and intermediate locations at an interval of about 80 km. At the receiving end, the slurry is filtered and water is removed. Coal is used as fuel in boilers and water is used as cooling water for condenser (Fig. 2.9).

2.20 COAL BLENDING

Coal blending is a process of mixing two coals of different reactivities. Alternatively coal derived ash and/or mineral matter coal are added to a less reactive coal. Coals differ in their liquefaction reactivity due to rank, petrography and concentration of mineral matter. Addition of a highly reactive coal to a

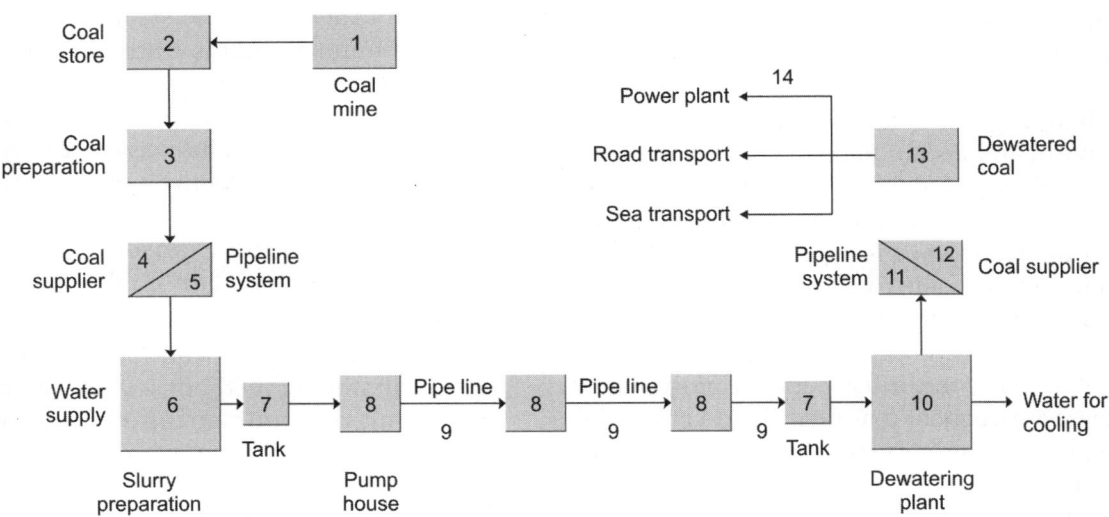

Fig. 2.9: Flow diagram of coal slurry pipeline system

low reactive coal can increase conversion and lower operating severity in a liquefaction plant. Coal blending is beneficial in a way that this decreases the severity of plant operating conditions.

2.21 COAL FROM INDIAN PERSPECTIVE

Coal is the predominant energy source (58%) in India, followed by oil (27%), natural gas (7%), liquate (4%), hydro power (3%) and nuclear power (0.22%).

2.22 COAL RESERVES AND MINING

India has a long history of commercial coal mining covering nearly 230 years starting from 1774 by M/s Summer and Heatly of East India Company in the Raniganj, coal field along the western bank of river Damodar. Major coal fields in India are found in Jharkand, Bihar, West Bengal, Madhya Pradesh, Maharastra, Assam, Andhra Pradesh, Odisha, Tamil Nadu and Kashmir. Jharia (Jharkand) and Raniganj (West Bengal) are the biggest and best coalfields of the country. Unlike the coals in Europe and America, Indian coals have high percentage of mineral matter, most of which is finely disseminated and intimately mixed with the coal substance. Jharkand came into existence on 15th November, 2000 as the 28th state of India as a result of the bifurcation of Bihar state. Nearly 32.98 % of coal deposits of India is in Jharkand. Its mines are in Jharia, Chandrapure, Bokaro, Ramgarh, Kamapur, Charhi and also in Rajmahal and Daltonganj area. Raniganj coal field is the largest coalfield in India, belonging to the Gondwana Super group (Gondwana is a geological term which refers to as certain rock system which is about 200 million years old. Most of the Indian coals belong to this group). Mining in this region dates back to the British period. Unfortunately there are frequent coal-fires reported from this region and India is loosing good quality coal prior to its exploitation by spontaneous combustion. Hence, there is a need for detection and monitoring of coal-fires in coal fields in

order to control them effectively. Lignite is found mainly at Neyveli in Tamil Nadu. Minor coal fields exist in Andhra Pradesh, Kashmir and Assam. Assam coals have very high sulphur content (3–8%). Kashmir coals are artificial anthracite converted from lignite deposits. Coal deposits of the Tertiary era (60 million years old) are found in Assam, Rajasthan and Jammu.

2.23 COAL MINING

The coal beds may be located deep underground or near ground surface with a hot so deep overburden of soil. The surface tarrain may be flat or with sleep sloper. Mining is the technique of extracting minerals as fossils from the Earth. Mine is an opening or excavation, or a pit in the Earth from which fossils, mineral, metallic ores extracted by digging.

Two distinct types of mining are:

- Surface mining, (strip mining, placer mining, open cot mining, open pit mining) in which the coal beds are near the ground surface with little overburden of soil. (Depth <30 m).
- Underground mining. In which the coal beds are located at depths.

1. *Surface mining* of coal involves removal of overburden from the steam of coal. Surface mining includes: Strip mining, placer mining, open out mining, open-pit mining etc. Depths are less than 30 m. About 75% mining in India is by surface mining. Surface mining involves stripping off the overburden of soil and rock and then loosening the exposed coal by digging, blasting. The coal is then excavated.

Strip mining is a type of surface mining. The strip mining methods include: area stripping, contour stripping, auger mining. Area stripping is employed in flat or slightly rolling terrains where coal beds cover continuous large area near ground surface (depth <30 m). Recovery rates are 80 to 90%. *Contour strip* mining is employed in hilly or mountainous regions. Bench cuts are made in sloping hill sides at the level of coal beds. Overburden is removed and dumped down the slope. The depth of the bench out on hill side is limited by the height of high wall on upslope side, recovery rates are low and vary widely. *Auger mining*: This procedure is used very rarely, a very few mines (3%) in the world employ this method. Auger is a rotating boring tool having spiral head along extendable shaft. Typical diameters are 0.5 m to 2 m; shaft length upto 70m.

2. *Underground mining* (shaft mining): The underground mining is used for attracting coal from coal beds at depths more than 50 m, deep mining involves use of cutting tools which penetrate deep into the coal bed, conveyers which collect the coal and transport it to the mine mouth; roof supports for overlying rock above the excavation, shafts for lighting the coal. Communication agents such as methanol, sodium hydroxide, ethylamine etc. are forced in drill hole to dissolve hard minerals and beat-up the coal rapidly. This reduces time of drilling and economizes the deep coal mining.

The methods of deep mining according to direction of access to deep coal beds include:

- Vertical—Shaft mining
- Drift mining
- Slope mining

In vertical shaft mines the main shaft and ventilation shaft are vertical. The coal is filled through the main shaft.In slope mine, the main shaft for lifting the coal is inclined.

2.24 COAL MINING AND THE ENVIRONMENT

Coal mining, particularly surface mining, requires large areas of land to be temporarily disturbed. This raises a number of environmental challenges, including soil erosion, dust, noise and water pollution, and impacts on local biodiversity. Steps are taken in modern mining operations to minimize impacts on all aspects of the environment. By carefully pre-planning projects, implementing pollution control measures, monitoring the effects of mining and rehabilitating mined areas, the coal industry minimizes the impact of its activities on the neighbouring community, the immediate environment and on long-term land capability.

Land Disturbance

In best practice, studies of the immediate environment are carried out several years before a coal mine opens in order to define the existing conditions and to identify potential problems. The studies look at the impact of mining on surface and ground water, soils, local land use, native vegetation and wildlife populations. Computer simulations can be undertaken to model impacts on the local environment. The findings are then reviewed as part of the process leading to the award of a mining permit by the relevant government authorities.

Mine Subsidence

Mine subsidence can be a problem with underground coal mining, whereby the ground level lowers as a result of coal having been mined beneath. A thorough understanding of subsistence patterns in a particular region allows the effect of underground mining on the surface to be quantified. The coal mining industry uses a range of engineering techniques to design the layout and dimension of its underground mine workings, so that surface subsidence can be anticipated and controlled. This ensures safe, maximum recovery of a coal resource, while providing protection to other land uses.

Water Pollution

Mine operations work to improve their water management, aiming to reduce demand through efficiency, technology and the use of lower quality and recycled water. Water pollution is controlled carefully by separating the water runoff from undisturbed areas, which contains sediments or salt from mine workings. Clean runoff can be discharged into surrounding water courses, while other water is treated and can be reused such as for dust suppression and in coal preparation plants.

Acidmine Drainage

Acid mine drainage (AMD) can be a challenge at coal mining operations. AMD is a metal-rich water formed from the chemical reaction between water and rocks containing sulphur-bearing minerals. The runoff formed is usually acidic and frequently comes from areas where ore- or coal mining activities have exposed rocks containing pyrite, a sulphur-bearing mineral. However, metal-rich drainage can also occur in mineralised areas that have not been mined. AMD is formed when the pyrite reacts with air and water to form sulphuric acid and dissolved iron. This acid run-off dissolves heavy metals such as copper, lead and mercury into ground and surface water.

There are mine management methods that can minimise the problem of AMD, and effective mine design can keep water away from acid generating materials and help prevent AMD occurring. AMD can be treated actively or passively.

- Active treatment involves installing a water treatment plant, where the AMD is first dosed with lime to neutralise the acid and then passed through settling tanks to remove the sediment and particulate metals.

- Passive treatment aims to develop a self-operating system that can treat the effluent without constant human intervention.

Dust and Noise Pollution

Dust at mining operations can be caused by trucks being driven on unsealed roads, coal crushing operations, drilling operations and wind blowing over areas disturbed by mining.

Dust levels can be controlled by spraying water on roads, stockpiles and conveyors. Other steps can also be taken, including fitting drills with dust collection systems and purchasing additional land surrounding the mine to act as a buffer zone. Trees planted in these buffer zones can also minimise the visual impact of mining operations on local communities.

Noise can be controlled through the careful selection of equipment and insulation and sound enclosures around machinery.

Rehabilitation

Coal mining is only a temporary use of land, so it is vital that rehabilitation of land takes place once mining operations have stopped. In best practice a detailed rehabilitation or reclamation plan is designed and approved for each coal mine, covering the period from the start of operation until mining has finished.

Where the mining is underground, the surface area can be simultaneously used for other uses—such as forests, cattle grazing and growing crops – with little or no disruption to the existing land use.

Mine reclamation activities are undertaken gradually – with the shaping and contouring of spoil piles, replacement of topsoil, seeding with grasses and planting of trees taking place on the mined-out areas. Care is taken to relocate streams, wildlife, and other valuable resources.

As mining operations cease in one section of a surface mine, bulldozers and scrapers are used to reshape the disturbed area. Drainage within and off the site is carefully designed to make the new land surface a stable and resistant to soil erosion as the local environment allows. Based on the soil requirements, the land is suitably fertilised and revegetated. Reclaimed land can have many uses, including agriculture, forestry, wildlife habitation and recreation.

Companies carefully monitor the progress of rehabilitation and usually prohibit the use of the land until the vegetation is self suppor-ting. The cost of the rehabilitation of the mined land is factored into the mine's operating costs.

Using Methane from Coal Mines

Methane (CH_4) is a gas formed as part of the process of coal formation. It is released from the coal seam and the surrounding disturbed strata during mining operations. Methane is a potent greenhouse gas, with a global warming potential of 23 times than that of carbon dioxide. While coal is not the only source of methane emission – agricultural activities are major emitters – methane from coal seams can be utilized rather than released to the atmosphere with a significant environ-mental benefit.

2.25 CONCLUSION

The time will inevitably come when there will be no more coal and no more petroleum for the rate at which the reserves are being consumed. Before the disappearance of coal and petroleum from every day life, mankind must develop a new source of power or perish. Coal is the soild fossil fuel containing several hydrocarbons and minerals. Coal mines are either surface mines or deep mines. Coal is mined and then prepared. The transportation is by land, sea and pipeline slurry. Coal is used as a fuel. Types of coal in descending order of heating values are anthracite, subanthracite, bituminous, sub-butuminous and lignite. Coal conversion technologies are gasification, liquefaction and carbonization etc.

3

Petroleum, Natural Gas and Refinery Products

3.1 INTRODUCTION

Petroleum is oily, flammable, thick dark brown or greenish liquid that occurs naturally in deposits, usually beneath the surface of the Earth; it is also called as crude oil. Petroleum means rock oil, (Greek: petra – rock, elaion – oil, Latin: oleum – oil), the name inherited for its discovery from the sedimentary rocks. It is used mostly for producing fuel oil, which is the primary energy source today. Petroleum and natural gases are very important energy sources. The crude oil and raw natural gas are obtained from production wells and refined in refineries. Several useful liquid and gaseous fuels, byproducts and chemicals are obtained. In India, oil and natural gas exploration, processing and distribution has been established through ONGC, Gas Authority of India, IOCL, HPL, BPCL etc. The important products obtained from petroleum and natural gas refineries are fuel gases, liquefied petroleum gas, liquefied natural gas, aviation fuel (car fuel), car turbine fuel (naphtha), furnance oils (diesel), kerosene (fuel oils) etc. The energy route of petroleum and natural gas are similar. The steps include:

Exploration—drilling—discovery—production wells—production of crude oil/natural gas—refining—transport and distribution—utilization.

3.2 PETROLEUM AND NAPHTHA

Petroleum is also the raw material for many chemical products, including solvents, fertilizers, pesticides and plastics. For its high demand in our day-to-day life, it is also called as 'black gold'. Petroleum and natural gas are important fossil fuels. Petroleum resources, both liquid and gaseous, have become the major energy sources because of their availability and convenience as fuels for IC engines as turbines, stationary and mobile engine, combined cycle power plants and land, air – transportation, military devices, etc. Petroleum is often considered the *lifeblood* of nearly all other industry. For its high energy content and use, petroleum remains as the primary energy source. Petroleum items (naphthas) are crucial in chemical industry, paints, varnishes and so forth, petroleum assets happen in very nearly all parts of the world, however, the significant economically extractable stores exist in moderately few areas of the world. The interest for petroleum has expanded quickly with the development of force segment and transportation division. Petroleum costs have expanded. Petroleum has picked up a significant position in the energy situation of the world. Crude oil (crude) is a mixture of hydrocarbons as present in nature before refining or transforming. Petroleum is a fossil fuel shaped by deterioration and bacterial anaerobic responses on busied vegetation and creature masses under favourable temperatures and weights and surroundings over a few million centuries. Crude petroleum is a complex mixture of various hydrocarbons and other usable substances. Petroleum naphtha is a generic term applied to refined, partially refined, unrefined petroleum and liquid products from natural gas. Petroleum spirits are refined petroleum distillates which are suitable for use as thinners. Petroleum tar is a black/brown product obtained from petroleum refining. Natural gas is a gaseous fossil fuel usually found in association with petroleum. Refining is the process of separation of component parts from the crude oil and to manufacture petroleum products. This process includes: distillation, cracking, chemical treating and solvent extraction.

3.3 ORIGIN OF PETROLEUM

Biogenic Theory

From geologists perspective, unrefined petroleum, for example coal and characteristic gas, is the result of compression and warming of old vegetation over geological time scales. Consistent with this hypothesis, it is structured from the rotted stays of ancient marine creatures and plants. Over numerous hundreds of years this natural matter, blended with mud, is covered under thick sedimentary layers of material. The resulting abnormal amounts of hotness and weight cause the remaining parts to transform, first into a waxy material known as kerogen, and afterward into liquid and gaseous hydrocarbons in a procedure known as *catagenesis*. These then move through adjoining rock layers until they get trapped underground in permeable rocks called *repositories*, framing an oil field, from which the liquid could be concentrated by boring and pumping. 150 m is for the most part acknowledged the "oil window". Despite the fact that this relates to diverse profundities for distinctive areas as far and wide as possible, a "regular" profundity for an oil window could be 4–5 km. Three circumstances must be available for oil stores to structure: a rich source rock, a relocation course, and a trap (seal) that structures the store. The responses that generate oil and common gas are frequently demonstrated as first request

breakdown responses, where kerogen breaks down to oil and natural gas by an alternate set of responses.

3.4 HOW ARE PETROLEUM FUELS DERIVED?

Crude petroleum is subjected to a variety of refining and blending processes at a petroleum refinery during production. Only a very small quantity of petroleum is utilised without refinement, directly to provide power during production. The principal liquid fuels derived from crude petroleum (crude oil) are produced by fractional distillation. Desulphurisation, hydrogenation, cracking, and other refining processes may be performed on selected fractions before they are blended and marketed as fuels. Usually gases, dirt and water are removed from the crude petroleum before transportation to the refinery, where the aim is to produce fractions or batches of different hydrocarbons, boiling within certain pre-determined temperature ranges, for various applications. The petroleum products are obtained by separation (e.g. distillation and stabilisation), conversion (e.g. cracking and reforming, alkylation and isomerisation). Figure 3. 1, shows the typical distillation scheme of an oil refinery.

3.5 PRINCIPAL PRODUCTS DERIVED FROM PETROLEUM

Approximately 84% of petroleum products include: refinery gases, LPG, aviation fuel, motor car fuel, kerosene, diesel, fuel oils, etc. Other products include naphtha, solvents, lube oil, bitumen, paraffin wax etc.

1. *Petrochemicals*: Many products derived from refining crude oil – such as ethylene, propylene, butylenes, and isobutylene – are primarily intended for use as petro-chemical feedstocks in the production of plastics, synthetic fibers, synthetic rubber, and other products. Some are also used as solvents, including benzene, toluene, and xylene.

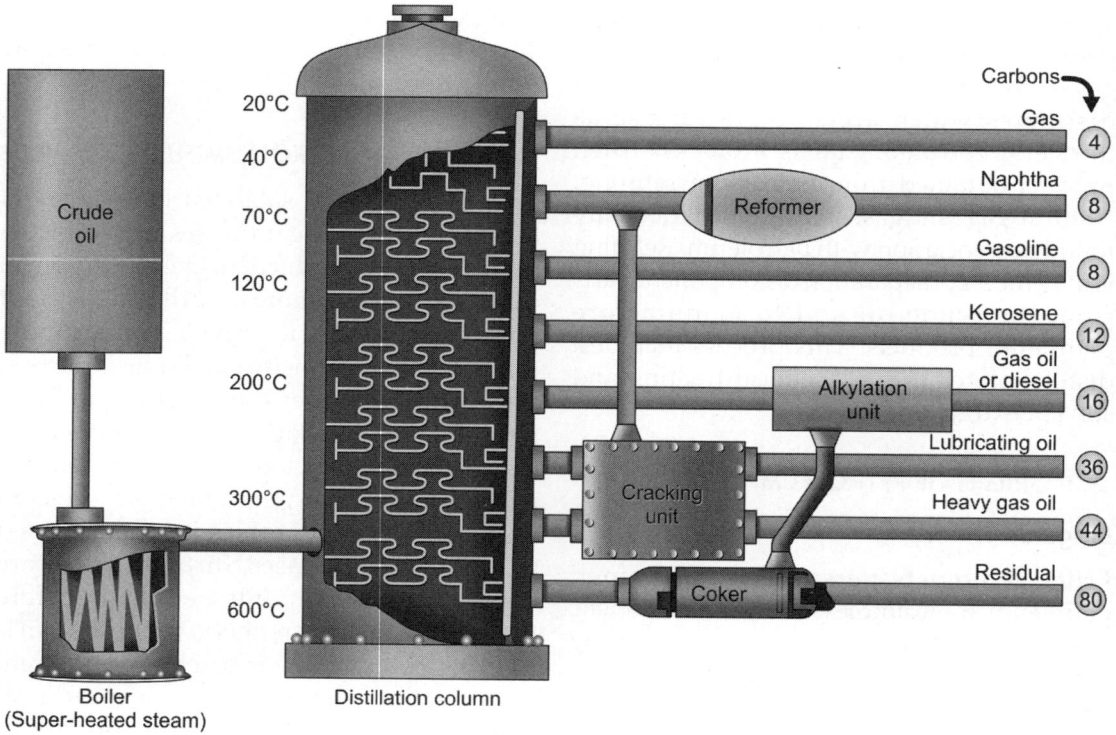

Fig. 3.1: Distillation of crude oil to derive petroleum products

2. *Liquid fuels* blending (producing automotive and aviation grades of gasoline, kerosene, various aviation turbine fuels, and diesel fuels, adding dyes, detergents, antiknock additives, oxygenates, and anti-fungal compounds as required). Shipped by barge, rail, and tanker ship. May be shipped regionally in dedicated pipelines to point consumers, particularly aviation jet fuel to major airports, or piped to distributors in multi-product pipelines using product separators called pipeline inspection gauges ("pigs").

3. *Lubricants* produces light machine oils, motor oils, and grease, adding viscosity stabilizers as required, usually shipped in bulk to an offsite packaging plant.

4. *Paraffin wax*, used in the packaging of frozen foods, among others. May be shipped in bulk to a site to prepare as packaged blocks.

5. *Gaseous fuels* such as propane, stored and shipped in liquid form under pressure in specialized railcars to distributors.

6. *LPG*: Liquefied petroleum gas, consisting primarily of propane and butane, is used as a fuel and as an intermediate in manufacturing petrochemicals.

7. *Gasoline*: Different gasoline blends are produced as regular or premium grades in both summer and winter formulations. Additives are often used to enhance performance and provide protection against oxidation and corrosion.

8. *Naphtha*: A low-octane gasoline product used as a feedstock by the chemicals industry, as a feedstock for catalytic reforming, and in the production of hydrogen.

9. *Middle distillates*: Middle distillates are diesel fuel, heating oil, and kerosene.

10. **Fuel oil:** Many ships, power plants, commercial buildings, and industry use fuel oil or combinations of fuel oil and distillate fuel.

11. **Bitumen:** For this low-value residual product of crude oil vacuum distillation is used primarily for road surfaces and in roofing materials.

3.6 PROPERTIES OF PETROLEUM FUELS

The eight most commonly used properties of liquid fuels are shown in Table 3. 1 below.

3.7 DISTILLATION OF CRUDE OIL

The products obtained by distillation of crude oil do not consist of single hydrocarbon, except in the case of simple gases such as ethane and propane. Each product fraction contains many hydrocarbon compounds boiling within a certain range and these can be broadly classified in order of decreasing volatility into gases, light, middle and heavy distillates and residues.

The gases consist chiefly of methane, ethane, propane and butane. The first two are utilised as fuel or petrochemical feedstocks. Propane and butane may also be liquefied by compression and marketed as liquefied petroleum gas (LPG). Butane may to some extent, be added to motor gasoline. The light distillates comprise fractions, which may be used directly in the blending of motor and aviation gasolines, or as catalytic reforming and petrochemical feedstocks; these fractions are sometimes referred to as tops or naphtha. The heavier, higher boiling-point fractions in this range are the feedstocks for reforming processes lighting, heating and jet engine fuel. Heavier distillates are used as gas oil and diesel fuel and also for blending with residual products in the preparation of furnace fuels. The residue is used for the manufacture of lubricating oils, waxes, bitumen, and feedstocks for vacuum distillation and cracking units, and as residual fuel oil.

Whilst considered 'basic' in the sense of being well established, these properties are not basic in the fundamental scientific sense. They invariably facilitate categorisation of fuels through very well defined experimental procedures.

Moreover, several are interdependent, which becomes apparent if one tries to vary these properties independently.

1. The *relative density* is often used as a broad indication of liquid fuel type and fuel storage capacity: it is defined as the mass of sample occupying unit volume at a specified temperature.

2. The *viscosity* a liquid is a well-defined measure of its internal resistance to flow, and decreases with temperature, whereas the *pour point* is used to characterise the freezing characteristics of fuels.

3. *Vapour pressure* provides a measure of fuel volatility. Whereas an individual hydrocarbon would exhibit a single boiling point, commercial fuel blends boil over a range of temperature. The distillation process or characteristic of a fuel blend facilitates a broad indication of the volatilities of the component fuels.

4. *Flash point* is the most widely used indicator of a liquid fuel's flammability, often used for safety purposes, whereas the spontaneous ignition temperature indicates the minimum temperature to which the fuel

Table. 3.1 Properties of petroleum fuels

No.	Property	Characterisation
i.	Relative density—(formerly specific gravity)	Specifying properties of the matter
ii.	Pour point	
iii.	Shear viscosity	
iv.	Vapour pressure	Indication of mass transfer characteristics
v.	Distillation	
vi.	Flashpoint	Indication of reactivity and energy characteristics
vii.	Spontaneous ignition temperature	
viii.	Calorific value	

must be heated in the presence of air to promote ignition spontaneously, i.e. in the absence of an 'external' source of ignition such as a spark.

5. The calorific value is the quantity of energy released as heat per unit mass of fuel burned under prescribed conditions.

3.8 OVERVIEW OF PETROLEUM FUELS

Liquid fuel products derived from petroleum are generally categorised in a number of broad categories (Fig. 3. 1). Within each category, there are various subdivisions or 'classes' for specific applications:

1. Gasolines are colourless blends of volatile fractions, which boil within the temperature range of about 20°–200 °C. For overall average properties, gasolines are often approximated to octane. The major application for gasoline is the spark-ignition reciprocating-piston engine used for transport, where the anti-knock rating, e.g. research octane number (RON) is very important as it governs the proportion of energy that can be extracted from the fuel in SI engines. 'Unleaded' gasoline is now widely used due to health concerns.

2. Kerosines are colourless blends of relatively involatile petroleum fractions, which boil between about 150°–250 °C, and have a relative density of about 0.8. The average properties of kerosene and high-flash kerosene are very roughly equivalent to dodecane and tridecane respectively. Depending on fuel 'cut', applications include domestic heating, cookers, camping stoves, some heavy SI engine applications and most notably, aviation fuel.

3. Gas oils are brownish-coloured petroleum fractions comprising distillates boiling between 180°–360 °C, with relative density of about 0.84. Net calorific value is typically in the region of 42.5 MJ/kg, viscosity does not exceed 6 cSt (at 37.8 °C) and the flashpoint minimum is 55 °C. Primary uses

are for high-speed diesel engines for transport and relatively small static installations, small heating applications, furnaces, food-processing and agricultural drying. They are sometimes dyed for brand identification.

4. Diesel fuels are darkish-brown petroleum fractions with relative density of about 0.87, net calorific value typically 41.9 MJ/kg, maximum viscosity of about 14 cSt (at 37.8 °C) and minimum flash-point of about 60 °C. Applications include heavy, large engines employed in marine and stationary electricity generating installations, operating at low rotational speeds, and which are less reliant on fuel quality. Industrial heating, hot-water boilers and drying processes are others applications, and minimum temperature for handling is 10 °C due to the relatively high pour point.

5. Residual fuel oils are brownish-black petroleum fractions with relative density typically 0.95. Net calorific values are typically 40 MJ/kg. Viscosity is the critical property for these fuels, with viscosities ranging from 30–500 cSt (at 82.2 °C) for some classes, hence necessitating pre-heating. Minimum flashpoints are 66 °C. High sulphur content – up to 3.5% can be prohibitive in terms of corrosion. Applications include heating, and steam-raising in ships, industrial process heating and power generation.

Another example of a modern industrial fuel derived from petroleum is orimulsion which is an emulsion of natural venezuelan bitumen and water, which has been fired as a "liquid" fuel, for example, in power station boilers designed for oil firing. Finally petroleum coke must be listed as a petroleum-derived industrial fuel, which is used, for example, in the firing of cement kilns. The subject of the firing of industrial fuels and hazards associated with storage and handling.

3.9 COMPOSITION OF PETROLEUM

Petroleum is a blend of vapourous fluid and strong mixtures of numerous alkanes, it comprises mainly of a mixture of hydrocarbons, with hints of different nitrogenous and sulphurous compounds. Gaseous petroleum composed of legates hydrocarbons with abundant methane substance and is termed as *natural gas*. Liquid petroleum composed of liquid hydrocarbons as well as incorporates breaks up gases, waxes (solid hydrocarbons) and bituminous material. Strong petroleum, heavier hydrocarbons and this bituminous material is typically alluded to as bitumen or as plant. Along these, petroleum additionally holds smaller measures of nickel, vanadium and different components. Vast stores of petroleum have been found in broadly diverse parts of the world and their chemical composition differs incredibly. Thus the elemental composition of petroleum vary significantly from raw petroleum to crude petroleum and Table 3.2 represents the composition of petroleum.

Table 3.2 Overall composition of petroleum

Element	% Composition
Carbon	83.0–87.0
Hydrogen	10–14
Nitrogen	0.1–0.2
Sulphur	0.05–6.0
Oxygen	0.05–1.5

3.10 ENERGY ROUTES TO PETROLEUM

The energy routes of petroleum are shown in (Fig. 3.2). Petroleum differs from coal in a manner of its conversion and use. Crude petroleum must be first refined and its components must be separated. The principle steps in the route are:

1. Geological survey
2. Drilling of petroleum wells and gas wells
3. Production of crude oil and natural gas in refineries
4. Storage and transport of crude oil and natural gas
5. Refining of crude oil
6. Storage of petroleum product
7. Transportation of petroleum product
8. Distribution of liquid petroleum and liquid/gaseous petroleum products
9. Supply to consumers

3.11 PRODUCTION OR EXTRACTION OF PETROLEUM

Locating an oil field is the first obstacle to overcome. Today, petroleum engineers use instruments such as gravimeters and magnetometers in the search for petroleum. Generally the first stage in the extraction of crude oil is to drill a well into the underground reservoir. Often many wells (called multilateral wells) are drilled into the same reservoir, to ensure that the extraction rate will be economically

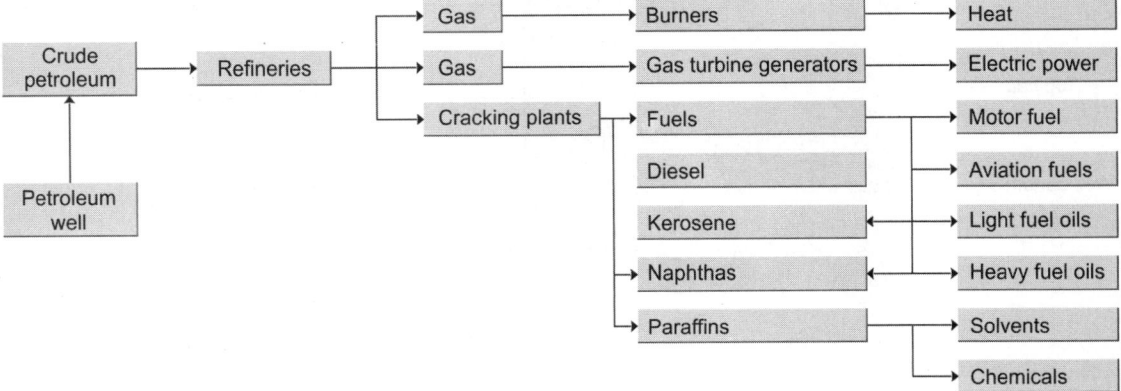

Fig. 3.2: Energy routes to petroleum

liable. Also, some wells (secondary wells) may be used to pump water, steam, acids or various gas mixtures into the reservoir to occur or mountain the reservoir pressure, and so maintain an economic extraction rate.

3.12 EXPLORATION OF PETROLEUM RESOURCES

Deposits of unrefined petroleum and common gas are accepted to have been structured by decay of covered natural vegetation and where organisms dwell, under favourable marine conditions. Profundity say such stores shift from 30 m to a few of thousand meters. in chronicled times such deposits could have happened in shallow sea which were later covered under dry area. Also, waterways and ice sheets could have dumped natural matters from trees, soil, salt and creatures in valleys which could have got covered via avalanches and geographical annoying methodologies.

Favorable high temperature, weight, high-impact bacterial assimilation over a large number of years is important to change the natural mass to fossil fuels.

Petroleum stores for the most part happen under sedimentary rocks and are structured throughout past 100 million years. Oil investigation action is directed in territories having sedimentary rocks. Suitable destinations for investigation are permeable grain like sedimentary rocks with sand, coarseness, limestone, dolomites and shales and here and there changeable rocks (Fig. 3.3). The store rocks are dome formed and sandwitched between impermeable sedimentary cap rock on top and impermeable base rock at bottom. The base water pushes weight on the petroleum oil and gas. Natural gas is normally gathered above the unrefined petroleum hold.

3.13 DRILLING

Drilling for oil and natural gas exploration is done from drilling platforms.

The oil recovery are of following types:

1. On–shore (land)
2. Off–shore (sea)

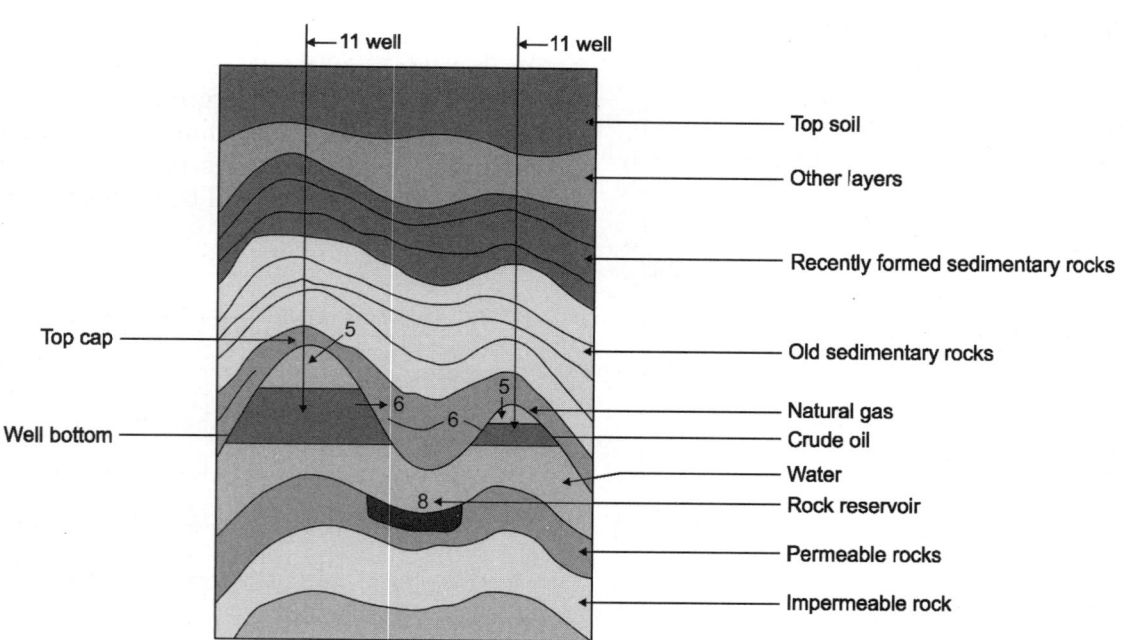

Fig. 3.3: Exploration of oil resources

Off-shore drilling has more prospects than land drilling. About 500 off-shore wells are in operation in the world (1992).

The types of drilling platforms are:

1. Drilling platform for land.
2. Drilling platform for sea:
   i. Shallow depths
   ii. Medium depths
   iii. Large depths.

In off-shore oil recovery, two different stages are distinguished as:

• Drilling stage and production stage.

Today, petroleum engineers use instruments such as gravimeters and magnetometers in the search for petroleum. Generally, the first stage in the extraction of crude oil is to drill a well into the underground reservoir. Often wells (called multilateral wells) are drilled into the same reservoir, to ensure that the extraction rate will be economically liable. Also, some wells (secondary wells) may be used to pump water, steam, acids or various gas mixtures into the reservoir to raise or maintain the reservoir pressure, and so maintain an economic extraction rate.

3.13.1 Drilling Platforms

An oil well is a vertical pipeline structured by screwed funnel joints. The pipeline achieves the oil supply. The funnel line packaging is solidified in the Earth–gap to prevent impact of water on channel material and falling of stones and detached earth in the well. Drilling platform is made of following components:

1. *Derrick*: It is a tall strong fabricated steel galvanized structure, which supports a load of a hundred tons and its height is in between 50 to 100 m (Fig. 3.4).
2. *Drilling platform*: It provides a support for installing the entire machinery of the drilling rig.
3. *Support legs*: Used in shallow off shore drilling, where the depth of ocean bed is about 100 m.

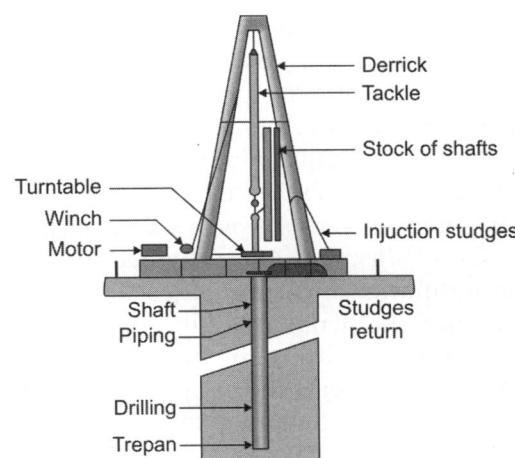

Fig. 3.4: Drilling platform and well on land

4. *Components on the platform include*: Turn table, winch, motor-gear, drilling rig, drill pipes, drilling-shafts with rotary drilling bit at the bottom end.

3.14 PETROLEUM REFINING

The petroleum business might be isolated into two general assemblies, upstream makers (investigation advancement and handling of raw petroleum or regular gas) and downstream transports (tankco, pipeline transports, refiners, retailers and shoppers. Crude oil or transformed raw petroleum is not extremely handy in the structure it escapes the ground. It needs to be broken down into parts and refined before utilization in a strong material, for example, plastics and froths, or as petroleum fossil energizes as on account of car and plane motors. An oil refinery is a streamlined procedure plant where unrefined petroleum is handled in three courses.

i. *Separation–separates crude oil into various fractions*: Oil might be utilized as a part of such a large number of different ways on the ground that it holds hydrocarbons of shifting atomic masses and lengths, for example, paraffins, aromatics, naphthenes (or cycloalkanes), alkenes, dienes, and alkynes. Hydrocarbons are atoms of fluctuating length and made of hydrogen and carbon

Fig. 3.5. The trap in the detachment of diverse streams in oil refinement methodology is the contrast in breaking points between the hydrocarbons, which implies they might be divided by refining. Figure 3.5 shows the typical distillation scheme.

ii. *Conversion–conversion to seleable products by skeletal alteration*: Once divided and any contaminants and impurities have been evacuated, the oil might be either sold without any further handling, or littler atoms, for example, isobutene and propylene or butylenes could be recombined to meet specified octane number prerequisites by methods, for example, alkylation or less usually, dimerization. Octane number prerequisite can likewise be enhanced by synergist transforming, which strips hydrogen out of hydrocarbons to generate aromatics, which have higher octane evaluations. Transitional products, for example, gas oils can even be reprocessed to break heavy, long-chained oil into a lighter short-chained one, by various forms of cracking such as fluid catalytic cracking, thermal cracking, and hydrocracking. The last venture in gas handling is the mixing of fills

Table 3.3 Common process units in an oil refinery unit process

Unit process	Function
Atmospheric distillation unit	Distills crude oil into fractions
Vacuum distillation unit	Further distills residual bottoms after atmospheric distillation
Hydro-treater unit	Desulfurizes naptha from atmospheric distillation, before sending it to a catalytic reformer unit
Catalytic reformer unit	Reformate paraffins to aromatics, olefins, and cyclic hydrocarbons, which are having high octane number
Fluid catalytic cracking	Break down heavier fractions into lighter, more valuable products—by means of catalytic system
Hydro-cracker unit	Break down heavier fractions into lighter, more valuable products—by means of steam
Alkylation unit	Produces high octane components by increasing branching or alkylation
Dimerization unit	Smaller olefinic molecules of less octane number are converted to molecules of higher octane number by dimerization of the smaller olefins
Isomerization unit	Straight chain normal alkanes of less octane number are isomerized to branched chain alkane of higher octane number

with diverse octane evaluations, vapour weights, and different properties to meet product specification.

iii. *Finishing*: Purification of the product streams.

3.15 DETAILS OF UNIT PROCESSES

3.15.1 Hydro-treater

A hydro-treater utilizes hydrogen to immerse aromatics and olefins and to uproot undesirable mixes of components, for example, sulphur and nitrogen. Basic significant components of a hydro-treater unit are a warmer, a

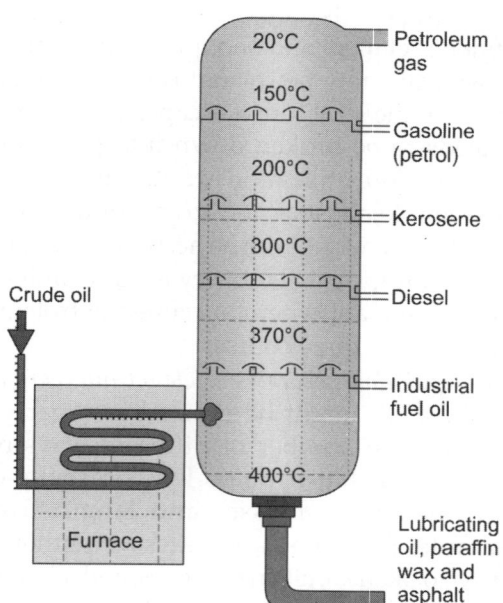

Fig. 3.5: Schematic of the distillation of crude oil

fixed bed catalytic reactor and a hydrogen compressor. The catalyst promotes the response of the hydrogen with the sulphur mixes, for example, mercaptans to transform hydrogen sulphide, which is then normally bled off and treated with amine in an amine treater. The hydrogen additionally soaked hydrocarbon double bonds which helps raise the strength of the fuel.

3.15.2 Catalytic Reforming

A catalytic changing procedure changes over a food stream holding paraffins, olefins and naphthenes into aromatics to be utilized either as an engine fuel mixing stock, or as a source for particular aromatic compounds, specifically benzene, toluene and xylene for utilization in petrochemicals production. The product stream of the reformer is by and large alluded to as a reformate. Reformate generated by this methodology has a high octane rating. Noteworthy amounts of hydrogen are likewise processed as side effect. Reactant improving is typically encouraged by a bifunctional catalyst that is fit for revising and breaking long-chain hydrocarbons and also evacuating hydrogen from naphthenes to transform aromatics. This methodology is not quite the same as steam changing which is additionally a reactant process that prepares hydrogen as the fundamental item.

3.15.3 Cracking

In an oil refinery cracking processes allow the production of light products (such as LPG and gasoline) from heavier crude oil distillation fractions (such as gas oils) and residues. Fluid catalytic cracking (FCC) produces a high yield of gasoline and LPG while hydrocracking is a major source of jet fuel, gasoline components and LPG. Thermal cracking is currently used to upgrade very heavy fractions or to produce light fractions or distillates, burner fuel and/or petroleum coke. Two extremes of the thermal cracking in terms of product range are represented by the high-temperature process called *steam cracking* or *pyrolysis* (750–900 °C

or more) which produces valuable ethylene and other feedstocks for the petrochemical industry, and the milder-temperature delayed coking (500 °C) which can produce, under the right conditions, valuable needle coke, a highly crystalline petroleum coke used in the production of electrodes for the steel and aluminum industries.

3.15.4 Fluid Catalytic Cracking

It is the most important conversion process used in petroleum refineries, earlier usage were dependent upon a low action alumina catalyst and a reactor where the catalyst particles were suspended in climbing stream of fluid hydrocarbon in a fluidized bed. In advanced plants, breaking happens utilizing an exceptionally dynamic zeolite-based catalyst in a vertical or upward sloped pipe called the "riser" for a short period. Preheated feed is showered into the base of the riser through feed spouts where it contacts amazingly hot fluidized catalyst at 665° to 760 °C. The hot catalyst vapourizes the feed and catalyzed the breaking responses that break down the high atomic weight oil into lighter parts including LPG, fuel, and diesel. The catalyst hydrocarbon mixture streams upward through the riser for simply a couple of seconds and after that the mixture is divided by means of violent winds. The catalyst free hydrocarbons are directed to a principle fractionator for partition into fuel gas, LPG, gasoline, light cycle oils utilized within diesel and plane fuel, and substantial fuel oil. The reactant splitting methodology includes the vicinity of corrosive catalyst (generally solid acids, for example, silica-alumina and zeolites) which advertises a heterolytic (asymmetric) breakage of securities yielding sets of particles of inverse charges, typically a carbocation and the exact unstable hydride anion.

During the tripup, the riser, and the cracking catalyst is "spent" by reactions which deposit coke on the catalyst and greatly reduce activity and selectivity. The "spent" catalyst is disengaged from the cracked hydrocarbon vapours and sent to a stripper where it is contacted with steam to remove hydrocarbons remaining in the catalyst pores. The "spent" catalyst then flows into a fluidized-bed regenerator where air (or in some cases air and oxygen) is used to burn off the coke to restore catalyst and also provide the necessary heat for the next reaction cycle, cracking being an endothermic reaction. The "regenerated" catalyst then flows to the base of the riser, repeating the cycle.

3.15.5 Hydrocracking

Hydrocracking is a catalyst breaking proce-dure supported by the presence of a raised partial weight of hydrogen. The results of this methodology are soaked hydrocarbons; contingent upon the response conditions (temperature, weight, catalyst movement) these items range from ethane, LPG to heavier hydrocarbons involving generally of isopraffins. Hydrocracking is regularly encouraged by a bifunctional catalyst that is fit for revising and breaking hydrocarbon chains and in addition adding hydrogen to aromatics and olefins to handle naphthenes and alkanes. Major products from hydro-cracking are plane fuel, diesel, moderately high octane rating gas parts and LPG. All these products have a quite low substance of sulphur and contaminants.

3.15.6 Steam Cracking

Steam cracking is a petrochemical process in which soaked hydrocarbons are broken down into more modest, regularly unsaturated, hydrocarbons. It is the central mechanical technique for processing the lighter alkenes (regularly olefins), including ethane (ethylene) and propene (propylene). In steam splitting, a vapourous or fluid hydrocarbon feed like naphtha, LPG or ethane is weakened with steam and after that quickly warmed in a heater (clearly without the presence of oxygen). Regularly, the response temperature is extremely hot; around 850 °C, however the response is just permitted to happen quite

quickly. In cutting edge breaking heaters, the living arrangement time is even decreased to milliseconds (bringing about gas speeds arriving at rates past the velocity of sound)/ so as to enhance the yield of desired products. After the splitting temperature has been arrived at, the gas is immediately extinguished to stop the response in an exchange line exchanger.

The products produced in the reaction depend on the composition of the feed, the hydrocarbon to steam ratio and on the cracking temperature and furnace residence time. Light hydrocarbon feeds (such as ethane, LPGs or light naphtha) give product streams rich in the lighter alkenes, including ethylene, propylene, and butadyene. Heavier hydrocarbon (full range and heavy naphtha as well as other refinery products) feeds give some of these, but also give products rich in aromatic hydro-carbons and hydrocarbons suitable for inclusion in gasoline or fuel oil. The higher cracking temperature (also referred to as severity) favours the production of ethane and benzene, where as lower severity produces relatively higher amounts of propene, C4-hydrocarbons and liquid products.

The thermal cracking process follows a hemolytic mechanism, that is, bonds break symmetrically and thus pairs of free radicals are formed. The main reactions that take place include:

Initiation reactions, where a single molecule breaks apart into two free radicals. Only a small fraction of the feed molecules actually undergo initiation, but these reactions are necessary to produce the free radicals that drive the rest of the reactions. In steam cracking, initiation usually involves breaking a chemical bond between two carbon atoms, rather then the bond between a carbon and a hydrogen atom.

$$CH_3CH_3 \rightarrow CH_3{}^\bullet + CH_3{}^\bullet$$

Hydrogen abstraction, where a free radical removes a hydrogen atom from another molecule, turning the second molecule into a free radical.

$$CH_3{}^\bullet + CH_3CH_3 \rightarrow CH_4 + CH_3CH_2{}^\bullet$$

Radical decomposition, where a free radical breaks apart into two molecules, one an alkene, the other a free radical. This is the process that results in the alkene products of steam cracking.

$$CH_3CH_2{}^\bullet + CH_2 = CH_2 \rightarrow CH_3CH_2CH_2CH_2{}^\bullet$$

Radical addition, the reverse of radical decomposition, in which a radical reacts with an alkene to form a single, larger free radical. These processes are involved in forming the aromatic products that result, when heavier feedstocks are used.

$$CH_3CH_2{}^\bullet + CH_2 = CH_2 \rightarrow CH_3CH_2CH_2CH_2{}^\bullet$$

Termination reactions, occur when two free radicals react with each other to produce products that are not free radicals. Two common forms of termination are recombination, where the two radicals combine to form one larger molecule, and disproportionation, where one radical transfers a hydrogen atom to the other, giving an alkene and an alkane.

$$CH_3{}^\bullet + CH_3CH_2{}^\bullet \rightarrow CH_3CH_2CH_3$$
$$CH_3CH_2{}^\bullet + CH_3CH_2{}^\bullet \rightarrow CH_2 = CH_2 + CH_3CH_3$$

The process also results in the slow deposition of coke, a form of carbon, on the reactor walls. This degrades the effectiveness of the reactor, so reaction conditions are designed to minimize this. Nonetheless, a steam cracking furnace can usually only run for a few months at a time between de-cokings.

3.15.7 Alkylation

Alkylation is the transfer of an alkyl group from one molecule to another. The alkyl group may be transferred as a alkyl carbocation, a free radical or a carbanion. In a standard oil refinery process, alkylation involves low-molecular-weight olefins (primarily a mixture of propylene and butylenes) with isobutene in the presence of a catalyst, either sulphuric acid or hydrofluoric acid. The product is called *alkylate* and is composed of a mixture of high-octane, branched-chain paraffin hydrocarbons. Alkylate is a premium gasoline blending stock because it has exceptional antiknock properties and is clean burning. Most crude oils contain only 10%–40% of their hydrocarbon constituents in the gasoline range, so refineries use cracking processes, which convert high molecular weight hydrocarbons into smaller and more volatile compounds. Polymeriation converts small gaseous olefins into liquid gasoline-size hydrocarbons. Alkylation processes transform small olefin and iso-paraffin molecules into larger iso-paraffins with a high octane number. Combination cracking, polymerization, and alkylation can result in a gasoline yield, representing 70% of the starting crude oil.

3.15.8 Isomerization

Isomerization is a process by which straight chain alkanes are converted to branched chain alkanes that can be blended in petrol to improve its octane rating (in presence of finely dispersed platinum on aluminium oxide catalyst).

3.16 PRODUCTION OF CRUDE OIL

Oil could be utilized as a part of such a large number of different ways on the grounds that it holds hydrocarbons of differing atomic masses and lengths, for example, paraffine, aromatics, naphthenes, dienes and alkynes. Hydrocarbons are atomic of fluctuating length and multifaceted nature made of hydrogen and carbon. The trap in the partition of diverse streams in oil refinement procedure is the distinction in breaking points between the hydrocarbons, which implies they might be divided by refining .The natural gas and crude petroleum store are typically under weight of the water head. The oil or common gases are discharged because of interior weight and no pumping is vital, if weight is sufficiently high. However, in the event that the weight is low, pumping of oil and natural gas get important.

3.16.1 Crude Gathering Station and Separation of Crude and Natural Gas

The crude-oil extracted by various production wells in the field is collected in a crude gathering station. The crude oil, natural gas and water are separated and the process is carried out in two or three stages with different range of temperature and pressure. First stage separation is carried out by high pressure (HP), where most of the natural gas is separated from the mixture in this stage. The remaining mixture is sent to a separator at atmospheric pressure. Water is removed by settling and the crude oil and natural gas from the separators, and are stored in separate tanks above ground.

3.17 TRANSPORTATION OF CRUDE OIL AND NATURAL GAS

Crude oil can be transported by the following alternative modes:

Land transport

Batch transport:

- Barrels
- Rail wagon
- Tankers
- Road tankers
- Trucks

Continuous transport: Pipeline (land, sea)

- Barges
- Tankers
- Super tankers

3.17.1 Electrical Transport

Conversion of chemical energy to electrical energy, then electrical transport.

3.17.2 Transport of Natural Gas and other Gaseous Hydrocarbons

Land transport: Pipeline gas, liquefied gas in cylinders, tankers, liquefied gas in road/rail tankers.

Ocean transport: Liquefied natural gas, tankers, liquefied gas in cyclinders.

Electrical transport: Conversion of chemical energy to electrical energy, then electrical transport.

3.18 STORAGE

Crude petroleum and liquid petroleum is stored in steel tanks or concrete tanks, above ground. Natural gas is stored in gaseous form in pressurized tanks. Alternatively, it is liquefied and stored in tanks above ground or underground. Low pressure water sealed cylindrical bell piston gas holders having storage volume of 5,00,000 to 7,00,000 cubic meter and dry type piston type gas holders are in use. High pressure gas holders are with pressure of 4 to 155 bar and with capacities 1000 to 15,000 cubic meter. Spherical gas holders of 30,000 cubic meter capacity are also in use.

3.19 NATURAL GAS

Natural gas is acknowledged to be the most eco-accommodating fuel dependent upon accessible data. Financially regular gas is more productive since just 10% of the produced gas wasted before utilization and it doesn't have to be created from different energizes. Additionally natural gas is utilized within its typical state. The natural gas is lighter and gets gathered in the upward arched depression in the cap rock. The unrefined petroleum stays at the bottom. Natural gas is not difficult to handle and helpful to utilize and vitality equal groundwork, it has been value price controlled below its competitor oil. Subsequently natural gas can substitute oil in both areas in particular powers (industry and domestic) and chemicals (compost petrochemicals and natural chemicals). Natural gas prospecting, investigation, penetrating are just about like those for petroleum. The significant distinction in extraction is: the natural gas stores can happen at greatly high weights (500 bar, 400 bar in a few cases) and must be permitted to develop discharge from the well.

3.20 NATURAL GAS OCCURRENCE AND PRODUCTION

Natural gas was formed from the remains of tiny sea animals and plants that died 200–400 million years ago. The ancient people of Greece, Persia, and India discovered natural gas many centuries ago. About 2,500 years ago, the Chinese recognized the natural gas could be put to work. The Chinese piped the gas from shallow wells and burn it under large pans to evaporate sea water from salt.

3.21 SOURCES OF NATURAL GAS

Natural gas can be hard to find since it can be trapped in porous rocks deep underground. However, various methods have been developed to find out natural gas deposits. The methods employed are as follows:

1. Looking at surface rocks to find dues about underground formations
2. Setting of small explosions or drop heavy weights on the surface and record the sound waves as they bounce back from the rock layers underground and
3. By measuring the gravitational rock masses deep within the Earth.

Coal beds and landfills are other sources of natural gas, however only 3% of the demand is achieved.

3.22 PHYSICAL PROPERTIES OF NATURAL GAS

Natural gas is a mixture of light hydrocarbons including methane, ethane, propane, butanes and pentanes. Different mixes found in characteristic gas incorporates CO_2, helium, hydrogen sulphide and nitrogen. The structure of natural gas is never consistent, be that as it may, the essential segment of common gas is methane (regularly no less than 90%). Methane is highly inflammable, burns easily and almost completely; and its less polluting. Natural gas is neither corrosive nor toxic, its ignition temperature is high, and it has a narrow flammability range, making it an inherently safe fossil fuel compared to other fuel sources. Like crude petroleum oil, the raw natural gas cannot be used in the form as obtained from the well. It must be refined to eliminate hydrogen sulphide, liquids and other unwanted components such as water, nitrogen, rare gases, and carbon dioxide. Thereafter, what remains are essentially methane (CH_4) and some heavier hydrocarbons like ethylene, propane, butane.

3.23 CLASSIFICATION OF NATURAL GAS

In terms of occurrence, natural gas is classified as non-associated gas, associated gas, dissolved gas and gas cap.

3.23.1 Non-associated Gas

There is non-associated natural gas which is found in reservoirs in which there is no or, at best, minimum amounts of crude oil. It is usually richer in methane but is markedly leaner in terms of the higher paraffinic hydrocarbons and condensate material. It could be kept underground as long as required.

3.23.2 Associated Gas

Natural gas found in crude oil reservoirs and produced dung, the production of crude oil is called *associated gas*. It exists as a free gas (gas cap) in contact with the crude petroleum and also as a 'dissolved natural gas' in the crude oil. Associated gas is usually leaner in methane than the non-associated gas but rich in higher molecular weight hydrocarbons. Non-associated gas can be produced at a higher pressure, whereas, associated gas (free or dissolved gas) must be separated from petroleum at lower separator pressures, which usually involves increased expenditure for compression.

3.24 NATURAL GAS PRODUCTS

Natural gas and/or its constituent hydrocarbons are marketed in the form of different products, such as lean natural gas, wet natural gas like compressed natural gas (CNG), natural gas liquids (NGL), liquefied petroleum gas (LPG), natural gasoline, natural gas condensate, ethane, propane, ethane-propane fraction and butanes.

3.25 NATURAL GAS REFINERY

The procedure is choosen on the support of structure of crude natural gas accessible from wells and final results of business significance possible from the methodology (Fig. 3.6). The crude natural gas, if initially dried with glycol, the glycol absorbs moisture from the crude characteristic gas. The glycol is recovered by vaccum drying and after that reused. Next, the heavy constituted gases are retained in poor oil under weight. The poor oil gets advanced by consumed gases. The gases are separated

and discharged in steps by expansion in every progressive rectifying section.

3.26 TRANSPORTATION

Natural gas arriving at the purchaser, closes regularly through pipeline which is typically made of steel funneling and measured between 20 and 42 inches. Since gas is transported at high weights, there are compressor stations along the pipeline with a specific end goal to administer the level of weight required. Contrasted with other vital sources, natural gas transportation is exceptionally effective in light of the fact that the allotment of vitality rundown from cause to goal is low.

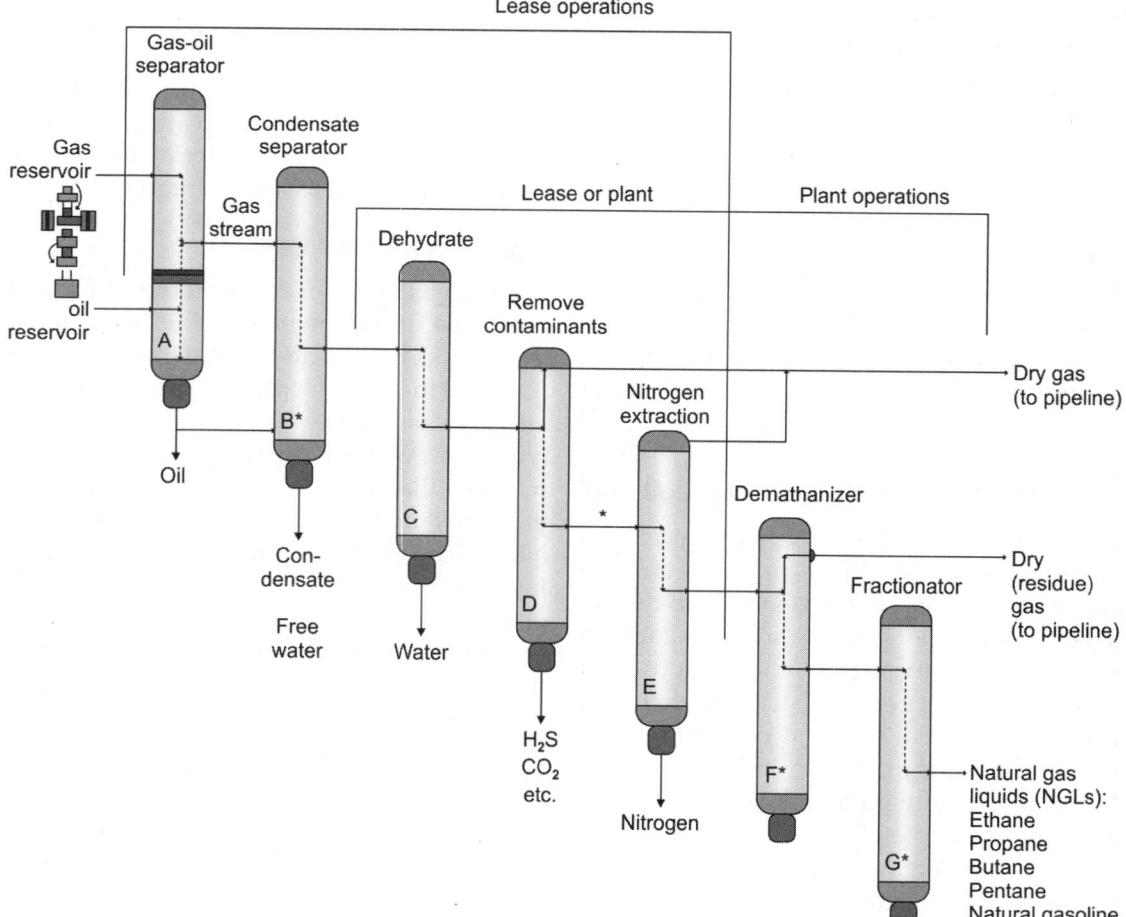

Fig. 3.6: Diagram of natural gas processing

3.26.1 Transported as LNG

Natural gas can also be transported by sea. In this case, it is transformed into liquefied natural gas (LNG). The liquefaction process retains oxygen, CO_2, sulphur compounds and water. A full LPG chain consists of a liquefaction plant, low temperature and pressurized transport ships and a regasification terminal.

3.27 SECTOR-WISE EXPLOITATION OF NATURAL GAS

3.27.1 Residential Usage

Natural gas is used in cooking, washing, drying, water warming and air-conditioning. Operating costs of natural gas equipment are generally lower than those of other energy usage.

3.27.2 Commercial Use

The flow diagram for commercial use is shown in Fig. 3.7.

3.27.3 Industrial Utilization of Natural Gas

Production of mash and paper, metals, chemicals, stone, dirt, glass and to process certain sustenance, different fields in which common

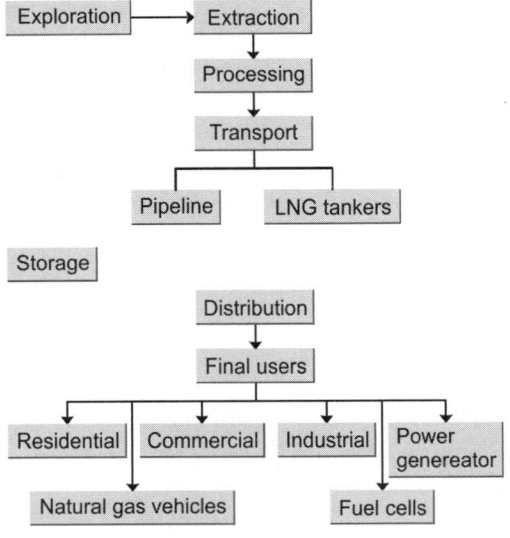

Fig. 3.7: Utilisation of natural gas

gas is successfully used. Gas is additionally used to treat waste materials, for incineration, drying, dehumidification, warming and cooling, and CO generation. It is additionally a suitable substance feedstock for the petrochemical industry. Natural gas has an incalculable number of mechanical utilization, including giving the base fixings to such shifted items as plastic, compost, liquid catalyst, and fabrics. Infact, industry is the biggest customer of natural gas.

3.28 POWER GENERATION

Natural gas works more efficiently and emits less pollution than other fossil fuel power plants. Due to economic, environmental, and technological changes, natural gas has become the fuel of choice for new power plants. Steam generation units, centralized gas turbines, microturbines, combined cycle limits and distributed generation are the other examples where natural gas is utilized.

3.29 TRANSPORTATION

Natural gas can be used as a motor vehicle fuel in two ways: as compressed natural gas (CNG), which is the most common form, and as liquefied natural gas (LNG). Cars using natural gas are estimated to emit 20% less greenhouse gases than gas alone or diesel cars. In many countries NGVs are introduced to replace buses, taxis and other public vehicle fleets. Most natural gas vehicles operate using compressed natural gas (CNG).

3.30 FUEL CELLS

Natural gas is one of the multiple fuels on which fuel cells can operate. Fuel cells became an increasingly important technology for the generation of electricity. They are like rechargeable batteries, except using an electric recharger; they use a fuel, such as natural gas, to generate electric power even when they are in use. Fuel cells for distributed generation systems offer a multitude of benefits and are an exciting area of innovation and research for distributed generation applications. One of the major technological innovations with regard to electric generation, whether distributed or centralized, is the use of combined heat and power (CHP) systems. These systems make use of heat that is normally wasted in the electric generation process, thereby increasing the energy efficiency of the total system.

3.31 IMPORT OF NATURAL GAS TO INDIA THROUGH TRANSNATIONAL GAS PIPELINES

Iran–Pakistan–India (IPI) pipeline project.
Myanmar–Bangladesh–India gas pipeline project.
Turkmenistan–Afghanistan–Pakistan (TAP) pipeline.

3.32 LIQUEFIED NATURAL GAS

Natural gas at 161°C transforms into liquid. This is done for easy storage and transportation since it reduces the volume occupied by gas by a factor of 600. LNG is transported in specially built ships with cryogenic tanks. It is received at the LNG. Deceiving terminals and is degasified to be supplied as natural gas to the consumers. Dedicated gas field development and production, liquefaction plants, transportation in special vessels, *regassification* plant and transportation and distribution to the gas consumer are various steps involved in the production and distribution of LNG.

3.33 NATURAL GAS AND THE ENVIRONMENT

All the fossil fuels, coals, petroleum, and natural gas release pollutants into the atmosphere, when burnt to provide the energy we need. The list of pollutants they release are carbon dioxide, carbon monoxides, reactive hydrocarbons, nitrogen oxides, sulphur oxides, and solid particulates (ash or soot). The good news is that natural gas is the most environmental friendly fossil fuel. It is a cleaner-burning fuel than coal or petroleum because it contains less carbon than its fossil fuel cousins. Natural gas also has less sulphur and nitrogen compounds and it emits less ash particulates into the air when it is burnt than coal or petroleum fuels.

Section 3

Renewable Energy Resources

4

Wind Energy

4.1 INTRODUCTION

Because of the present day energy demand and developing ecological awareness, there is a need to supplement energy demand with clean and renewable sources of energy. Humanity has utilized the wind as a source of energy for many years. Together with hydro-power wind was the most used energy hotspot for quite some time. With the improvement of electrical designing and interest for power, towards the conclusion of the nineteenth century the first tests were completed on use of windmills for creating power. In the conti-nuous search of clean, safe and renewable energy sources, wind power, certainly is the most attractive solution. Wind power was used earlier for several centuries for propelling ships, driving wind mills, pumping water, irrigating fields and numerous other purposes. Wind is one of the potential renewable energy sources which can be harnessed in a commercial way and a detailed knowledge of wind characteristics is required for efficient planning and implementation of any wind engineering project. Wind blows due to the warming and cooling of the Earth's atmos-phere and the changes in temperature. In extreme cases the wind energy can be destruc-_tive in nature. Fortunately, most regions of the world experience moderate range of wind speeds that can allow human to extract energy from the wind. Wind results from air in motion. Air in motion arises from a pressure gradient. On a global basis one primary terracing function causing surface wind from the poles towards the equator is corrective circulation. Solar radiation heats the air near the equator, and this low density heated air is bayed up. At the surface it is displaced by cooler, more dense higher pressure air flowing from the poles. In the upper atmosphere near the equator the air thus tend to flow back towards the poles and away from the equator. The net result is a global corrective circulation with surface winds from North to South in the northern hemisphere.

The wind is basically caused by the solar energy predicting the Earth that is why wind utilization is considered a part of solar techno-logy. A large wind farm may consist of several hundred individual wind turbines which are connected to the electric power transmission network. Off-shore wind farms can harness more frequent and powerful winds than are available to land based installations and have less visual impact on the landscape but cons-truction costs are considerably higher. Small on-shore wind facilities are used to provide electricity to isolated locations and utility companies increasingly produced by small domestic wind turbines. Wind energy is consi-dered to be a very clean, cheap important renewable energy source particularly for rural areas, farms, remote on-shore and off-shore installations away from main electrical grid.

The development of wind power in India began in 1990s, and has significantly increased in the last few years. Although a relative newcomer to the wind industry compared with Denmark or the United

51

States, India is the fifth largest installed wind power station in the world. In 2009–10 station India's growth rate was highest among the other top 10 countries. As of 31 March, 2011 the installed capacity wind power in India was 16078 MW, mainly spread across Tamil Nadu (6007 MW), Maharashtra (2310.70 MW), Gujarat (2175.60 MW), Karnataka (1730.10 MW), Rajasthan (1524.70 MW), Madhya Pradesh 275.50 MW), Andhra Pradesh (200.20 MW), Kerala (32.8 MW), Odisha 2 MW, (West Bengal (1.1 MW) and other states (3.20 MW). It is estimated that 600 MW of additional wind power capacity will be installed in India by 2012.

The wind power industry is one of the fastest expanding industries as a result of the rapid growth of installed capacity. The wind power over the last 20–30 years has become a competitive technology for clean energy production. There is no reason why wind power should not become as important to the world's future energy supply as nuclear power is today. The question that has to be asked is how wind power will affect the whole electrical grid, in particular the distribution network to which it is usually connected to. Now-a-days, wind power is a fully established branch of the electricity market. When making decisions about new wind turbines placement, the energy gain is not the only criteria to be considered during the planning phase; cost efficiency, the impact on the environment and the impact on the electric grid are some of the most important issues.

Unit ratings of wind turbine generators cover a wide range from 0.5 kW to 14 kW. The broad classification is as follows:

1. Small: 1 to 15 kW
2. Medium: 15 to 200 kW
3. Large: 250 to 1000 kW
4. Very Large: 1000 kW to 6000 kW

Wind energy is a manifestation of the solar energy. Wind is the air in motion. Energy in the wind is converted into rotary mechanical energy by the wind turbine. The rotary mechanical energy is used for several applications such as:

• Pumping water
• Grinding flour
• Driving generator rotors to produce electrical energy.

Energy chains of wind energy are:

i. Wind energy → Mechanical energy at wind turbine shaft → Mechanical energy utilization

ii. Wind energy → Mechanical energy at wind turbine shaft → Electrical energy by generator → Electrical energy utilization

Conversion of the kinetic energy (i.e. energy of motion) of the wind into mechanical energy that can be utilized to perform useful work, or to generate electricity. Most machines for converting wind energy into mechanical energy consist basically of a number of sails, vanes, or blades radiating from a hub or central axis. Several types of wind turbines have been installed and are being operated successfully. These are broadly classified into two main categories, *horizontal shaft wind turbine* and *vertical shaft wind turbine*. The horizontal-axis turbine has a three-blade vertical propeller that catches the wind face-on. The generator–turbine unit is mounted on a tall tower. The vertical turbine has a set of blades that spins around a vertical axis and the units are mounted on ground level. Wind energy conversion devices are commonly known as wind turbines because they convert the energy of the windstream into energy of rotation and the component which rotates is called the rotor. Wind energy offers another source for pumping as well as electric power generation. India has a potential to generate over 20,000 MW power and ranks as one of the promising countries for tapping this source.

4.2. WIND FARMS

A wind farm is a collection of windmills or turbines which are used to generate electrical power through their mechanical motions as they are pushed by the wind. The energy generated by a wind farm can be fed directly into the general energy grid after passing through transformers. As a potentially large source of renewable energy, wind farms are particularly popular in nations which are focusing on alternative energy. Other types of renewable energy include wave power and solar arrays. All of these technologies take advantage of already existing energy, converting it into a usable form. Since a wind farm does not actively deplete resources as it generates power, it is considered a form of "green" energy. As a general rule, economic wind generators require windspeed of 16 km/h (10 mph) or greater. An ideal location would have a near constant flow of non-turbulent wind throughout the year, with a minimum likelihood of sudden powerful bursts of wind.

4.3 USES AND HISTORICAL BACKGROUND

Wind energy has been in use for several centuries in the past for:

1. Agricultural and rural applications such as grinding flour mills, wood cutting saw, stone crushes, mixers, water pumps, irrigation facility etc.
2. Ocean transport by sails for ships. An application of wind energy for producing electrical energy was introduced first in 1985. Several units were installed in Europe during early part of 20th century. Wind turbine generators have been developed on commercial scale and have received more importance after 1980.

The wind has played a long and important role in the history of human civilization. The first known use of wind dates back 5,000 years to egypt, where boats used sails to travel from shore to shore. The first true windmill, a machine with vanes attached to an axis to produce circular motion, may have been built as early as 2000 B.C. in ancient Babylon. By the 10th century A.D., windmills with wind-catching surfaces as long as 16 feet and as high as 30 feet were grinding grain in the area now known as eastern Iran and Afghanistan. Since

early recorded history, people have been harnessing the energy of the wind. Wind energy propelled boats along the Nile River as early as 5000 B.C. By 200 B.C., simple windmills in China were pumping water, while vertical-axis windmills with woven reed sails were grinding grain in Persia and the Middle East. In the 1930s and 1940s, hundreds of thousands of electricity producing wind turbines were built in the US. They had two or three thin blades which rotated at high speeds to drive electrical generators. These wind turbines provided electricity to farms beyond the reach of power lines and were typically used to charge storage batteries, operate radio receivers and power a light bulb or two. By the early 1950s, however, the extension of the central power grid to nearly every American household, via the Rural Electrification Administration, eliminated the market for these machines. Wind turbine development lay nearly dormant for the next 20 years.

The wind-turbine generator units have become commercially successful after 1988, and are being accepted for rural, remote and other suitable areas as important source of electrical power supply. The wind-turbine, gears, generator together form a wind-turbine generator unit. A wind farm has several such units which operate in parallel electrical and feed electrical energy to isolated load or the electrical grid. The first propeller type wind mill to drive electrical generator was built by Prof. P. La Cour of Denmark in 1985.It consisted of a four-blade propeller driving a gear-train and shaft. First large wind turbine generator unit (rated 1.25 MW) was installed in USA, around 1941 and operated for four years successfully. After 1973 oil crisis, the interest in wind turbine generator was revived all over the world.

4.4 WIND DIRECTION AND SPEED

To work properly, the horizontal-axis turbine needs the wind to flow at a right angle to the blades. If it blows from a different direction than the blades are facing, the turbine gets much less energy from the wind. To accommodate changes in wind direction, the turbine has a yaw drive that rotates the unit's direction. However, the drive adapts slowly to changing directions because it must turn the entire turbine and propeller assembly. By contrast, a vertical turbine runs well regardless of wind direction, making it better-suited to urban areas with tall buildings where wind turbulence is a given. The vertical-axis design allows it to operate on lower wind speeds than is possible with the horizontal turbine.

4.5 BASIC PRINCIPLES OF WIND ENERGY CONVERSION

4.5.1 Nature of the Wind

The circulation of air in the atmosphere is caused by the non-uniform heating of the Earth's surface by the sun. The air immediately above a warm area expands; it is forced upwards by cool, denser air which blows in from surrounding areas causing a wind. The nature of the terrain, the degree of cloud cover and the angle of the Sun in the sky are all factors which influence this process. In general, during the day the air above the land mass tends to heat up more rapidly than the air over water. In coastal regions this manifests itself in a strong on shore wind. At night the process is reversed because the air cools down more rapidly over the land and the breeze therefore blows off-shore. Flow or air (wind) results in flow of mass. Flowing mass has kinetic energy (KE). Hence wind is a natural source of kinetic energy (KE). Energy in the wind can be converted into useful mechanical energy by means of wind turbine, wind-mill, sails of ships etc.

Winds are natural phenomena in the atmosphere and has two different origins:

1. Planetary winds are brought about by day by day rotation of the Earth around its polar axis and unequal temperatures between polar regions and tropical areas (Fig. 4.1).

The primary planetary winds are brought about similarly. Cool surface air encompasses down from the shafts constraining the warm air over the tropics to ascent yet the direction of these monstrous air developments is influenced by the revolution of the Earth and the net impact is a huge clockwise flow of winds in the northern half of the globe, and clockwise course in the southern side of the equator. The quality and heading of these planetary winds change with the seasons due to sun's orientation.

2. Local winds are brought on by unequal warming and cooling of ground surfaces and ocean/lake surfaces throughout day and night (Fig. 4.2). Average wind velocities are more terrific in rocky and coastal zones than they are inland. The winds additionally have a tendency to blow more constantly and with more amazing quality over the

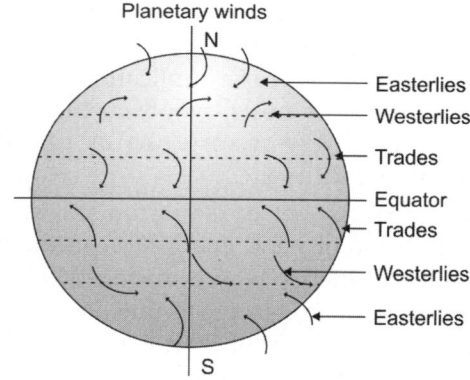

Fig. 4.1: Planetary winds

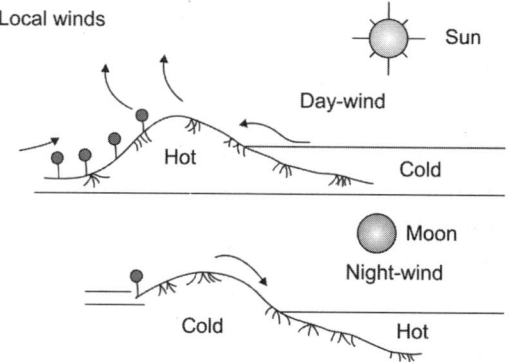

Fig. 4.2: Local day and night winds

surface of the water, where there is a less surface drag. Wind rates increases with tallness. They have customarily been the measure information standard tallness of ten meters where they are found to be 20–25% greater than near the surfaces. At a stature of 60 m they may be 30–60% higher in view of the decrease in the dray impact of the world's surface.

4.5.2 The Power in the Wind

Wind is the movement of air from an area of high pressure to an area of low pressure. In fact, wind exists because the Sun unevenly heats the surface of the Earth. As hot air rises, cooler air moves in to fill the void. As long as the sun shines, the wind will blow. And as long as the wind blows, people will harness it to power their lives. Wind possesses energy by virtue of its motion. Any device capable of slowing down the mass of moving air, like a sail or propeller, can extract part of the energy and convert it into useful work. Three factors determine the output from a wind energy converter

i. The wind speed
ii. The cross-section of wind swept by rotor; and
iii. The overall conversion efficiency of the rotors, transmission system and generator or pump.

The power in the wind can be computed by using the concept of kinetics. The wind mill works on the principle of converting kinetic energy of the wind into mechanical energy. We know that power is equal to energy per unit time. The energy available is the kinetic energy of the wind.

4.6 FORCES ON THE BLADES AND THRUST ON TURBINES

There are two types of forces which are acting on the blades of a propeller type wind turbine. One is *circumferential force* acting in the direction of wheel rotation that provides the torque and other is the *axial force* acting in the direction of the wind stream that provides an axial thrust that must be counteracted by proper mechanical design.

4.7 WIND VELOCITIES AND HEIGHT FROM GROUND AND SITE-SELECTION

4.7.1 Site Selection Process

The power available in the wind increases rapidly with the speed hence wind energy conversion machines should be located especially in those areas where the winds are strong and persistent. Site selection for large wind turbine is an important matter which requires consideration of a comprehensive set of factors and balancing of multiple objectives in determining the suitability of a particular area for a defined land use. The selection of suitable project area involve a complex array of critical factors drawing from physical, demographical, economic policies, and environmental disciplines. Location plays a vital part in the performance and efficiency of a wind turbine, so get it wrong and it could be disastrous but get it right than turbine will have a long, happy and profitable future together.

The following guidelines have been designed to install a wind turbine:

• Turbines work at the best when on high, exposed sites. Coastal sites are especially good.
• Town centers and highly populated residential areas are usually not suitable sites for wind turbines.
• Avoid roof mounted turbines as there is no guarantee that these devices will not damage your property through vibration.
• The further the distance between turbine and power requirement, the more power will lose in the cable.
• Turbulence disrupts the air flow which can wear down the blades and reduce the lifecycle of the turbine, i.e. during the installation of a wind turbine the distance between the turbine and the nearest obstacle is more than twice the height of the turbine, or the height of the turbine is more than twice the height of the nearest obstacle.

- Smaller turbines require an average wind speed of over 4.5 m/s to produce an efficient level of electricity.
- Require planning permission before you can install your wind turbine.
- If the location is next to a listed building, in a conservation area, or in an area of natural beauty, might experience difficulties when trying to obtain planning permission, so contact your local authority at the very early stages of project to avoid disappointment.
- Wind is the fuel that drives a wind turbine. A windmill needs to be placed where the wind is, i.e. putting it on too short, a tower is like installing solar photovoltaic panels in the shade. Neither will work very well. Not just any wind will do, a wind turbine needs air that moves uniformly in the same direction. The rotor cannot extract energy from turbulent wind, and the constantly changing wind direction due to turbulence causes excessive wear and premature failure of your turbine. This means that turbine must be placed high enough to catch strong winds, and above turbulent air. Since the tower price goes up quickly with height there is a limit to what is practical and affordable.

4.8 WIND ENERGY CONVERSION SYSTEM

The natural wind, as an energy source, is extremely variable. In order to predict the energy output of a wind energy conversion system (WECS), there has been considerable interest in finding a suitable statistical model of wind speed frequency distribution in the last few years. Traditional windmills were used extensively in the Middle Ages to mill grain and lift water for land drainage and watering cattle. Wind energy converters are still used for these purposes today in some parts of the world, but the main focus of attention now lies with their use to generate electricity. Differential heating of the Earth's surface by the sun causes the movement of large air masses on the surface of the Earth, i.e. the wind. Wind energy conversion systems convert the kinetic energy of the wind into electricity or other forms of energy. Wind power generation has experienced a tremendous growth in the past decade, and has been recognized as an environmentally friendly and economically competitive means of electric power generation. More and more countries are ratifying the 1997 Kyoto protocol, and wind power has become one of the most effective ways to reach its goals. The Kyoto protocol sets targets for participating countries to reduce greenhouse gas emissions to at least 5% below the 1990 level in the commitment period of 2008 to 2012. According to the US energy information administration, world electricity consumption will increase from 12,833 TWh in 1999 to 22,230 TWh in 2020, mainly driven by developing countries, where two billion people are still without access to electricity. The fuel mix for the world's electricity generation in 1999, as presented in Fig. 4.3, indicates that fossil fuels accounted for 62% while renewable sources including hydropower, wind and solar etc. accounted for 20.2%. The term wind mill is still widely used to describe wind energy conversion systems. Wind energy conversion systems are more correctly referred to as WECS, aerogenerators, wind turbine generators, or simply wind turbines. By addition of the electrical generator and other blocks, the wind energy is converted into electrical energy and is used locally for various purposes.

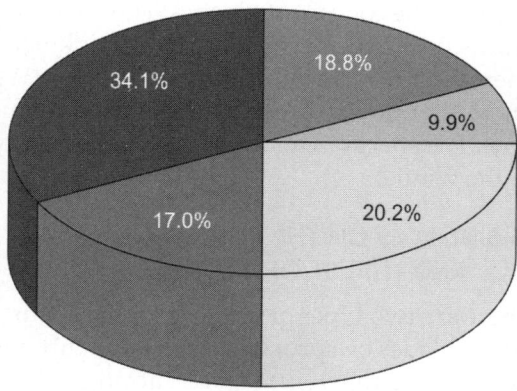

■Coal ■Natural gas ■Nucleas ■Oil □Renewables

Fig. 4.3: Fuel mix for world electricity generation in 1999

4.8.1 Lift and Drag

The basis for wind energy conversion is lift and drag forces. The extraction of power, and hence energy, from the wind depends on creating certain forces and applying them to rotate a mechanism (Fig. 4.4). There are two primary mechanisms for producing forces from the wind.

Drag forces, in the direction of the airflow, and lift forces, perpendicular to the airflow. Either or both of these can be used to generate the forces needed to rotate the blades of a wind turbine. Lift forces are produced by changing the velocity of the air stream flowing over either side of the lifting surface. In other words, any change in velocity generates a pressure difference across the lifting surface. This pressure difference produces a force that begins to act on the high pressure side and moves towards the low pressure side of the lifting surface which is called an airfoil. A good airfoil has a high lift/drag ratio, in some cases it can generate lift forces perpendicular to the air stream direction that are 30 times as great as the drag force parallel to the flow. For efficient operation, a wind turbine blade needs to function with as much lift and as little drag as possible because drag dissipates energy.

4.8.2. Drag-based Wind Turbine

In drag-based wind turbines, the force of the wind pushes against a surface, like an open

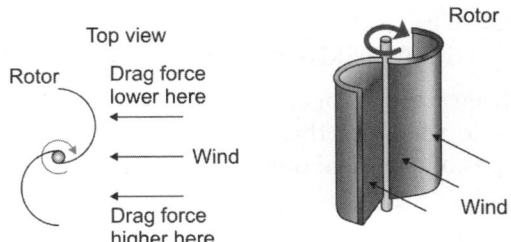

Fig. 4.5: Drag-based wind turbine concept; rotation created by difference in drag forces on the convex and concave surfaces of the rotor

sail. (Fig. 4.5). In fact, the earliest wind turbines, dating back to ancient Persia, used this approach. The Savonius rotor is a simple drag-based windmill that you can make at home. It works because the drag of the open, or concave, face of the cylinder is greater than the drag on closed or convex section.

4.8.3 Lift-based Wind Turbines

More energy can be extracted from wind using lift rather than drag, but this requires specially shaped airfoil surfaces, like those used on airplane wings. The airfoil shape is designed to create a differential pressure between the upper and lower surfaces, leading to a net force in the direction perpendicular to the wind direction. Rotors of this type must be carefully oriented (the orientation is referred to as the rotor pitch), to maintain their ability to harness the power of the wind as wind speed changes. (Fig. 4.6).

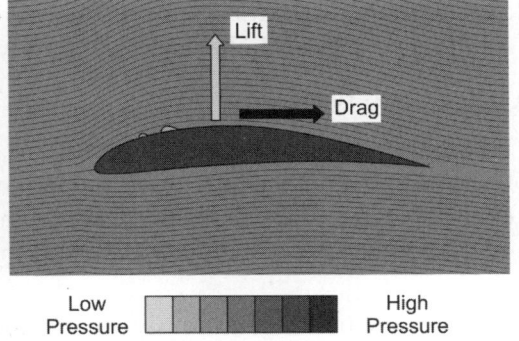

Fig. 4.4: Lift and drag force; flow field around airfoil with pressure field (colours) and aerodynamic forces

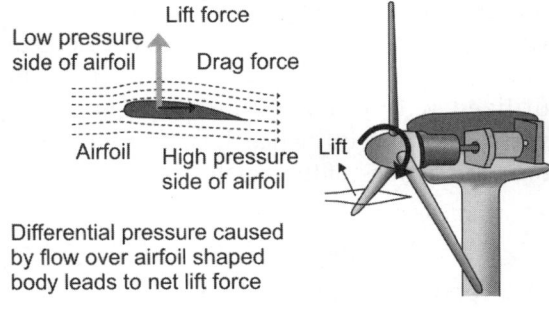

Fig. 4.6: Lift-based wind turbine concept

4.9 WIND TO ELECTRICAL ENERGY CONVERSION ALTERNATIVES

Three basic aspects should be considered while selecting the system. They are three types of electrical output:

- DC
- Variable frequency AC
- Constant frequency AC

Wind turbine rotor speed—Features:

- Constant speed, with pitch control and gears
- Nearly constant speed with simpler pitch control
- Variable speed with fixed pitch blades utilization aspects
- Stand alone or grid connected load
- Energy storage requirements for unfavorable wind periods

4.9.1 Wind Data and Energy Estimation

The wind speed data are by and large accessible in time arrangement position, in which every data represents an immediate specimen wind speed or an average of wind rate taken at short interval of time. The regular and also quick changes in winds both as to magnitude and direction need to be well comprehended to make the best utilization of them in wind plant outlines. Winds are known to vacillate by an element of 2 or all the more within seconds and therefore bringing on the ability to vary by a variable of 8 or more. The information on wind conducts is gathered by different courses relying upon the utilization it is expected to be put into. The hardly mean wind speed as gathered by the meteorological perceptions is the essential information utilized within a windmill designes. The variables which influence the way of the wind near the surface of the Earth, are:

   i. Latitude of the place

  ii. Altitude of the place

 iii. Topography of the place

 iv. Scale of the hours, month or year

Large-scale wind turbines have already proved themselves as cost competitive electricity generators in locations where wind resource is good enough. Winds' being an unsteady phenomenon, the scale of the periods considered is an important set of data required in the design. The hourly mean velocity provides the data for establishing the potential of the place for trapping the wind energy. The scale of the month is useful to indicate whether it is going to be useful during particular periods of the year on storage if necessary is to be provided for. The location, height above ground level at which wind is measured and the nature of the surface on Earth have an influence on the velocity of wind at any given time.

4.9.2 Wind Surveys

Typical wind measurements at potential sites for wind machines usually require the following:

1. **Instrumentation:** It consists of 3 cup anemometer and wind direction sensors. Height of instruments; 10 m (3 feet) for preliminary data; 15–45 m (50–150 feet) for long time data.

2. **Data recording systems:** It contains two part, strip chart and magnetic tap.

3. **Type of data:** Can be determined by wind speed and directional hourly averages.

4. **Data reporting:** Can be done by wind frequency curves, daily, weekly, monthly.

4.9.3 Energy Estimation

The quantity of energy that can be captured by a wind turbine depends upon the power versus wind speed characteristic of the turbine and the wind speed distribution at the turbine site.

The basic wind data of hourly mean wind velocity is recost into number of hours in the year for which the speed equals or exceeds each particular value.

4.10 BASIC COMPONENTS OF A WECS (WIND ENERGY CONVERSION SYSTEM)

Greenhouse gas reduction has been one of the crucial and inevitable global challenges, especially for the last two decades as more evidences on global warming have been reported. This has drawn increasing attention to renewable energies including wind energy, which is regarded as a relatively mature technology. It recorded 159 GW for the total wind energy capacities in 2009, which is the highest capacity among the existing renewable energy sources with excluding large-scale hydro power generators. Also, its annual installation growth rate marked 31.7% in 2009 with its growth rate having been increasing for the last few years, which indicates that wind energy is considered one of the fastest growing and attractive renewable energy source. The increasing price-competitiveness of wind energy against other conventional fossil fuel energy sources such as coal and natural gas is another positive indication on wind energy. Therefore, a vast amount of researches on WECS have been and is being undertaken intensively. The main components of a WECS are shown in Fig. 4.7. WECS consists of three major aspects; aerodynamic, mechanical and electrical. The electrical aspect of WECS can further be divided into three main components, which are wind turbine generators (WTGs), power electronic converters (PECs) and the utility grid. The main components of a WECS are shown in Fig. 4.7. Summary of the system operation is as follows, aeroturbines convert energy in moving air to rotary mechanical energy. Pitch control and yaw control are require only in case of horizontal or wind axis machines for proper operation.

4.10.1 Yaw Control

Yaw is the angle of rotation of the nacelle around its vertical axis. Efficient yaw control is essential to ensure that wind turbines always face directly into the wind. The rotor can be in a fixed orientation with the swept area perpendicular to the predominant wind direction. Such a machine is said to be yaw fixed. Most wind turbines, however, are yaw active, that is to say, as the wind direction changes, a motor rotates the turbine slowly about the vertical axis so as to face the blades into the wind. In the small turbines, yaw action is controlled by a tail vane, similar to that in a typical pumping windmill. The purpose of the controller is to sense wind speed, wind direction, shafts speeds and protect the system from extreme conditions brought upon by strong winds electrical faults and the like.

The sub-component of wind mill are:
1. Wind turbine or rotors
2. Wind mill head
3. Transmission and control
4. Supporting structure

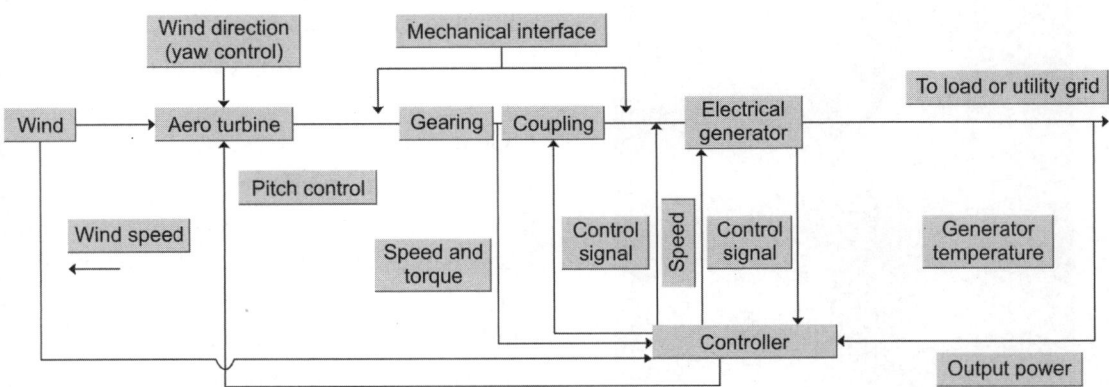

Fig. 4.7: The main components of a WECS

4.10.2 Rotors

Rotors are mainly of two types:
  i. Horizontal axis rotor and
  ii. Vertical axis rotor

One advantage of vertical axis machines is that they operate in all wind directions and thus need no yaw adjustment. The rotor is only one of the important components. For an effective utilization, all the components needs to be properly designed and matched with the rest of the components.

4.10.3 The Windmill Head Supports

The windmill head supports the rotor, housing the rotor bearings (Fig. 4.8). It also control the pitch of the blades for safety devices and tail vane to orient the rotor to face the wind.

4.10.4 Transmissions

The rate of rotation of large wind turbine generators operating at rated capacity or below is conveniently controlled by varying the pitch of the rotor blades, but it is low, about 40 to 50 revolutions per minute. Because optimum generator output requires much greater rates of rotation, such as 1800 rpm, it is necessary to increase greatly the low rotor rate of turning.

4.10.5 Generator

The generator of choice is the synchronous unit for large aerogenerator systems because it is very versatile and has an extensive data base.

Fig. 4.8: The windmill head

4.10.6 Controls

The modern large wind turbine generator requires a reliable control system to perform the following functions:

1. The orientation of the rotor into the wind
2. Startup and cut-in of the instrument
3. Generator output monitoring
4. Protection for the generator

It involve the following components for the proper functioning of the equipment like sensor, decision elements (relays, logic modules, microprocessor) and actuators (electric and pneumatic).

4.10.7 Towers

Four types of supporting towers are:

1. The reinforced concrete tower
2. The pole tower
3. The built up shell tube tower
4. The truss tower

Out of these four types of towers, the truss tower is favoured because widely adaptable, cost is low, parts are readily available. The minimum tower height for a small WECS is about 10 m and the maximum height is estimated to be roughly 60 m.

4.11 TYPES OF WIND TURBINE–GENERATOR UNITS

A wind turbine generator unit consists of the following major parts:

1. A wind turbine with vertical or horizontal axis
2. Gear chain
3. An electrical generator
4. Associated civil works, electrical and mechanical auxiliaries, control panels etc.

The wind turbine generator units convert wind power into electrical power.

  i. Darrius wind turbine with configuration.
  ii. VAWT with H configuration

4.12 CLASSIFICATION OF WEC SYSTEMS

WEC systems are classified into two categories:

4.12.1 Horizontal Axis Machines

Horizontal axis machines are manufactured very widely. The axis of rotation is horizontal and the aeroturbine plane is vertical facing the winds are manufactured very widely, the axis of rotation is horizontal and the aero-turbine plane is vertical, facing the wind the three blade version is the most essential, all over the world for unit ratings from 15 kW to 3 MW. Figure 4.9 shows the horizontal axis wind turbine with three blades. Horizontal axis wind turbines (HAWT) have their axis of revolution even to the ground and just about parallel to the wind stream. Most of the commercial wind turbines fall to this class. Horizontal axis machines have some unique preferences, for example, low cut-in wind pace and simple rolling. All in all, they indicate generally high power coefficient. However, the generator and gearbox of these turbines are to be set over the tower which makes its plan more unpredictable and expensive. An alternate impediment is the requirement for the tail or yaw head to turn the turbine towards wind.

The rotor of the horizontal axis wind turbine turns around a horizontal axis, and

Fig. 4.9: Horizontal axis wind turbine with three blades

throughout meeting expectations, the turning plane is vertical to the wind direction. The cutting edges of the wind turbine are introduced perpendicularly to the rotating axis, and structure a certain heavenly attendant. The amount of the sharpened pieces of steels depends upon the capacity of the wind turbine. The wind turbine with additional sharpened pieces of steels is frequently called as the low-speed wind turbine, and when it works at low speed, it will pick up a high proportion of use of the wind and with a high torque. The wind turbine with fewer blades is frequently named as the high-speed wind Turbine, and when it works at high speed , it will pick up a high proportion of use of the wind, yet the beginning wind speed should to be high. Due to its fewer sharpened pieces of steels (blades), in the same state of force, the rotor of the low-speed is much lighter, so it is suitable to create power.

Contingent upon the diverse relative position of the rotor and tower, the horizontal axis wind turbine might be isolated into the upwind wind turbine and the downwind wind turbine. The rotor turns before the tower confronting the wind is regarded as the *upwind wind turbine* and the rotor is introduced on the tower accompanying the wind known to be the downwind wind turbine. The upwind wind turbine must have a certain sort of directing establishment to guarantee the rotor to face the wind throughout working, though the *downwind wind turbine* can confront the wind immediately to spare the inconvenience of introducing the controlling apparatus. The horizontal axis wind turbine can likewise be separated into the lift type wind turbine and the resistance type wind turbine, and the previous has a high pivoting rate, while the last has a low turning velocity. As to create power with wind, the lift-sort wind turbine is all the more habitually received. A large portion of the horizontal axis has the directing unit, and can turn with the wind. To the little estimated wind turbine, the directing mechanism utilizes the tail vane, while to the large sized wind turbine, it frequently embraces the apparatus comprising of wind sensors and servo engine.

There are many types of the horizontal axis wind turbine:

1. Rotor with reversal blades
2. Several rotors installed on one tower to reduce the cost of tower on the condition of a certain output power
3. Tapered hood to ensure that when the gas flow goes through the horizontal axis in a centralized or decentralized way in order to increase or decrease the speed
4. Some with whirlpool shaped by the horizontal axis wind turbine around the rotor to centralize the wind and raise the speed of the gas flow.

The innovation of the horizontal axis wind turbine is more developed, and it is not difficult to process high-power wind turbines, yet the structure is complicated. The degree of the use of the wind force is not so high as the vertical axis wind turbines. Not just must be distinct parts, for example, the rotor, transmission, generator, and tower, should be as effective as could reasonably be expected, however these segments must work successfully in combination. A portion of the principle outline contemplations will be acknowledged later, Fig. 4.10 shows the types of horizontal axis wind turbines. Contingent upon the amount of cutting edges, level hub wind turbines are further considered single bladed, two bladed, three bladed and multi bladed. Single bladed turbines are shabbier because of investment funds on sharpened

steel materials. The drag losses are likewise least for these turbines. On the other hand, to adjust the sharpened steel, a counter weight must be set inverse to the center point. Single bladed outlines are not extremely well known because of issues in equalizing and visual adequacy. Two bladed rotors additionally have these impediments, however to a lesser degree. The majority of the present business turbines utilized for power generation have three blades.

4.12.1.1 Horizontal Axis Using Two Aerodynamic Blades

In this type of design, rotor drives a generator through a step up gear box. The blade rotor is usually designed to be oriented downwind of the tower. The components are mounted on a bed plate which is attached on a pintle at the top of the tower because of the high cost of the blade, rotors with more than two blades are not recommended. Rotors with more than two, say 3 or 4 blades would have slightly higher power coefficiency.

4.12.1.2 Horizontal Axis Propeller Type Using Single Blade

In this type of wind mill, a long blade (above say 60 m) is mounted on a rigid hub and induction generator and gear box are also placed in this turbine. To reduce rotor cost use of low cost counter weight is recommended which balance long blades centrifugally.

4.12.1.3 Horizontal Axis Multibladed Type

This type of wind turbine is made from sheet metal or aluminum. The rotors have high strength to weight ratio and operating speed is 60 km/hr. This wind mill has an excellent power efficiency and is of low cost.

4.12.1.4 Horizontal Axis Wind Mill–Dutch Type

This is a very old design wind mill, whose surfaces are made from wooden slats which function at very high speed.

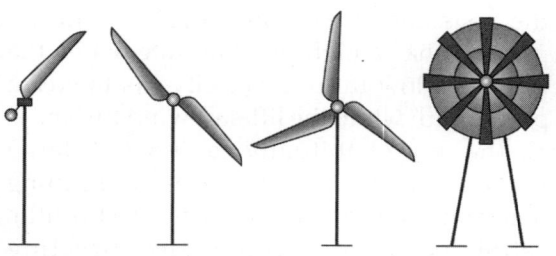

Fig. 4.10: Types of horizontal axis wind turbine

4.12.1.5 Sail Type

Sail type is a recent wind mill made from cloth, nylon, or plastics arranged as mast and pole or sailwings.

Design Consideration of Horizontal Axis Machines

Background: For thousands of years, the wind has provided energy to mankind. The wind has allowed ships to sail from place to place, and it has turned windmills which allowed the grinding of grain and the pumping of water. The first windmills were primitive and caught as much air as they could pointing in only are direction and using simple, straight, non-aerodynamic blades; none the less, they did what no other machines had done before.

However, actual power generation using windmills did not occur until the end of the 19th century. The advancement of this technology was slow, especially with the rapid industrialization taking place in Europe and the United States. As oil prices fell in the 20th century, wind energy became a less attractive energy source. As such a new technology, wind turbines were expensive to develop and build, and even though the wind itself was free, wind power was still too unreliable to take the place of fossil fuels.

Not until the oil crisis of the 1970s, when gasoline and other petroleum products were very limited and expensive, did serious interest in wind turbine technology once again emerged. Since then, wind power has become one of the fastest growing energy sources in the world. Constant improvements in aero-dynamics, structures, and generators have led to more efficient wind turbines outputting more power than ever.

It is in these large; commercial utility wind turbines that much technological improve-ment has taken place. In addition to individual turbine improvements, changes in the wind turbine size and the total number being used have increased dramatically over recent years. Today's large wind turbines are sometimes more than a hundred meters in diameter and could be only one of tens or hundreds of wind turbines, all being used together to form a "wind energy power plant".

Size is only one design consideration for wind turbines. The number of turbine blades varies between wind machines, along with the blade's aerodynamic properties, such as taper, twist, and cross sectional shape. Other advances in wind turbine technology continue today, leading to more powerful and more efficient wind turbines, which will be used to power more of the world in coming years. Some of the main design considerations are outlined below:

1. *Rotor*: The blades of wind turbine rotor are made from wood, metal or composites of several materials. At present carbon fibres are used which are light weight and flexible with immense strength. In common type of machine the rotor is mounted on a hori-zontal shaft and connected either to a generator or some mechanical device. For better performance, the tower should be taller, so that the wind speed increases with height.

2. *No. of blades*: Wind turbines have been built with upto six propellers type blades but two and three bladed propellers are most common.

3. *Blade design*: Wind turbine blades have an airfoil type cross-section and a variable pitch. Better performance of wind mill is obtained when the blades are narrower at the tip than at the root.

4. *Yaw control*: The area of the windstreams swept by the wind turbine is maximum when blades face into the wind. This is achieved by control arrangement, in which when the wind direction changes a motor rotates the turbine slowly about the vertical axis so as the face the blades into the wind.

5. *Turbine tower system*: The horizontal wind turbines are mounted on towers and there are wind forces on the tower.

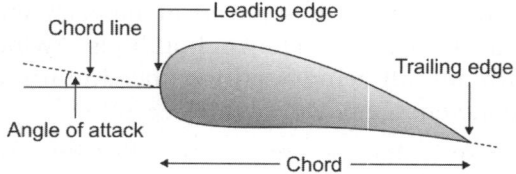

Fig. 4.11: Important parameter of an airfoil

Airfoils: Wind turbine cutting edges have an airfoil sort cross segment and a variable pitch. The real characteristics of such an airfoil are demonstrated in Fig. 4.11. For the proficient energy extraction, cutting edges of current wind turbine are made with airfoil sections. The airfoils utilized for the prior days, wind turbines were the flight airfoils under the NACA (national advisory committee for aeronautics) arrangement. NACA details the characteristics of the airfoil by numbers. For example in a four digit particular, the first number means the greatest camber of the airfoil at the chord line (percent of chord), the second number gives the area of the purpose of most extreme camber from the heading edge and the third and fourth numbers show the greatest thickness (in one percent of the chord). In a couple of gadgets, the razor sharp edges had a consistent chord length, i.e. steady separation from edge to the other. Better performance of wind mill is acquired when the blades are narrower at the tip than at the root.

Fig. 4.12, shows the strengths following up an airfoil when it is put in airstream. The point when an airfoil is put in a wind stream, air passes through both upper and lower levels of the blade. Due to the typical curvature of the blade, air passing over the upper side has to travel more distance per unit time than that passing through the lower side. In this manner the air particles at the upper layer move quicker. As per the Bernoulli's hypothesis, this may make a low-weight area at the highest point of the airfoil. This pressure difference between the upper and easier surfaces of the airfoil will bring about an force *F*. The segment of this power perpendicular to the course of the undisturbed stream is known as the lift force *L*. The power towards the undisturbed stream is known as the drag energy *D*. The horizontal axis by and large have better execution. They have been utilized for different provisions, yet the two significant areas of interest are electric power generation and pumping water. The recent brings some many-sided quality into the outline as the mechanical energy must be transmitted over a distance. Likewise in a few cases the rotor movement must be changed over to responding movement.

4.12.2 Vertical Axis Machines

The axis of rotation is vertical and the blades may also be vertical and are built commercially by a few manufactures. The sails or blades may also be vertical, as on the ancient Persian wind mills, or nearly so, as on the modern, darrieus rotor machine. Two types of designs are commercially successful. Different types of vertical axis panemones have been created in the past that utilization drag powers to turn rotors of distinctive shapes. These incorporate those panemones, heat utilization plates, measures, or turbines as the drag apparatus, and also the savonius. Formed cross-segment rotors which really give some lift energy, are still prevalently drag devices. Such apparatus have moderately high beginning torques contrasted with lift devices in view of their higher solidity, however, have generally low tip-to-wind speeds and more level force yields for every given rotor size, weight and expense. A vertical axis rotor could

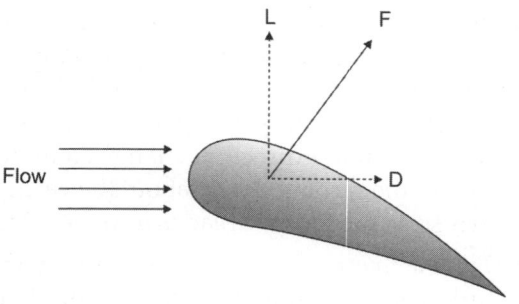

Fig. 4.12: Airfoil lift and drag

be either drag or lift based. The cup anemometer is an illustration of a drag-based, vertical axis wind apparatus. The drag on a cup is greater when its concave side faces the wind which causes the device to rotate. Lift also plays a small part, the cups crossing the wind experience a small lift because their convex surfaces deflect the wind and causes a pressure reduction. The primary righteousness of the glass anemometer is that it has a tendency to pivot inside a restricted reach of TSRS under all conditions, so its rotational rate is nearly relative to wind speed. However, it cannot carry a load with any efficiency, it has never been constructed on a large scale for use as a wind turbine. The wind turbines mounted with the axis of rotation in vertical position have the accompanying notable favorable circumstances like:

1. They are omnidirectional. They do not need the complete yaw control. This type of wind mill captures wind in any direction because their operation is independent of wind direction.

2. They require less structural support because heavy components like gear generators are located at the base level.

The following are two types of vertical wind rotor:

4.12.2.1 Savonius Rotor

Savonius wind turbines are vertical axis wind turbines, or VAWT, and are used for wind force conversion into torque through the rotation of the main shaft (Fig. 4.13). Savonius wind turbines mainly operate on the drag of the aerofoils by their opposing directions, and their interaction is with the wind movement and works like a cup anemometer. This wind turbine was invented by Engineer Sigurd Johannes Savonius in the year of 1920. This machine has become popular, since it requires relatively low velocity winds for operation.

It consists of two half-cylinders confronting inverse headings in such a path as to have just about a S-molded cross area. These two semi

Fig. 4.13: Vertical axis wind turbines

round drums are mounted on a vertical axis perpendicular to the wind direction with a gap at the axis between the two drums. Independent of the wind course the rotor turns, for example, to make the arched sides of the pails head towards the wind. The primary movement of the wind is exceptionally straightforward; the energy of the wind is more excellent on the measured face than on the adjusted face. A few aerofoils of the savonius wind turbine make the state of a S, if observed arially and hold one of the least difficult wind turbine outlines ever planned. Essentially they exploit their curvaceous shape keeping in mind the end goal to endure less grinding in development and consequently build pivoting velocity with the fueling power of the wind. The wind bending around the posterior of the measured face pushes a lessened weight much as the wind does over the highest point of an air-foil and this serves to drive the rotation, the air whip around inside the forward-moving measured face subsequently pushing both towards the revolution. The sharpened steel outline additionally prompts an inquisitive conservation of wind force, because of their real utilization of drag fueled development. This is created by the bending state of the shaft plan which is fabricated to assimilate an insignificant measure of wind power with a specific end goal to capacity; despite the fact

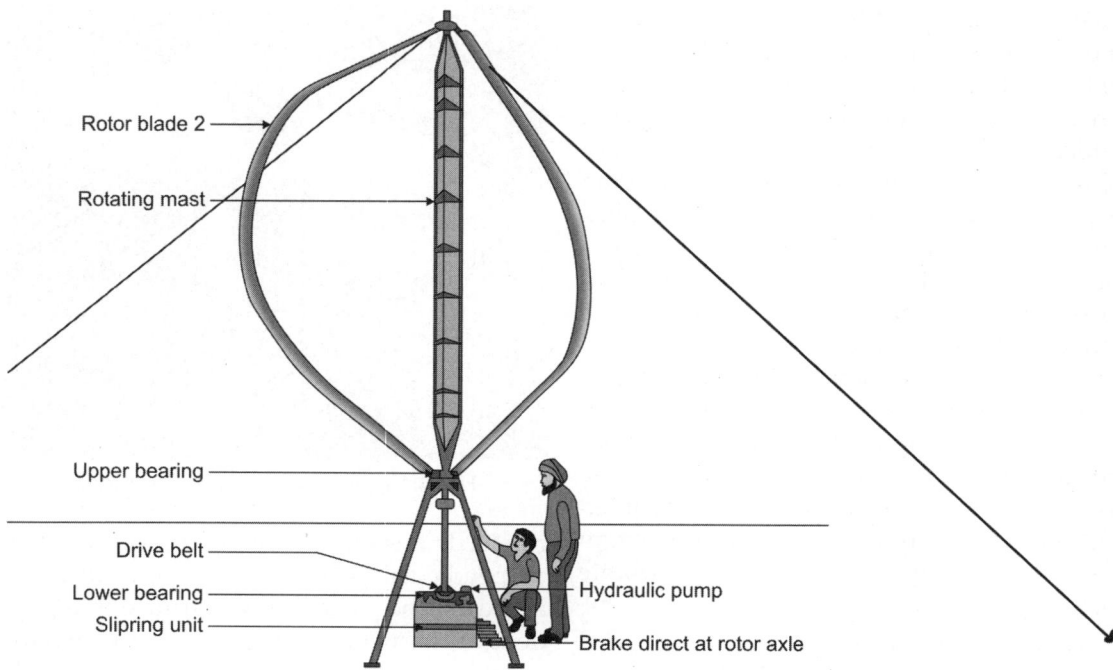

Rotor blade 2

Rotating mast

Upper bearing

Drive belt

Lower bearing

Slipring unit

Hydraulic pump

Brake direct at rotor axle

Fig. 4.14: Darrieus rotor wind turbine

that, this protection of wind force happens at the expense of losing speed. The traditional lower setting positioning of the shaft of savonius wind turbines cause them to be significantly less effective than wind turbines set up on higher posts. The higher mounting allows them to take advantage of the increase in wind speed with height.

The turbines are self starting and have been widely used as deep-water buoys in order to generate electric power, due to their low speed, simple design, and minimal maintenance. In addition, they move independently and change direction in order to adapt the speed and direction of the wind. Savonius wind turbines additionally been utilized as anemometers as a part of request to assess and measure the pace and the course of the wind. Savonius wind turbines can additionally work or turn in both a vertical direction and an even one. The point when the turn of the wind turbines is flat, the turbine can utilize the concentrated energy as a part of request to create noise, heat or power. The fundamental

characteristic of this turbine is that, it is self beginning, has low speed and low productivity.

4.12.2.2 Darrieus Rotor

This machine was invented originally and patented in 1925 by G.J.M. Darrieus, a French engineer and his concept has recently been given serious consideration once again (Fig. 4.14).

The Darrieus rotor is a vertical axis wind turbine (VAWT) with two or more cutting edges having an aerodynamic airfoil (Fig. 4.15). The blades are typically twisted into a chain and are associated with the center at the upper and easier side, both the closures of sharpened

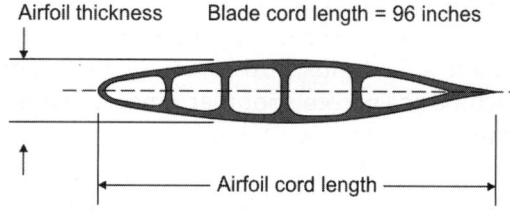

Airfoil thickness   Blade cord length = 96 inches

Airfoil cord length

Fig. 4.15: Diagram of airfoil's geometry

pieces of steels are joined to a vertical shaft. Hence, the energy in the sharpened steel because of wind rotation is pure tension. This gives a solidness to help withstand the wind strengths it encounters, and the sharpened pieces of steels are made lighter than in the propeller sort. However, additionally Darrieus rotors with straight cutting edges (H-Darrieus) have been produced which along these lines have vast centers furnished with spokes. The energy is taken from the wind by a part of the lift energy L working toward turn. Consequently the Darrieus rotors are lift gadgets, portrayed by bended edges with air foil. The same guideline is utilized for a horizontal axis wind turbine (HAWT). There are additionally VAWT's for which the force is gained by the contrast in drag D of the blade moving towards the wind and the edge moving against the wind. A cup anemometer, utilized for measuring wind rates, is an illustration of such a rotor. There are additionally VAWT's, for example the Savonius rotor, which are working a shot at a blending of lift and drag. One of the points of interest of utilizing the lift rule is that much higher power coefficients, Cp might be acknowledged than for windmills utilizing the drag guideline. An alternate point of interest is that the tip speed degree, could be much higher bringing about a much higher rpm for the same rotor distance across and in utilizing much lesser material. The Darieus rotor guideline is dependent upon the rule by which a Bermuda–rigged sailing ship can cruise over the wind at paces more terrific than that of the wind, much like the blade of a horizontal axis wind mill. Lift causes a turning movement stronger than the drag constrains that block it as long as the TSR is sufficiently high to avoid blades from stailing as they cross the wind. This notion was patented by a Frenchman, Darrieus.

Characteristics of Darrieus rotor:

1. Self starting
2. High speed

3. High efficiency
4. Potentially low capital cost.

Advantages of Darrieus rotor:

1. The rotor shaft is vertical. Therefore it is possible to place the load, like a generator or a centrifugal pump at ground level. As the generator housing is not rotating, the cable to the load is not twisted and no brushes are required for large twisting angles.
2. The rotor can take wind from every direction.
3. The visual acceptation for placing of the wind mill on a building might be larger than for an horizontal axis wind mill.
4. Airfoil rotor fabrication costs are expected to be reduced over conventional rotor blade costs.
5. The major advantage of this design is that the rotor blades can accept the wind from any direction.

Disadvantages of Darrieus rotor:

1. The angle of attack varies strongly and therefore the lift, the drag varies strongly too. During the parts of the revolution where the blade moves about parallel to the wind there is only drag and the distribution to the torque is negative. Therefore the maximum power coefficient Cp is about 0.35 for big rotors (but only 0.2–0.3 for small rotors).
2. Because the blade load is fluctuating strongly, this results, especially for a two bladed rotor, in strong fluctuating loads on the tower and the foundation.
3. The starting torque coefficient is zero and at low tip speed ratios it is even negative. Therefore a special motor is required to start the rotor.
4. It is difficult to protect the rotor against high wind speeds. Turning the rotor out of the wind is not possible. Pitch control is only possible for Darrieus rotors with straight blades but pitch control requires a complex construction with many turning points. Systems with air brakes appear not to work in practice. Braking the vertical shaft mechanically is the only possibility, but this is not really safe because the rotor will turn unloaded if the brake fails for some reason.

5. A very large bending moment is created in the rotor shaft, if it is not supported at the top because support at the top requires a large, wide guiding.
6. For a traditional Darrieus rotor with blades bent into a chain line, the upper and lower parts of the blade don't contribute to the torque or even have a negative torque because the local radius and therefore the tip speed ratio is too small.
7. For a Darrieus rotor with straight blades (H-Darrieus), the blade is loaded very strongly by bending because of the centrifugal force in the blade. A H-Darrieus requires long spokes to connect the blades to the hub and to minimise drag, these spokes should have an aerodynamic airfoil.
8. The pipe which is placed in between the upper and the lower side of the rotor creates, especially for large diameters, a turbulent wake which has an unfavourable influence on the aerodynamics of the blade.

9. At the same tip speed ratio and the same swept area, the required amount of material for a Darrieus rotor is very much larger than for an horizontal axis windmill (Fig. 4.16).

4.12.2.3 Analysis of Aerodynamic Forces Acting on the Blade

An airfoil is characterized as the cross segment of a form that is submitted in an airstream in request to process service flight optimized drive in the most productive way conceivable. The cross segments of wings, propeller edges, windmill sharpened pieces of steels, compressor and turbine cutting edges in a plane motor, and hydrofoils are illustration of air foils. The fundamental geometry of an airfoil is indicated in Fig. 4.17. The most paramount characteristics of airfoil geometry are the chord, camber, and thickness. The straight line joining the heading and trailing edges is chord line, the separation measured between the

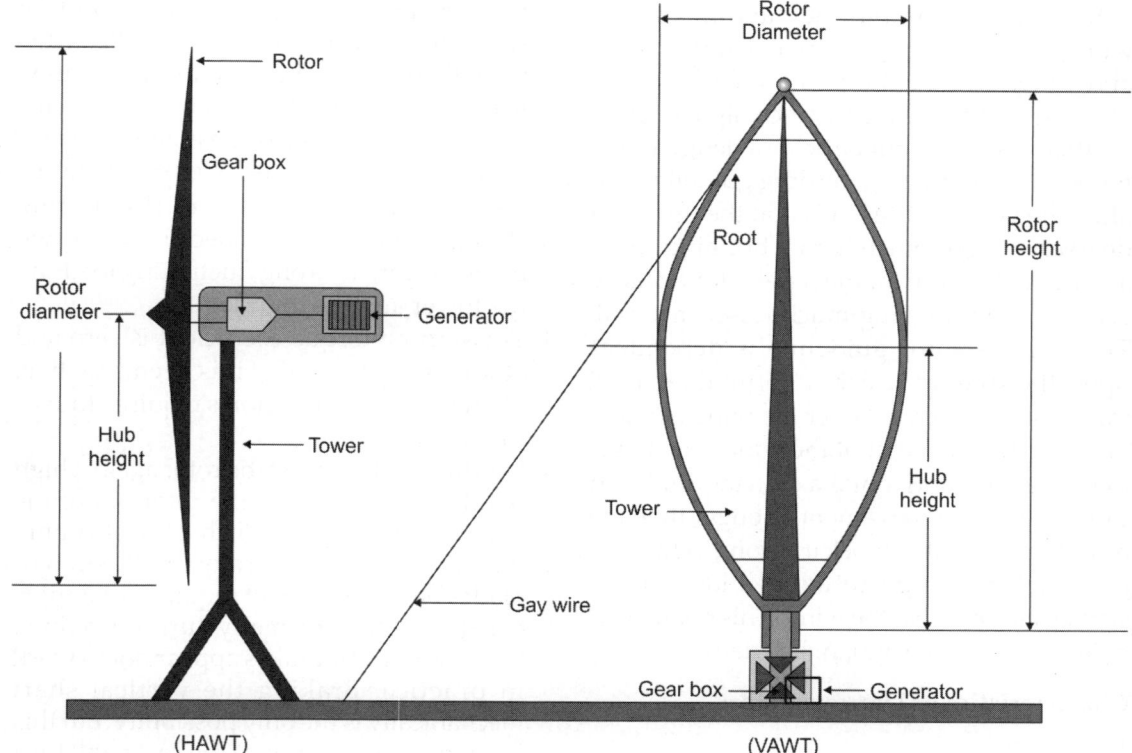

Fig. 4.16: Comparison between HAWT and VAWT

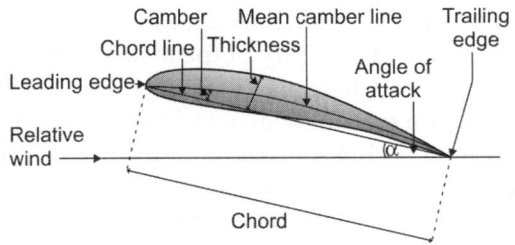

Fig. 4.17: Diagram of airfoil's geometry

trailing and heading edge along the chord line is the chord of an airfoil. The line of focus that are partly between the upper and lower surfaces is the mean camber line as measured perpendicularly from the chord line. The thickness of an airfoil is the distance from the upper and lower surfaces as measured perpendicularly to the chord line, and shifts in separation along the chord line. Camber is the greatest separation that happens between the mean camber line and the chord line. Greatest camber and thickness, and additionally where they happen along the chord line, are essential segments for airfoils, and are utilized within the characterization of airfoils. Lift is characterized as the part of air element power perpendicular to the relative wind current.

The theory most commonly found in text books and pilot training manuals utilizes Bernoulli's principle which states that for a liquid or gas, areas with high relative velocity create lower pressure systems, and areas with low relative velocity create high pressure systems. The theory states that airfoils are shaped so that the upper surface is longer than the lower surface; therefore, when air molecules are separated by the leading edge of the airfoil, they have a greater distance to travel as they cross the upper surface than along the lower surface. Thus, for the air molecules to meet at the trailing edge at the same time, the molecules traveling along the upper surface must be traveling faster than the air molecules along the bottom. Since the airflow on the upper surface is faster. According to Bernoulli's principle a lower,

pressure system is created. The difference between the low pressure above the airfoil and the high pressure below causes lift to occur. However, there are no principles of fluid dynamics stating that two free moving air particles must meet at a single point beyond an obstacle once separated by the obstacle. The correct theory of lift generation is known as the flow turning theory. It states that the airfoil bends the direction of the airflow around it as the airflow passes over the upper surface, and creates a vertical velocity of airflow past the trailing edge. The effect of the airflow bending is due to the viscosity of a fluid and the *Coanda* effect. As the airfoil bends the airflow near the upper surface, it pulls on the air above it and causes an acceleration of that air down to the airfoil. The pulling of the air causes a low pressure system to form over the airfoil creating a net force that is lift. The other aerodynamic force that affects an airfoil in a wind tunnel is perpendicular to the lifting force, called drag. The airfoil experiences a drag force that opposes the relative motion of the airfoil and has the direction parallel to the airfoil. Skin friction drag is the friction that occurs between the air molecules and the surface of the airfoil. The drag is dependent on the overall shape of the air foil, and pertains to the pressure distribution about the airfoil's surface. As with the lifting force, the airfoil changes the local momentum of the air around it, affecting the velocity and pressure. The resulting pressure distribution produces a force that acts on the airfoil

4.13 ENVIRONMENTAL ASPECTS

Wind turbines are not without environmental impact and their operation is not entirely risk-free. Following are the main issues pertaining to a wind turbine are as follows:

1. *Bird and bat deaths*: Turbines are proven to kill birds and bats. Bird and bat deaths are one of the most controversial biological issues related to wind turbines. Turbines are proven to disrupt wild animals and their natural habitats.

2. *Weather and climate change*: Wind farms may affect weather in their immediate vicinity. Spinning wind turbine rotors generate a lot of turbulence in their wakes like the wake of a boat. This turbulence increases vertical mixing of heat and water vapour that affects the meteorological conditions downwind. Overall, wind farms lead to a slight warming at night and a slight cooling during the day time.

3. *Noise*: Like all mechanical systems, wind turbines produce some noise when they operate. Most of the turbine noise is marked by the sound of the wind itself, and the turbines run only when the wind blows. Revolving blades generate noise which can be heard in the immediate vicinity of the installation.

4. *Other concerns*: Unlike most other generation technologies, wind turbines do not use combustion to generate electricity, and hence don't produce air emissions. The only potentially toxic or hazardous materials are relatively small amounts of lubricating oils and hydraulic and insulating fluids. Therefore, contamination of surface or ground water or soils is highly suspected. The primary health and safety considerations are related to blade movements and the presence of industrial equipment in areas potentially accessible to the public.

5

Hydroelectricity

5.1 INTRODUCTION

Water flowing from higher level to lower level has kinetic energy which can be converted to mechanical energy and then electrical energy. A generating station which utilizes the potential energy of water at a high level for the generation of electrical power is known as hydroelectric power station. The hydro energy in flowing water is a function of weight of water and the head through which the water falls. Hydropower is considered as one of the most economic and non-polluting source of energy. Hydroelectricity means electricity generated by hydropower or from the use of the gravitational force as falling or flowing water. This is one of the most common forms of power generation since this form of energy neither produces any direct waste matter nor is subjected to exhaustion. Hydropower is energy that comes from the force of moving water. The head of water fall and movement of water in the form of kinetic or potential energy is called as water cycle or hydropower cycle. Kinetic energy of water is a function of mass and velocity while potential energy is a function of the differences in level of water between two points. Prior to the widespread availability of commercial electric power, hydropower was used for irrigation, and operation of various machines, such as water mills, textile machines, saw mills, dock cranes, and domestic lifts. A Hydroelectric power station is used to supply electrical energy to consumers where water resources available.

In hydroelectric power plants the energy of water is utilized to drive the turbine which in turn to the generator to produce electricity. Hydropower or water power is important only next to thermal power. Nearly 20% of the total power of the world is met by hydropower stations. In India in 1897, a unit was initiated near Darjeeling. After independence, a substantial growth in hydropower has occurred with the commissioning of large multipurpose projects like Damodar valley corporation (DVC), Bhakra Nangal, Hirakund, Nagarjunasagar, Mettur, Koyna, Rihand and so on. The potential for hydroelectric potential in terms of installed capacity in India is estimated to be about 1,48,700 MW, out of which a capacity of 30,164 MW (20.3%) has been developed so far and 13,616 MW (9.2%) of capacity is under construction.

Hydropower is a renewable energy resource because it uses the Earth's water cycle to generate electricity. Water evaporates from the earth's surface, forms clouds, precipitates back to earth, and flows toward the ocean. The movement of water as it flows downstream creates kinetic energy that can be converted into electricity and about 2700 TWh is generated every year. Hydropower supplies at least 50% of electricity production in 66 countries and atleast 90% in 24 countries. Out of the total power generation installed capacity in India of 1,76,990 MW (June, 2011), hydropower contributes about 21.5%, i.e. 38,106 MW. A capacity addition of 78,700 MW is envisaged from different conventional sources during 2007–2012 (the 11th plan), which includes 1,5,627 MW from large hydro projects. In addition to this, a capacity addition of 1400 MW was envisaged from small hydro plants, up to 25 MW station capacity. The total hydroelectric power potential in the country is assessed at about 1,50,000 MW, equivalent to 84,000 MW at 60% load factor. The potential of small hydro power projects is estimated at about 15,000 MW.

5.2 HYDROPOWER

Mainly the hydropower depends on the rain. The rain defined as, energy from the Sun evaporates water from oceans and rivers and draws it upward as water vapour. When the water vapour reaches the cooler air in the atmosphere, it condenses and forms clouds. The moisture eventually falls to the Earth as rain or snow, replenishing the water in the oceans and rivers. The storage of rainy water is completely used for a year or more than a year. The hydro power depends on the power requirement and can be defined as base load station and peak load station. In base load station the load requirement is constant throughout the year and cost of the power/ unit is comparatively less. In peak load station is used when the demand of power is more and cost of the power/unit will be more. The time that a peak plant operates may be many hours a day or as little as a few hours per year, depending on the condition of the region's electrical grid. Most of the time electricity will be used during the day hours as the power is required for watermills, textile machines, sawmills etc. The initial capital cost for the hydro power stations is more and running cost is less. Hydro plants are more energy efficient than most thermal power plants too, that means they waste less energy to produce electricity. In thermal power plants, a lot of energy is lost as heat. Hydro plants are about 95% efficient in converting the kinetic energy of the moving water into electricity.

5.3 TECHNOLOGY

A hydroelectric power plant consists of a high dam that is built across a large river to create a reservoir, and a station where the process of energy conversion to electricity takes place. The first step in the generation of energy in a hydropower plant is the collection of run-off of seasonal rain and snow in lakes, streams and rivers, during the hydrological cycle. The run-off flows to dams downstream. The water falls through a dam, into the hydropower plant and turns a large wheel called a turbine. The turbine converts the energy of falling water into mechanical energy to drive the generator. After this process has taken place electricity is transferred to the communities through transmission lines and the water is released back into the lakes, streams or rivers. This is entirely not harmful, because no pollutants are added to the water while it flows through the hydropower plant.

5.4 HYDROPOWER GENERATION IN THE WORLD

The inherent technical, economic and environmental benefits of hydroelectric power make it an important contributor to the future world energy mix, particularly in the developing countries. These countries have a great and ever-intensifying need for power and water supplies and they also have greatest remaining hydro potential. When highlighted hydropower in world, was found that hydropower now supplies about 888.8 GW or above 25% of world electricity. Still the construction of large dams are going on, specially the worlds largest is the *Three Gorges Dam* on the third longest river in the world, the Yangtze river. Annual installed capacity surged during 2004 mainly due to the rise in new installations in China. In addition, rising interest in the sector has led to increased government support policies which will derive installations in many countries. Hydropower, or hydro-electric power generation, is used around the world to generate electricity. Hydropower has been used for thousands of years to convert the energy of water into mechanical work for irrigation, and later to power machinery in the factories of the industrial revolution. By the year 2050, the world population is expected to increase by 50%, from 6 to 9 billion. Energy consumption per inhabitant per year is generally in correlation with the standard of living of the population, which is characteristic of welfare from an economic, social and cultural point of view. Today the less developed countries in the world, with 2.2 billion inhabitants, have an annual per capita consumption of primary energy which is 20 times less than those of the industrialized countries (with 1.3 billion inhabitants), and per capita electricity consumption which is 35 times less. Hydroelectric power plants generally range in size from several hundred kilowatts to several hundred megawatts, but a few enormous plants have capacities near 10,000 megawatts in order to supply electricity to millions of people. According to the national renewable energy laboratory, world hydroelectric power plants have a combined capacity of 6,75,000 megawatts that produces over 2.3 trillion kilowatt-hours of electricity each year; supplying 24% of the world's electricity to more than 1 billion customers. Only 2,400 of the 80,000 dams in the United States are used for hydroelectric power. It is costly to construct a new hydroelectric power plant, and construction uses much water and land. In addition, environmental concerns have been voiced against their use. According to the US geological survey, the likely trend for the future is towards small-scale hydroelectric power plants that can generate electricity for single communities. A number of countries, such as China, India, Iran and Turkey, are undertaking large-scale hydro development programmes, and there are projects under construction in about 80 countries. According to the recent world surveys, conducted for the World Atlas and Industry Guide, published annually by hydropower and dams, a number of countries see hydropower as the key to their future economic development: Examples are Sudan, Rwanda, Mali, Benin, Ghana, Liberia, Guinea, Myanmar, Bhutan, Cambodia, Armenia, Kyrgyzstan, Cuba, Costa Rica, and Guyana.

World distribution of hydropower is given under:

- Hydropower is the most important and widely-used renewable source of energy.
- Hydropower represents 25% of total electricity production.
- China is the largest producer of hydroelectricity, followed by Canada, Brazil, and the United States (*Source*: Energy information administration).
- There are so many dams proposed which are considered to be in list of largest dams. Presently the Three Gorges Dam of China having capacity of 22.5 GW generation capacity stands in the first position and it will complete by the year 2011.

5.5 POTENTIAL IN INDIA

India is blessed with immense amount of hydroelectric potential and ranks 5th in term of exploitable hydro-potential on global scenario. As per assessment made by CEA, India is endowed with economically exploitable hydropower potential to the tune of 1,48,700 MW of installed capacity. The basin wise assessed potential is as given in Table 5.1.

In addition, 56 number of pumped storage projects have also been identified with

Table 5.1 Basin wise assessed potential

Basin/rivers probable	Installed capacity (MW)
Indus basin	33,832
Ganga nasin	20,711
Central Indian river system	4,152
Western flowing rivers of southern India	9,430
Eastern flowing rivers of southern India	14,511
Brahmaputra basin	66,065
Total	1,48,701

probable installed capacity of 94,000 MW. In addition to this, hydro-potential from small, mini and micro schemes has been estimated as 6,782 MW from 1,512 sites. Thus, in total India is endowed with hydro-potential of about 2,50,000 MW. The total installed capacity of India is **3,6,878 MW.**

5.5.1 Beginning of Hydropower in India

India's first major hydroelectric power installation, started generating electricity in 1902. This power installation was from Sivasamudram, which was island located in upper course of Cauvery river. This powerhouse was designed to generate three megawatts. And the power was transmitted to Kolar gold mines which were under control of British companies. At that time gold mining was a small business in South India. In the 1880's heavy machineries were required to carry mining business like heavy depth operations. Due to shortage of low calorific steam coal which was not available in South India, it was necessary to be brought at considerable expense from distant fields in Hyderabad and Bengal. Despite the chronic fuel problems, the mining companies were prospering and were investing their profits in new equipments to improve the productivity of the mines. Looking at the above problems the mine engineers at Kolar, thinking that hydroelectricity might meet the gold mines' requirements for a source of inexpensive electricity. The Niagara power company had devised a dual strategy for marketing electricity, first through attraction of power consuming electro-chemical and electro-metallurgical to the area.

The development and implement of hydropower at Niagara falls had determined an electrical frequency suitable for industrial processing, urban lighting had to develop power lines capable of overcoming recurring problems with lightning, switching and cable insulation. The machinery and technology developed first by *Westing house* and then by General Electric (GE) at Niagara falls and by 1898, developed into a reproducible power complex that would set the standards for other hydroelectric power installations. GE's work at Niagara falls was widely known, they had installed hydroelectric power equipments for gold mines in America and South America, and the company was intent upon expanding its sales throughout the world. The power station was named after the island of Sivasamudram, nearby the fall. Mysore retained one of General electric's engineers, Harry Parker Gibbs, as the chief electrical engineer of the state's new electrical department and sent four Indian members of the departmental staff to GE's headquarters in Schenectady, New York for training. Gibbs was later hired by Tata hydroelectric power company as general manager, to supply electricity to cotton textile mills of Bombay city. Later it was found that, the original agreement of 1900's between Mysore and the Madras Presidency had stipulated that "all water diverted from the river for the power works shall be returned to the river below the fall without being limitations in quantity." Plans in 1910 were made to build a reservoir across the Cauvery river just above Mysore city. This was to be one of the world's first multipurpose reservoirs, for developing more power at Sivasamudram and for irrigation. This brought the Sivasamudram power development into the context of a long standing dispute over the Cauvery river's water. By the end of the 19th century, the Cauvery river system had been fully utilized for irrigation. The new exploitation of the Cauvery river has increased interstate disputes over the projected uses of the river and remain to this day. Figure 5.1, shows the major hydropower plants in India.

5.6 HYDROPOWER PROJECTS IN INDIA

Hydro power is considered as one of the most economic and non-polluting sources of energy. Power generated from water is termed as hydroelectricity. Hydroelectricity means electricity generated by hydropower or from the use of the gravitational force of falling or

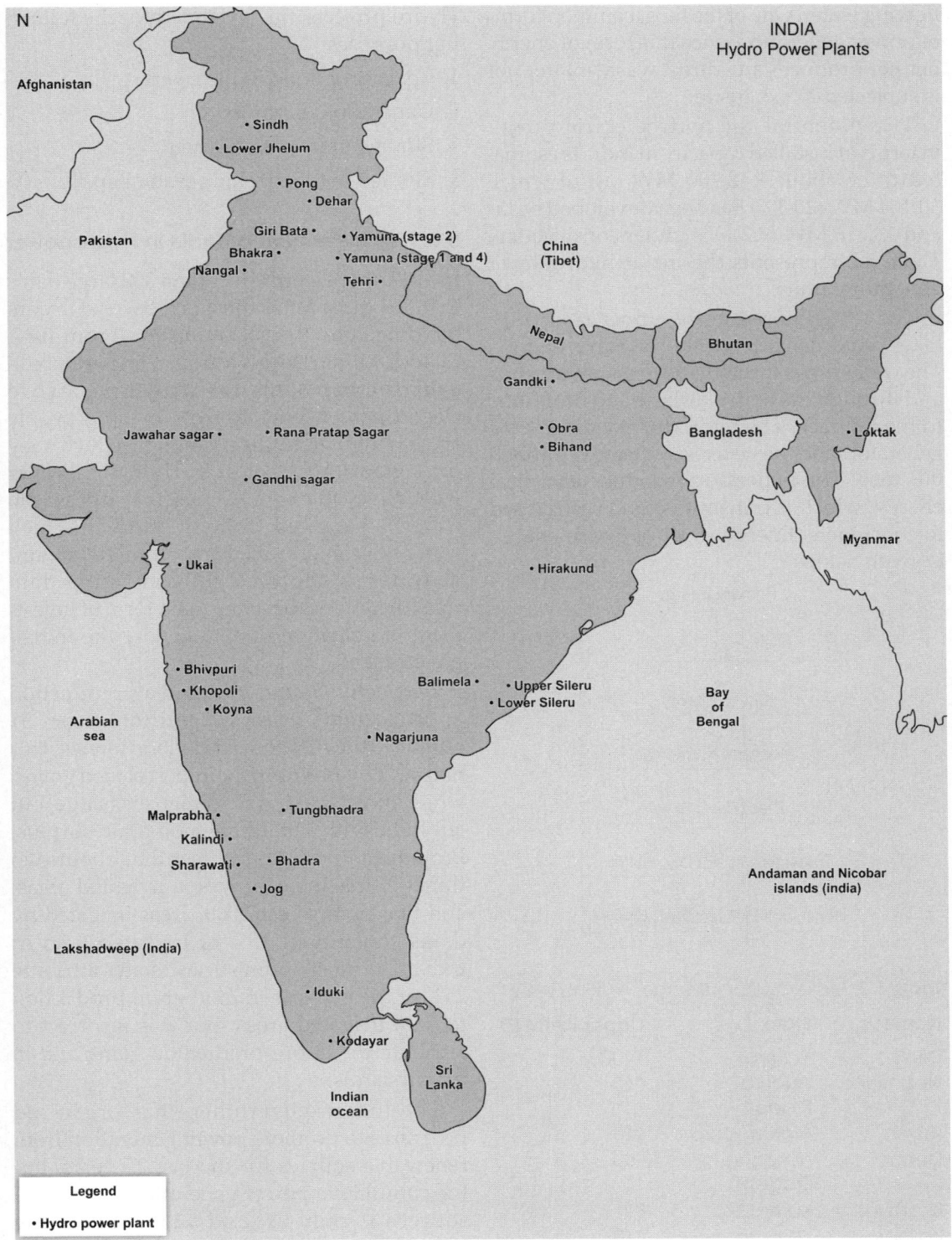

Fig. 5.1: The map showing the major hydropower plants in India

flowing water. One of the most common forms of power generation since this form of energy neither produces any direct waste matter nor is subjected to exhaustion.

The potential for hydroelectric power in terms of installed capacity in India is estimated to be about 1,48,700 MW out of which 30,164 MW (20.3%) has been developed so far and 13,616 MW (9.2%) is under construction. Table 5.2 represents the major hydropower generating units.

The water stored in the upper reservoir/lake/pond constitutes the primary energy. The water flows through the pressure pipeline and through the hydro turbine and draft tube to the tail race. The hydro turbines drive the generators and the water is discharged through tail race. The generators produce electrical energy, which is transmitted and distributed to consumers. Energy route of hydro-energy is given below:

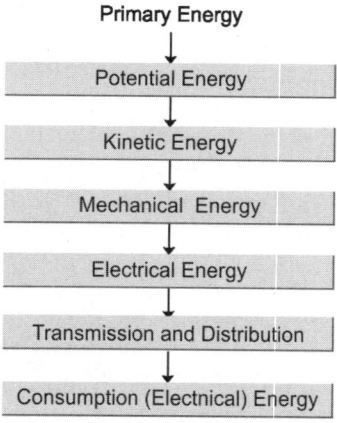

Primary Energy
↓
Potential Energy
↓
Kinetic Energy
↓
Mechanical Energy
↓
Electrical Energy
↓
Transmission and Distribution
↓
Consumption (Electrical) Energy

Table 5.2 Major hydropower generating units

Name	State	Capacity (MW)
Bhakra	Punjab	1100
Nagarjuna	Andhra Pradesh	960
Koyna	Maharashtra	920
Dehar	Himachal Pradesh	990
Sharavathy	Karnataka	891
Kalinadi	Karnataka	810
Srisailam	Andhra Pradesh	770

Hydro projects are developed for the following purposes:

1. To control flood in the rivers
2. Generation of power
3. Storage of irrigation water
4. Storage of the drinking water supply.

5.6.1 Achievement of India in Hydropower

In 1947, there were fewer than 300 large dams in India. India ranks third in the world in dam building, after the US and China. From 1947, till today India has given much importance in constructing dams for irrigation, hydro electric power generation etc. In India, people are having agricultural lands in which they grow grains, vegetables etc. The natural water will be available in rainy season. In dry season water is supplied through dams in which water flow is controlled and required amount of water is supplied. In fact, large dam construction has been the main form of investment in irrigation undertaken by the Indian government.

Between 1951 and 2000, India's production of food grains increased fourfold, from 51 million tonnes to about 200 million tonnes. This not only obvious from the import of food grains, which directs saving in foreign exchange, but left India with a marginal food grain surplus. Proponents point to the fact that about two thirds of this increase was in irrigated areas, and that by the year 2000, areas irrigated by dams constituted 35% of irrigated land in India. The most optimistic estimates attribute 25% of the increase in food grain production to dam irrigated areas. But it is incorrect to attribute the entire production gains *in dam irrigated areas.*

The Indian government has organised programs to promote power generation from renewable sources for the last 25 years, but the cumulative power generation from these sources is only around 12,000 MW. The Himachal Pradesh plant's availability factor– the amount of time a power plant can produce electricity over a certain period, divided by the amount of time in the period–for the month of July 2009 was 105.26 percent, with the cumulative factor for the station is at 102.88%. This marks the highest factor achieved by the central sector hydropower stations operating in the northern grid.

Large dam construction has been an important and expensive undertaking for the Indian government. While dams have enhanced agricultural productivity in India, there is no evidence that they have been very cost effective, and they have significantly adverse distributional implications. The case of large dams suggests strongly that distributional implications of public polices should be central to any evaluation. Clearly, the case of large dams suggests the need to understand the institutions, and power structures, which led to the implementation of these projects.

By 2012, the country will see three new projects of 1,000 MW and above. These are the Karcham Wantoo project (1,000 MW) in Himachal Pradesh, the Tehri pump storage scheme of 1,000 MW and the 2,000 MW plant at Subansiri in Arunachal Pradesh. Post independence, we have made lots of progress in dam and water reservoirs in India. Dams are basically used for power generation, water supply, to stabilize water flow/irrigation, flood prevention, land reclamation, water diversion, recreation and aquatic beauty. India is very *rich in Dams* and India is having some of the largest Dams and reservoirs. There are states in India, where dams have been established but Karnataka is very rich in having most of dams and reservoirs. Dams in Karnataka is very popular, serving the purpose of people of Karnataka and Bangalore. Karnataka is much enriched in terms of dams in South India. Table 5.3 summarizes the list of the major hydro power plants in India.

5.6.2 Hydropower Stations in India

In India, the conventional alternatives to hydroelectric power are diesel, coal or natural gas. After the full impact of chernobyl nuclear disaster became known, there is less talk of nuclear energy as a major alternative. Natural gas has the limitation as its reserves are low. Considering India's coal reserves and the fact that it imports petroleum, coal would rank equally with diesel. Though thermal plants using coal used to be highly polluting, modern technologies have helped to bring down pollution to very low levels. However, coal is ranked below oil in the West as it produces a lot of carbon dioxide, a greenhouse gas. A fast growing power sector is crucial to sustain India's economic growth. India has an assessed hydropower potential of 84,000 MW at 60% load factor, out of this and only about 20% has been developed so far. In the past, various factors such as the dearth of adequately investigated projects, environmental concerns, resettlement and rehabilitation issues, land acquisition problems, regulatory issues, long clearance and approval procedures, power evacuation problems, the dearth of good contractors, and in some cases, inter-state issues, law and order problems have contributed to the slow pace of hydropower development. There have been a long time and cost overruns in case of some projects due to geological surprises, resettlement and rehabilitation issues, etc. However, considering the large potential and the intrinsic characteristics of hydropower in promoting the country's energy security and flexibility in system operation, the Government is keen to accelerate hydropower development. The Central Electricity Authority (CEA) undertook reassessment of the hydropower resources of the country in 1980s. In this survey, theoretical and the economic hydro potential of the rivers was worked out. The potential was assessed by identifying specific suitable sites and water availability corresponding to a 90% dependable year. CEA had identified 845 economically feasible schemes in various river basins of the country.

Table 5.3 List of the major hydropower plants in India

Name	Location	Operator	Configuration	Important facts
Babail	Uttar Pradesh	Uttar Pradesh Jal Vidyut Nigam Ltd	2 × 1.5 MW Tube	The Babail minihydel project was approved in Sept. 1986 and was awarded to PGM in Sept. 1988 as a ₹ 6:22 cr turnkey project
Bhandardara-1	Maharashtra	Dodson-Lindblom Hydropower Pvt Ltd	1 × 14.4 MW Francis	This plant was acquired in 1996 from Maharashtra water resource dept. and overhauled in 1997/98 with assistance from AHEC
Belka	Uttar Pradesh	Uttar Pradesh Jal Vidyut Nigam Ltd	2 × 1.5 MW Tube	The Belka and Babail minihydel projects were approved in Sept. 1986 and Belka was awarded to FCC and PGM in July 1988, as a ₹ 5.66 crore project. Construction on Belka did not start until Dec. 1996 after delays in securing forest clearance and land acquisition
Chenani-1	Jammu and Kashmir	Jammu and Kashmir Power Development Corp	5 × 4.66 MW Pelton	The Chenani land-1 projects in Udhampur district were inaugurated in 1971 by Prime Minister Indira Gandhi. They were closed on 25 Feb. 2005 following a landslide, that damaged a 700 m diversion tunnel. The complete repairing process costs ~₹ 8 crore and the plants put back in service in June 2008
Bhatgar	Maharashtra	Maharashtra State Power Generation Co Ltd	1 × 16 MW Kaplan	The dam was part of the world's largest irrigation project, known as Lloyd Barrage. This was a multipurpose scheme which was initiated in 1923 by Sir George Ambrose Lloyd, then Governor of Bombay. The project was opened in Jan 1932
Indira Sagar	Madhya Pradesh	Narmada Hydroelectric Development Corp Ltd	8 × 125 MW Francis	NHDC is a joint venture of NHPC and the MP government set up on 1 Aug 2000
Little Ranjit	West Bengal	West Bengal State Electricity Distribution Co Ltd	2 × 1 MW Pelton	Operations commenced in 1970
Jammu Canal	Jammu and Kashmir	Jammu and Kashmir power development corp	2 × 500 kW Francis	This power station has been out of service since 1995

(contd.)

(contd.)

Name	Location	Operator	Configuration	Important facts
Matatila	Uttar Pradesh	Uttar Pradesh Jal Vidyut Nigam Ltd	3 × 10 MW Kaplan	This dam was built between 1952 and 1964 on the Betwa river in the Ganga basin
Salal	Jammu and Kashmir	National Hydro Power Corp Ltd	6 × 115 MW Kaplan	Built on the Chenab river, this power station has a 118 m high, 630 m long rockfill dam and a 113 m high, 450 m long concrete dam plus an 11m, 2,46 km tail-race tunnel. The reservoir is 33 km long. Project development started in 1970
Omkareshwar	Madhya Pradesh	Narmada Hydroelectric Development Corp Ltd	8 × 65 MW Francis	This 949 m long concrete gravity dam reaches a maximum height of 53 m. The annual production is expected to be 1.1 TWh
Samal	Odisha	Odisha Power Consortium Ltd	5 × 4 MW S-Turbine	OPCL is a power company promoted by VBC Ferro Alloys Ltd. Samal uses releases from Samal barrage reservoir on the Brahmani river

Table 5.4 shows that the percentage of electricity generated by thermal power plants are about 70%, 21% by hydroelectric power plants and 4% by nuclear power plants. Some of the measures announced by; Govt. of India have already been introduced, which include simplified procedures for transfer of techno-economic clearances, streamlining of clearance process and introduction of three-stage clearance approach for development of hydro projects in central sector/joint ventures, etc. The Indian government considers hydro-power as a renewable economic, non-polluting and environmentally benign source of energy. Experience of running hydropower stations in India has shown that even after careful project planning and good quality control measures from construction to commissioning, unforeseen problems do occur in service resulting in unplanned outages/low generation and load shedding etc. This causes disruption to consumers and reduced cash generation for the operator.

5.6.3 Major Steps Taken By Indian Government

1. The hydro-electric potential in terms of installed capacity is proposed to be about 1,48,700 MW out of which a capacity of 30,164 MW (20.3%) has been developed so far and 13,616 MW (9.2%) is under construction.

2. In addition, 6,782 MW in terms of installed capacity from small, mini and micro hydro schemes have been assessed.

3. Also, 56 sites for pumped storage schemes with an aggregate installed capacity of 94,000 MW have been identified.

4. The government expects to harness its full potential of hydropower by 2027 with a whopping investment of 5,000 billion rupees.

Table 5.4 Percentage of electricity generated

Hydropower station	Thermal power station	Nuclear power	Other
21%	70%	4%	5%

5. Additionally, India has committed massive funds for the construction of various nuclear reactors rather than hydropower plants which would generate at least 30,000 MW.

5.6.4 Hydroelectric Energy Resources in India

India's total estimated potential of hydro-electric power resources are 70,000 MW to 1,00,000 MW. (The difference, 30,000 MW in the estimates is due to assumption in limit of capital expenditure considered for liable schemes).

Resources in Northern and North-Eastern regions are yet to be exploited. The major difficulties in their exploitation are high installation and development costs of civil works, high cost of long transmission lines upto loud centers, long construction periods, lack of financial resources and environmental problems including earthquake possibility, deforestation and submergence of agricultural lands/villages.

5.7 MERITS OF HYDRO-ELECTRIC ENERGY

Hydro energy have some inherent advantages which makes it very attractive, like

1. Hydro energy is a clean and renewable energy. The hydroelectric power generation does not produce pollution. The hydro energy is renewed naturally by rain water and by melting of snow on high maintains during summer.

2. The running costs of hydropower installations are very low as compared to thermal or nuclear power stations.

3. The natural renewable energy is stored in the high level reservoir and used whenever necessary.

4. Being simple in design and operation , the hydroplants do not require highly skilled workers.

5. Hydropower stations does not produce any greenhouse effect.

5.8 DEMERITS OF HYDRO-ELECTRIC ENERGY

Demerits of the hydroelectric energy are as follows:

1. Hydropower projects are capital-intensive with a low rate of return.

2. Large hydro-power plants disturb the ecology of the area, by way of deforestation, destroying vegetation and uprooting people.

3. Power generation is dependent on the quantity of water available, which may vary from season to season and year to year.

5.9 HYDROELECTRIC POWER PLANTS

Factors affecting the selection of site for hydro electric station are:

1. **Amount of water available:** Water is the main heart of hydro power plants. Previous records of rainfall are studied and also minimum and maximum quantity of water available during the year are estimated. After allowing for losses due to evaporation and percolation the net volume of water available for power generation can be determined.

2. **Storage of water:** Wide variation of rainfall during the year makes it necessary to store water for continuous generation of power throughout the year, intend to provide the sufficient storage for one year so as to make available during the worst dry periods (Fig. 5.2).

3. **Head of water:** The available water head depends on topography of the area. This is the important factor and it decides the generation of power, low falls on unregulated streams are subject to wide variations which affect the net head, and may, infact, reduce it to abnormal low value, uneconomical for power generation.

4. **Transportation facilities:** The site selected for a hydro plant should be accessible by rail as well as by road.

5. **Distance of power site from power grid:** If the power site is near the power grid, the cost of power will be less. If it is far, the cost will be more.

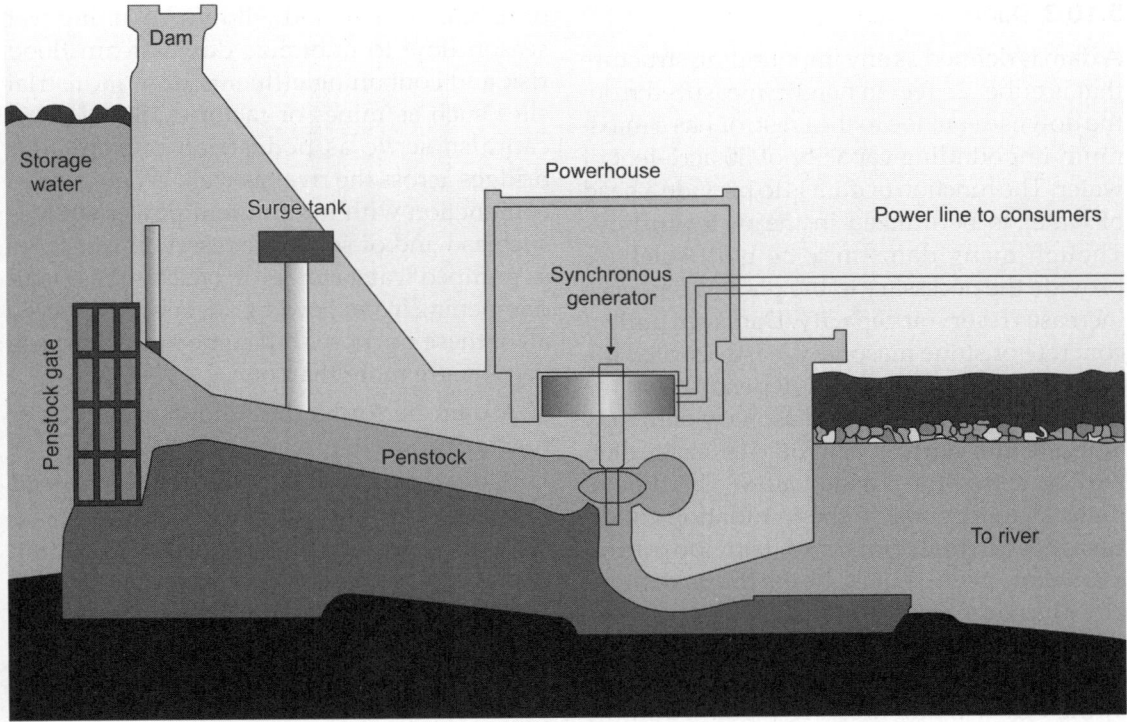

Fig. 5.2: Layout of hydropower station

5.10 COMPONENTS OF HYDROELECTRIC SCHEME

A typical hydroelectric scheme has the following essential components:

1. Catchment area
2. Reservoir
3. Dam
4. Spill ways
5. Penstock
6. Surge tanks
7. Draft tubes
8. Powerhouse

5.10.1 Catchment Area

The whole area behind the dam draining into a stream or river across which the dam has been constructed is called the catchment area. The catchment area of a hydro plant is the whole area behind the dam, draining into a stream or river across which the dam has been built at a suitable place. The characteristics of the catchment area include its size, shape, surface orientation, altitude, topography and geology. The bigger the catchment, steeper is the slope, higher is the altitude, and greater is the total runoff water.

5.10.2 Reservoir

Whole of the water available from the catchment area is collected in a reservoir behind the dam. The purpose of the storing of water in the reservoir is to get a uniform power output throughout the year. Water stored is not only used for power generation, but also for irrigation, flood control, water supply and navigation. A reservoir may be natural, like a lake to maintain or artificially built by erecting a dam across a river. Water held in upstream reservoir is called storage, whereas, water behind the dam at the plant is called poundage. A natural reservoir is a lake in high mountains and an artificial reservoir is made by constructing a dam across the runs.

5.10.3 Dams

A dam is defined as any impounding structure that is either 25 feet in height, measured from the downstream toe to the crest, or has a maximum impounding capacity of 50 acre-feet of water. The function of dam is to provide a head of water to be utilized in the water turbine. Though many dams may be built solely to provide the necessary to the plant, a dam also increases reservoir capacity. Dams are built of concrete or stone masonary, earth or rock fill. The type and arrangement depends upon the topography of the site. A masonary dam may be built in a narrow canyon. An earth dam may be suited for a wide valley. The type of dams also depends on the foundation conditions, local materials and transportation available, occurrences of earthquakes and other hazards. Structures that fail to meet these criteria but have the potential to cause significant property damage or pose a threat to life in the downstream area are regulated in the same manner as dams. All such structures except federal dams and those permitted by the division of mine reclamation and enforcement must be reviewed, and a stream construction permit must be issued by this office. Design criteria, hazard classification information and submittal requirements can be found in the publication "Design Criteria for Dams and Associated Structures."

Construction inspections are performed periodically and during critical stages of work. Upon completion of construction, the owner submits a notice of completion and as-constructed drawings. When as-constructed drawings are received, a final inspection is conducted. If all work is satisfactory, the owner is granted permission to impound water and the completed dam is placed on the inventory of dams maintained by the section.

Intended purposes include providing water for irrigation to a town or city water supply, improving navigation, creating a reservoir of water to supply industrial uses, generating hydroelectric power, creating recreation areas or habitat for fish and wildlife, retaining wet season flow to minimize downstream flood risk and containing effluents from industrial sites such as mines or factories. Some dams can also serve as pedestrian or vehicular bridges across the river as well. When used in conjunction with intermittent power sources such as wind or solar, the reservoir can serve as pumped water storage to facilitate base load dampening in the power grid. Few dams serve all of these purposes but some multi-purpose dams serve more than one.

A dam performs the following two basic functions:

1. It develops a reservoir of the desired capacity to store water and
2. It builds up a head for power generation dams, may be classified according to their structural materials such as: timber, steel, earth and rock filled and masonry. Timbers and steel are used for dams of height 6 m to 12 m only. Earth dams are built for larger heights, up to about 100 m to protect the dam from the wave erosion, a protecting coat of rock, concrete or planking must be laid at the water line. To protect the dam from the wave erosion, a protecting coat of rock, concrete or planting must be laid at the water line. The other exposed surfaces should be covered with grass or vegetation to protect the dam from rainfall erosion.

5.10.3.1 Failure of Dams

1. **Water leakage:** If required action is not taken at the right time, when water leakage is observed, leads to dam failure. Proper maintenance is required to maintain the dam for long life. It is necessary to anticipate any problems and action to be taken before structure fails.
2. **Poor maintenance:** If maintenance is not carried at regular intervals, it leads to dam failure.
3. **Spillway design error:** When river flow exceeds the storage capacity, the disfunctional spillway leads to dam failure.

4. **Poor survey:** Dams built on slopes must be properly engineered to avoid issues with instability or landslide. If it is still continued, leads to dam failure.

5. **Material:** Building material used for the dam should be of high quality. Low quality building material leads to dam failure.

6. **Maintenance:** Poor maintenance like when water is overflowing, if the gates are not working properly, leads to dam failure.

7. **Foundation:** Defects can occur in the foundation supporting the dam. If this weight is not properly taken into account in the engineering of the dam, the ground underneath can settle unequally and compromise the foundation. Similarly, any event causing the movement of a foundation, such as an earthquake, can also compromise a dam's foundation. The main cause of concrete dam failure is a problem with the foundation. High uplift pressures and uncontrolled foundation seepage can also compromise the dam's foundation. When the foundation of an earth fill dam is composed of fine silt, clay, or similar soft soil, the whole dam may slide due to water thrust.

8. **Seepage failure:** Seepage always occurs in the dams. If the magnitude is within design limits, it may not harm the stability of the dam. However, if seepage is concentrated or uncontrolled beyond limits, it will lead to failure of the dam. Following are some of the various types of seepage failure.

  i. **Piping through dam body:** When seepage starts through poor soils in the body of the dam, small channels are formed which transport material downstream. As more materials are transported downstream, the size of channels grow bigger and bigger which could lead to wash out of dam.

  ii. **Piping through foundation:** When highly permeable cavities or fissures or strata of gravel or boorish sand are present in the dam foundation, it may lead to heavy seepage. The concentrated seepage at high rate will erode soil which will cause increased flow of water and soil. As a result, the dam will settle or sink leading to failure and strength.

5.10.3.2 Disadvantages of Dams

1. **Human land loss:** Many poor people lost their lands as a result of dams. People who already lost agricultural lands, caused major unemployment in some countries like India.

2. **Failure of dam:** Dam is a major storage unit of water. The dam failure leads to flood and related catastrophies.

3. **Overflow of water:** Over topping of dam during rainy season leads to vacate the houses of many people in small towns.

5.10.3.3 Advantages of Dams

1. **Cheap:** If the dam is maintained properly, a hydroelectric power source in a given area is comparatively cheap.

2. **Reliable:** Once dam is constructed with design and by quality engineers then dam will have a long life and is reliable. It uses no fuel and with low escape risk, and as an alternative energy source it is cheaper than both nuclear and wind power.

5.10.3.4 Classification of Dams

Dams can be classified in various ways based on the following:

1. **Function:** Based on their functions dams can be classified into storage dams, diversion dams and detention dams. Storage dams are mainly for storing water and using it for various purposes such as hydro-power, irrigation and water supply. Diversion dams are constructed to raise the water level and to divert the river flow in another direction. Detention dams are used to store flood waters.

2. **Shape:** Dams can be either trapezoidal or arch to gauze structural function.

3. **Material of construction:** Can be constructed of earth fills, rock pieces, stone masonry, concrete, RCC and even of timber and rubber. Plain as well as steel-reinforced, earthern and rockfill dams are most popular.

4. **Hydraulic and structural design:** According to hydraulic design dams can be of non-overflow type, in which water is not allowed to flow over the top of the dam and the overflow type which allows water to flow over it. Based on the structural design there can be gravity dam, arch dam and buttress dam, where water thrust is resisted by gravity, arch action and buttresses.

5.10.3.5 Selection of Site and Choice of Dam

The selection of site depends on the function of the dam. Smaller the length of dam, less are the costs. The site has to be the one where the river valley has a neck formation. Quite often, the dam is located after the confluence of two rivers, so as to provide large storage capacity. Selection of site also depends on the geology, accessibility of materials by transport, economics and safety. Failure of dam is a big catastrophe and hence a dam has to satisfy tests of stability during high floods and shock loads, say due to earthquakes. Further considerations regarding the selection of site are the following:

i. **Geology of foundations:** The station should be located in a place where the land or the rock structure on which the dam will be built on is strong enough to hold the weight and the force of the water in the dam. The walls should have a capability of holding and sustaining both visible and invisible forces, whether man-made or natural. The rock structures should have the capability of withstanding an earthquake and it should not allow seepage of water, since this weakens the dam. The walls should be waterproof to avoid being weakened by water.

ii. **River path:** The best location for a hydro-electric power station should be along the path of a river. It should be at least at the river canyon or at the place where the river narrows. This enables the collection of the water or the diversion of the river. If the hydroelectric station aims to store maximum water on the dam, the volume of the basin located above the dam should be calculated to ensure that the dam does not suffer from the problem of insufficient water supply, which in turn, would affect the running of the turbines.

iii. **Raw material:** The materials used in the construction of dams determine whether it will last for long and effectively serve its purpose. The materials that are used to make the walls of the dam should be able to hold the force of the water. This means the site for the dam should be at a place where these materials, such as cement and ballast, can be easily found. It is crucial to use high-quality materials to prevent disasters, such as water flooding in areas near the dam.

iv. **Sufficient water:** The flow of the water to the place where the dam is located should be sufficient enough to fill the dam. Hydroelectric dams are usually big, and this makes them lose a lot of water through evaporation. The flow of the water from the river should be high enough to accommodate this loss of water without affecting the amount of electricity produced.

5.10.3.6 Types of Dams

1. **Gravity dams:** Gravity dams are constructed either in stone masonry or in concrete. The concrete dams can, however, be built faster with better quality control. It is a dam that curves upstream in a narrowing curve that directs most of the water against the canyon rock walls, providing the force to compress the dam. The gravity dam is a massive concrete or earth type structure. Gravity dams use their own weight to resist opposing forces. They rely on their great weight and size for stability. The gravity dam is the most commonly built dam in the world.

Gravity dams classified with respect to their structural height, are:

- Low, up to 10 feet
- Medium high, between 10–30 feet
- High, over 30 feet

2. **Earth dams:** For a small project of upto 70 m height, dams constructed are of earth fill or embankment. A large volume of material is required and it should be available in the vicinity. The dam construction varies with the height and the side slopes are flatter. It is cheaper than masonry dam, but has more seepage losses.

3. **Rockfill dams:** It is made of loose rock of all sizes and has a trapezoidal shape with a wide base, having a watertight section to reduce seepage.

5.10.4 Spillways

Spillways may be considered as a sort of safety valve for a dam. Passage for surplus water over or around a dam when the reservoir itself is full. Spillways are particularly important safety features for earth dams, protecting the dam and its foundation from erosion. They may lead over the dam or a portion of it or along a channel around the dam or a conduit through it. There are times when the river flow exceeds the storage capacity of the reservoir. Such a situation arises during heavy rain fall in the catchment areas. In order to discharge the surplus water from the storage reservoir into the river on the downstream side of dam, spillways are used. Spillways are constructed of concrete piers on the top of the dam. Gates are provided between piers and surplus water is discharged over the crest of the dam by the opening these gates.

There are two types of spillways:

  i. **Controlled spillway:** In this type gates are provided to regularize the flow of water. When water is overflowing the dam, gates are mainly used to control the pressure of water.

  ii. **Uncontrolled spillway:** In this type, gates are not provided. When the water rises above the "crest" of the spillway, it begins to release from the reservoir. The rate of discharge is controlled only by the depth of water within the reservoir. All of the storage volume in the reservoir above the spillway crest can be used only for the temporary storage of floodwater, and cannot be used as water supply storage because it is normally empty.

5.10.5 Penstock

Penstocks are open or closed conduits which carry water to the turbines in hydropower stations. Penstock is generally made of reinforced concrete or steel. Concrete penstocks are suitable for low heads less then 30 m. The steel penstocks are designed for any head. The thickness of penstocks increases with head or water pressure. When the distance from the forebay to the powerhouse is short separate penstocks are used for each turbine. In high head dams, the penstocks are provided with penstock gates or butterfly valves and air valves. The penstock gates are fixed and flow of water is controlled by operating penstock gates. In high head power plants the butterfly valve and air valve are provided in the powerhouses. Air valve maintains the air pressure inside the penstock equal to atmospheric pressure. When water runs out of penstock faster then it enters, a vacuum is created which may cause the penstock to collapse. Under such circumstances, air valve opens and admits the air in the penstock to maintain inside pressure equal to the outside. If the butterfly suddenly opens and water enters with high pressure there will be chances of penstock collapse. Hence, there is a water pressure valve fixed before the butterfly valve, which maintains the water pressure at both sides of penstock by bypassing butterfly valve. Before operating the butterfly valve, this water pressure valve operates and fills up the water at the other side of penstock. The setting of the system is such that the butterfly valve operates, when the water pressure at both sides of penstock are equal.

There are two types of penstock which are mainly used for power generation:

Low pressure penstock: The low pressure type consists of a canal commonly flume or a pipe.

High pressure penstock: The high pressure consists of steel pipes which can take water under pressure

5.10.6 Surge Tank

Surge tank is an additional storage near to turbine, usually provided in high head plants. A surge tank is located near the beginning of the conduit (penstock). When turbine is running at a steady load, there are no surges in the flow of water through the conduit, i.e. the quantity of water flowing in the conduit is just sufficient to meet the turbine requirements. As the load on the turbine decreases or during load rejection by the turbine the surge tank provides space for holding water. Similarly when load on turbine increases, it furnishes additional water. Thus the conduit (penstock) prevented from bursts. Hence, a surge tank overcomes the abnormal pressure in the conduit when load on the turbine falls and acts as a reservoir during increase of load on the turbine. Several designs of surge tanks have been adopted in power stations, the

important considerations being the amount of water to be stored, the amount of pressure to be release off and the separate space available at the site of construction.

The guide bearing is of pivoted pad type with self contained oil bath lubrication (Fig. 5.3). It consists of Babbitt lined pads arranged along the outer circumference of skirt of the shaft. Each pad is adjustable by means of studs. The pads remain pivoted against the spherical end of studs. The bearing housing is of split construction and is secured on the upper side and covered by studs. Four dowels are used for locating bearing housing on top cover. In stationary condition, the pads are kept immersed in oil bath up to a level slightly above the centre line of bearing. Rotation of shaft induces centrifugal force, due to which oil flows through the holes in shaft collar and rises along the journal surface, thus lubricating the pads. The oil returns through the pipeline bearing body and is led to the oil sump by gravity. The oil cooled in the oil sump which is directly placed on turbine top cover and cooling is carried out by main turbine discharge. The oil after passing through the tank flows back into the bearing through a pipeline provided on the bottom ring.

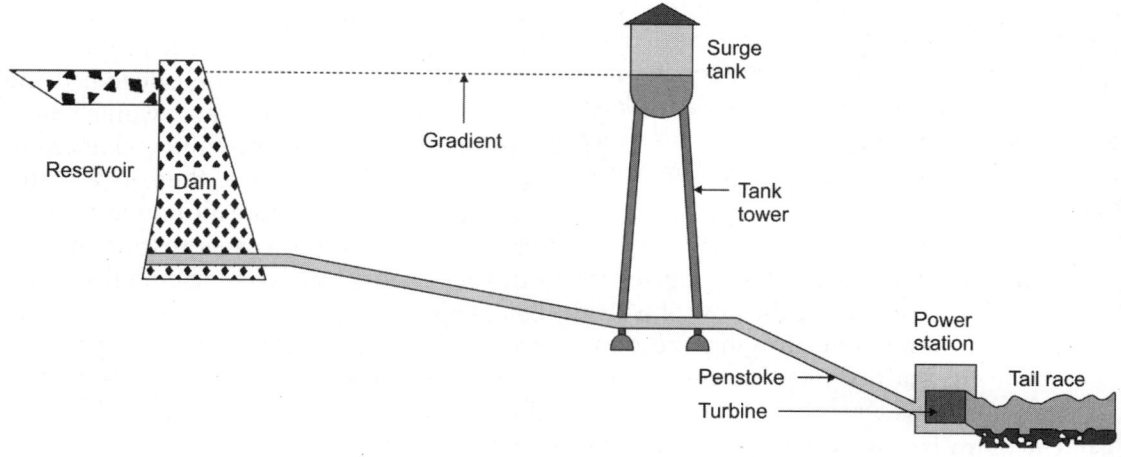

Fig. 5.3: Guide bearing

5.10.7 Draft Tube

The draft tube is of welded steel construction and consists of a cone and elbow liner. A water and air tight manhole access having clear opening with hinged door, bronze hinge pins, stainless steel bolts and jacking screws has been provided in cone. A test cock and a pressure gauge are provided. Slots are provided below the manhole for supporting the inspection platform. When water flows on the turbine there is sudden pressure difference existing between water in the turbine and atmosphere. Therefore turbines are completely enclosed. Hence it is necessary to connect the turbine outlet by means of a pipe or passage like conic shape upto tail race level. The simplest and most efficient, turbine draft tube is the conical shaped draft tube. It is usually vertical and is designed with a truncated cone similar to an inverted ice cream cone. Originally, turbines were designed without draft tubes. Therefore, work on the runner, stop logs were inserted into the tailrace training walls and the discharge pit was pumped out. It converts a large proportion of the velocity energy rejected from the runner into useful pressure head as it recuperates energy.

5.10.8 Powerhouse

A powerhouse should have a stable structure and its layout should be such that adequate space is provided around the equipment for convenient dismantling and repair. The equipment provided in the powerhouse includes: hydraulic turbines, electric generators, governors, gate valves, relief valves, water circulation pumps, air duct, switch board and instruments, storage batteries and cranes.

5.11 CLASSIFICATION OF HYDRO-ELECTRIC POWER PLANTS

In hydroplants, water is collected behind the dam. This reservoir of water may be classified as either storage or poundage according to the amount of water flow regulation they can exert. The function of the storage is to impound excess river flow during the rainy season to supplement the low rate flow during dry seasons. They can meet the demand of load fluctuations for 6 months or even a year. Poundage involves in storing water during low loads, so that this water can be utilized for carrying the peak loads during the week. They can meet the hourly as well as weekly fluctuations of load demand with poundage, the water level always fluctuates during operations. It rises at the time of storing water, fails at the time of drawing water, remains constant when the load is constant.

The hydropower plants can be classified as below:

1. **Based on the availability of head:**
   i. **High head plants:** Hydroelectric power station operating on a high head of water is generally located at high altitudes. (Fig. 5.4). They take advantage of the large difference of level between the dam which holds back the water and the plant where electricity is generated. These plants work under a head of 100 m and above. This plants works above 500 mtrs and Pelton wheel turbines are commonly used. In this plant, water is carried out from the main reservoir by a tunnel up to surge tank and then from the surge tank to the powerhouse in penstock. Mainly in these plants pressure tunnel is provided before the surge tank, which in turn connects the penstock. A pressure tunnel is taken off from the reservoir and water brought to the valve house at the start of the penstock. The penstocks are huge steel pipes which take large quantity of water from the valve house to the powerhouse. The valve house contains main sluice gates and in addition automatic isolating valves which come into operation when the penstock bursts, cutting further supply of water. Surge tank is an open tank and is built just in between the beginning of the penstock and the valve house. In absence of surge tank, the water hammer can damage the

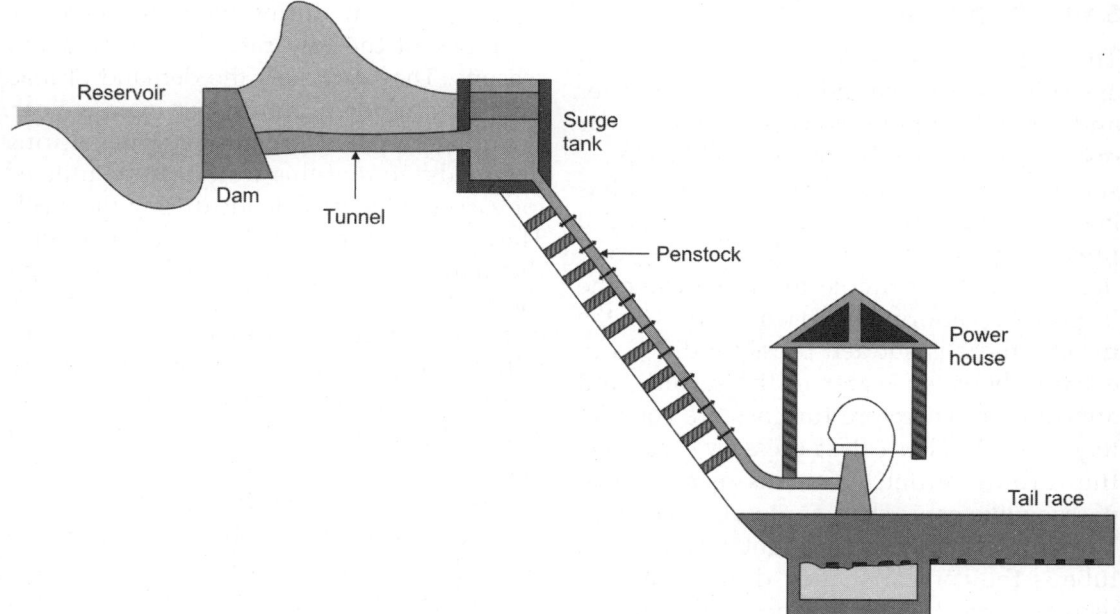

Fig. 5.4: High head plant

fixed gates. In majority of dams sluice gates are provided. The sluice gates are opened when dam level is below level and there is shortage of water for irrigation. Normally the high head plants are 500 m above and for heads above 500 m Pelton wheels are used.

ii. **Medium head plants:** These plants operate under heads varying from 30 m to 100 m. Hydroelectric power station operating on a medium head of water of approximately a few dozens of meters are located in mountainous regions of middling high. In general, the plant is located at the foot of the dam. Mainly forebay provided before the penstock, acts as a water reservoir for medium head plants. In this plants mainly water is carried through main reservoir to forebay and then to the penstock. The forebay acts as a surge tank for these plants. The turbines used will be Francis type of the steel encased variety (Fig. 5.5).

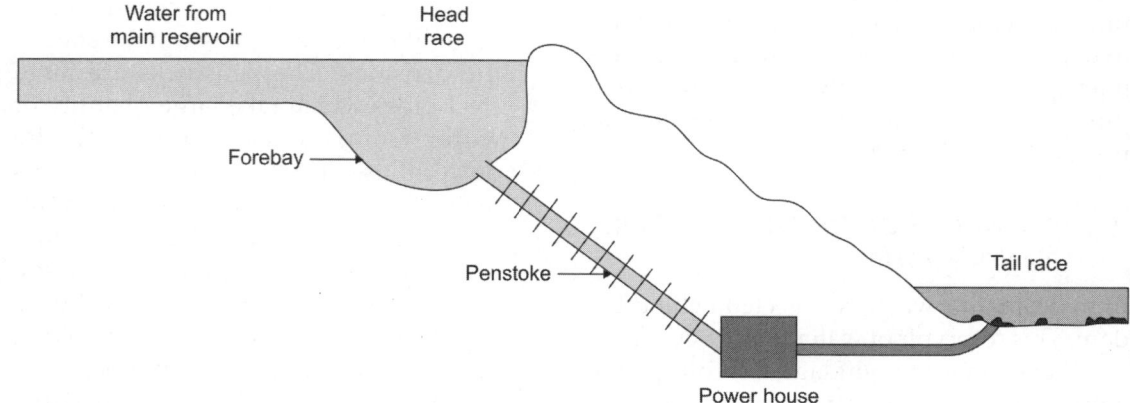

Fig. 5.5: Medium head plant

iii. **Low head plants:** A small dam is built across the river to provide the necessary head (Fig. 5.6). The excess water is allowed to flow over the dam itself. Francis, Propeller or Kaplan type of turbines are used for power generation. Also no surge tank is required. These plants are constructed where the water head available less then 30 m. The production of electricity will be less due to low head.

2. **Based on quantity of water available:** These plants can be classified as either without poundage or with poundage. A run-off river plant without poundage has no control over river flow and uses the water as it comes. These plants usually supply peak load. During floods, the tail water level may become high rendering the plant inoperative. A run-off river plant with poundage may supply base load or peak load power. At times of high water flow it may be base loaded and during dry seasons it may be peak loaded.

i. **Run-off plants without poundage:** As the name indicates this type of plant doesn't store water, the plant uses as water comes.

ii. **Run-off plants with poundage:** Poundage permits storage of water during the off-peak period and uses this water during peak periods.

iii. **Pumped storage power plants:** These plants are used when quantity of water available for generation is insufficient. If it is possible to pond at head water and tail water locations afterpassing through the turbine, is stored in the tail race pond from where it may be pumped back to the head water pond.

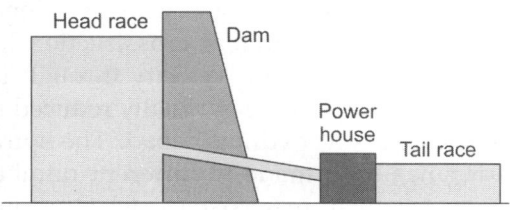

Fig. 5.6: Low head plant

iv. **Reservoir plants:** A reservoir plant is that, which has reservoir of such size as to permit carrying overstorage from wet season to the next dry season.

5.12 ROLE OF HYDROELECTRIC STATIONS IN POWER INDUSTRY

In power industry, three types of power plants exist: the first one being thermal (fossil fuel powered), the other is nuclear power plant (energy released in fission reaction). The third one is hydro-electric plant where power is generated from the generator coupled with the turbine being rotated by water energy. All types of power stations are linked together by HV/EHV transmission lines. Power generation must be used as soon as it is generated. Thus, it is obvious that at any time the power demand must by equal to generation. However, power demand not often remains constant. It varies with time of day and season. Thus generation is needed to be adjusted all through. Thermal power plants are put into service throughout the day though the generation capacity may to some extent vary. Nuclear plants supply the base loads and hydroelectric plants can be utilized to take care of the flexibilities in the generation and demand. Its power output can be altered very quickly. Hydroelectric plants can also be utilized as reserve capacity plants. In some hydel power stations the dam is constructed only for irrigation purpose. Power should be generated whenever water demand increased by farmers in irrigation dams. Irrigation can remarkably improved by routing the canals in the downstream of the hydel plants. Draught and flood both can be effectively controlled utilizing the dam water and discharge water of hydel plants. Though hydroelectric power can effectively serve major share in power generation still the role played by it cannot be fully evaluated by the share of hydro power in the overall production alone. Apart from purely power generating aspect, hydel stations have most commendable effect on environmental control and much lower man-power requirement. Hydroelecric power doesn't pollute the atmosphere and the unit cost of production of energy is obviously low as the main 'fuel' is only water. However, the hydel generation is to some extent dependent on weather variations.

5.13 PUMPED STORAGE HYDROPOWER PLANTS

In these plants mainly water is pumped back for more generation of power. Pumped-storage hydroelectricity is a type of hydro-electric power generation used by some power plants for load balancing. The method stores energy in the form of water pumped from a lower elevation reservoir to a higher elevation. Low-cost off-peak electric power is used to run the pumps. These plants are used when amount of water available for generation of power is otherwise inadequate. If it is possible to pond, at head water and tail water locations water after passing through the turbine is stored in the tail race pond where it may be pumped back to head water pond. The pumping back from tail race pond to the head water pond is done during off-peak period. During peak load period water is drawn from the head water pond through the penstock to operate the turbines.

Although the losses of the pumping process make the plant a net consumer of energy overall, the system increases revenue by selling more electricity during periods of peak demand, when electricity prices are highest. Pumped storage is the largest-capacity form of grid energy storage now available. (Fig. 5.7). Such plants can recover almost 70% of the power used in pumping the water. A recent development in this field is the use of rever-sible turbine-pump unit. Such unit can be used as a turbine while generating power and as pumping water to storage. The generator in this case works as a motor during reverse operation. The efficiency in such a case is high and is almost the same in both the operations. With the use of reversible-turbine pump sets,

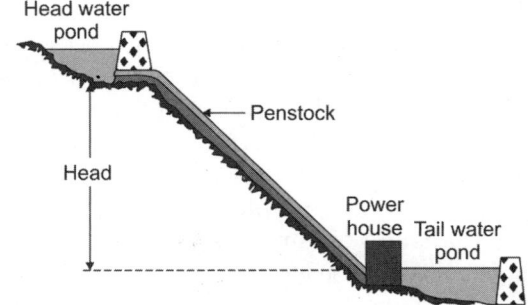

Fig. 5.7: Pumped storage plant

additional capital investment on pump and its motor can be saved and the scheme can be worked more economically. Such plants can be operated only in interconnected systems where other generating plants (steam, hydro etc.) are also available.

A pumped-storage plant has two reservoirs:

1. **Upper reservoir:** Like a conventional hydro-power plant, a dam creates a reservoir. The water in this reservoir flows through the hydropower plant to create electricity. Using a reversible turbine, the plant can pump water back to the upper reservoir. This is done in off-peak hours. Essentially, the second reservoir refills the upper reservoir. By pumping water back to the upper reservoir, the plant has more water to generate electricity during periods of peak consumption.

2. **Lower reservoir:** Water exiting the hydro-power plant flows into a lower reservoir rather than re-entering the river and flowing downstream.

5.14 SPIRAL CASING

The spiral casing and stay ring are fabricated from steel plates. The spiral casing is of loga-rithmic form and circular cross section for maintaining a constant velocity through its passage. The plates are gradually reduced in thickness to suit hydraulic load. The spiral casing has been made in different number segments which are welded to each other and with stay ring at site. Stay ring is made in two parts, joined by machined flanges and fasteners and are steel welded. The top cover and runner envelop are attached to stay ring flanges. Before 1849, in old turbines spiral casing was not there and water was directly following in the centre of the runner and was radially flowing outwards. At that time the American Engineer, James B. Francis, set out to improve upon the design of the few hydraulic turbines operating at the time in France and the US.

All investigations of Mr. Francis proved that water to enter the runner from the outside and to flow inward through the radial blades. Afterwards the design was improved so that the water was turned from a radial to an axial path within the runner, rather than outside it. Looking at the result it was found that the water pressure can be controlled and turbine was named as Francis. Mainly Francis turbines comes under reaction turbines. The reaction turbines designed as per velocity and the pressure of the water. Water pressure in the penstock is more and directly hitting on the runner may damage the runner blades. The reaction turbine controls the flow of water equally maintains the turbine speed.

5.15 INDIA TO CONSTRUCT TUNNEL FOR QUAKE STUDIES

An 8 km deep underground tunnel next to one of India's largest dams overlooking a pictures-que lake in southern Maharashtra may soon provide scientists with vital clues on how earthquakes can be predicted.

Through the tunnel, scientists will lower a string of instruments for setting up an under-ground seismic observatory, which will collect data to help understand the complex geo-physical processes leading to an earthquake.

"Koyna is the only place in the world for this experiment as quakes are happening within a small area of 20 × 30 km and even within that 600 sq km, they are concentrated within a core area of 5 sq km," Harsh Gupta, who success-fully predicted Koyna dam's potential to trigger earthquake more than three decades ago told the Deccan Herald on the sidelines of the 98th Indian Science Congress.

The first step, a 350-crore experimental project planned on that day with India signing a memorandum of understanding with the International Continental Drilling Programme ensuring participation of foreign scientists in that project. "By June, we would be able to finalise the scientific plans and go for the cabinet approval. The actual construction," said Shailesh Naik, secretary in the Ministry of Earth Sciences. An international workshop will be organised in March. Later, scientists will go to the site to decide the tunnel's exact location. Once the observatory is in place, the researchers will record hundreds of tiny quakes with magnitude one and more every year besides four to five quakes of magnitude 2 and at least five to six quakes of magnitude 3 and above. They also hope to get at least one magnitude 4 or 5 quake once in every two years.

The experiment has been initially planned for five years involving a large number of institutions and universities. There will be enough number of earthquakes to generate data on the stress build up and how the rock behaves under stress condition. This will help understand the physics of earthquake, using which a prediction model can be developed.

"With fresh information, our approach to earthquake prediction will be closer to reality," said Gupta who had once forecasted on quakes in the north-east and raised red flags on Koyna. Even though deep boreholes were dug to install underground observatories in Kobe (Japan) and Chi Chi (Taiwan), those tunnels were dug at the boundaries of tectonic plates and did not yield much information. The experiments can neither be conducted at the 1,300 km long San Andreas fault in California nor in the Himalayan range as there is no way to find out the exact region where the stress is being built up.

5.16 HYDRAULIC TURBINE

Hydraulic turbines convert the potential energy of water into shaft work, which, in turn, rotates the electric generator coupled to it in producing electric power. A water turbine is a *rotary engine* that takes energy from moving water. Water *turbines* were developed in the nineteenth century and were widely used for industrial power prior to electrical grids. Now they are mostly used for *electric power* generation. They are clean and *renewable energy* sources. Flowing water is directed on to the blades of a turbine runner, creating a force on the blades. Since the runner is spinning, the force acts through a distance. In this way, energy is transferred from the water flow to the turbine. Hydraulic turbines convert the potential energy of water into shaft work, which, in turn rotates the electric generator coupled to it in producing electric power. Hydraulic turbines transfer the energy from a flowing fluid to a rotating shaft. Turbine itself means a thing which rotates or spins. Hydraulic turbines have a row of blades fitted to the rotating shaft or a rotating plate. Flowing liquid, mostly water, when pass through the hydraulic turbine, strikes the blades of the turbine and makes the shaft rotate. While flowing through the hydraulic turbine the velocity and pressure of the liquid reduces, these result in the development of torque and rotation of the turbine shaft. There are different forms of hydraulic turbines in use depending on the operational requirements. For every specific use a particular type of hydraulic turbine provides the optimum output.

I. **Classification of hydraulic turbines:** *Based on flow path*—water can pass through the hydraulic turbines in different flow paths. Based on the flow path of the liquid hydraulic turbines can be categorized into three types:

   a. **Axial flow hydraulic turbines:** This category of hydraulic turbines has the flow path of the liquid mainly parallel to the axis of rotation. Kaplan turbines has liquid flow mainly in axial direction.

   b. **Radial flow hydraulic turbines:** Such hydraulic turbines has the liquid flowing mainly in a plane perpendicular to the axis of rotation.

   c. **Mixed flow hydraulic turbines:** For most of the hydraulic turbines used there is a significant component of both axial and radial flows. Such type of hydraulic turbines are called as mixed flow turbines. Francis turbine is an example of mixed flow type, in which water enters in radial direction and exits in axial direction.

   None of the hydraulic turbines are of purely axial flow or purely radial flow. There is always a component of radial flow in axial flow turbines and of axial flow in radial flow turbines.

II. **Classification of hydraulic turbines:** *Based on pressure change*—One more important criterion for classification of hydraulic turbines is whether the pressure of liquid changes or not, while it flows through the rotor of the hydraulic turbines. Based on the pressure change hydraulic turbines can be classified into two types:

   a. **Impulse turbine:** The pressure of liquid does not change while flowing through the rotor of the machine. In impulse turbines pressure change occur only in the nozzles of the machine. These types of turbines are mainly used in high head plants, water from high head received with high velocity and high kinetic energy impinges on the buckets of the wheel and the wheel rotates. In this turbine the entire pressure of water is converted into kinetic energy in a nozzle and the velocity of the jet drives the blades of turbine. The nozzle consist of a needle, and quantity of water jet falling on the turbine is controlled, this needle placed on the tip of the nozzle. If the load on the turbine decreases, the governor pushes the needle into the nozzle, thereby reducing the quantity of water striking the turbine.

Newton's second law describes the transfer of energy for impulse turbines.

Impulse turbines are most often used in very high (>300 m/984 ft) head applications.

Power: The power available in a stream of water is

$$P = \eta \cdot \rho \cdot g \cdot h \cdot \dot{q}$$

where, P = power (J/s or watt)

η = turbine efficiency

ρ = density of water (kg/m³)

g = acceleration of gravity (9.81 m/s²)

h = head (m). For still water, this is the difference in height between the inlet and outlet surfaces. Moving water has an additional component added to account for the kinetic energy of the flow. The total head equals the pressure head plus velocity head.

$\dot{q}$ = flow rate (m³/s)

Examples of impulse turbines are:

5.16.1 Pelton Wheel

A Pelton wheel, also called a Pelton turbine, is one of the most efficient type of water turbines available. It was invented by Lester Allan Pelton (1829–1908) in the 1870s, and is an impulse machine, meaning that it uses Newton's second law to extract energy from a jet of fluid. The Pelton wheel turbine is a tangential flow impulse turbine, water flows along the tangent to the path of the runner. Studying the high head plants the Pelton turbines are designed. The Pelton turbines are available in various sizes. Depending on that number of nozzles decided. Nozzles are arranged around the wheel such that the water jet emerging from a nozzle is tangential to the circumference of the wheel of Pelton turbine. Accordingly nozzles direct forceful streams of water against a series of spoon-shaped buckets mounted around the edge of a wheel. Each bucket reverses the flow of water, leaving it with diminished energy. The resulting impulse spins the turbine. The buckets are mounted in pairs, to keep the forces on the wheel balanced, as well as to ensure smooth, efficient momentum transfer of the fluid jet to the wheel. The Pelton wheel is most efficient in high head applications. For a constant water flow rate from the nozzles the speed of turbine changes with changing loads on it. For quality hydroelectricity generation the turbine should rotate at a constant speed. To keep the speed constant, despite the changing loads on the turbine water flow rate through the nozzles is changed. To stop the striking water jet to the turbine blades when load decreased on the turbine, the servo controlled spear valves are used in jets. In sudden reduction of load jets are made to stop by deflector plates so that over speed of turbine should not take.

In a Pelton turbine or Pelton wheel water jets impact on the blades of the turbine making the wheel rotate, producing torque and power. Learn more about design, analysis, working principle and applications of Pelton wheel turbine. Hydraulic turbines are being used from very ancient times to harness the energy stored in flowing streams, rivers and lakes. The oldest and the simplest form of a hydraulic turbine was the waterwheel used for grinding grains. Different types of hydraulic turbines were developed with the increasing need for power. Three major types are Pelton wheel, Francis and Kaplan turbine.

5.16.1.1 Design of Pelton Wheel Turbine

The Pelton turbine has a circular disk mounted on the rotating shaft or rotor. This circular disk has cup shaped blades, called as buckets, placed at equal distance around its circumference. Nozzles are arranged around the wheel such that the water jet emerging from a nozzle is tangential to the circumference of the wheel of Pelton turbine. According to the available water head (pressure of water) and the operating requirements the shape and number of nozzles placed around the Pelton wheel can vary.

Fig. 5.8: Pelton wheel turbine

5.16.1.2 Working Principle of Pelton Turbine

The high speed water jets emerging from the nozzles strike the buckets at splitters, placed at the middle of a bucket, from where jets are divided into two equal streams. These streams flow along the inner curve of the bucket and leave it in the direction opposite to that of incoming jet. The high speed water jets running the Pelton wheel turbine are obtained by expanding the high pressure water through nozzles to the atmospheric pressure. The high pressure water can be obtained from any water body situated at some height or streams of water flowing down the hills.

The change in momentum (direction as well as speed) of water stream produces an impulse on the blades of the wheel of Pelton turbine. This impulse generates the torque and rotation in the shaft of Pelton turbine. To obtain the optimum output from the Pelton turbine the impulse received by the blades should be maximum. For that, change in momentum of the water stream should be maximum possible. That is obtained when the water stream is deflected in the direction opposite to which it strikes the buckets and with the same speed relative to the buckets.

5.16.1.3 Pelton Turbine Hydroelectric Setup

A typical setup of a system generating electricity by using Pelton turbine will have a water reservoir situated at a height from the Pelton wheel (Fig. 5.9). The water from the reservoir flows through a pressure channel to the penstock head and then through the penstock or the supply pipeline to the nozzles, from where the water comes out as high speed jets striking the blades of the Pelton turbine. The penstock head is fitted with a surge tank which absorbs and dissipates sudden fluctuations in pressure.

For a constant water flow rate from the nozzles the speed of turbine changes with changing loads on it. For quality hydroelectricity generation the turbine should rotate at a constant speed. To keep the speed constant despite the changing loads on the turbine, water flow rate through the nozzles is changed. To control the gradual changes in load servo controlled spear valves are used in the jets to change the flow rate. And for sudden reduction in load the jets are deflected using deflector plates, so that some of the water from the jets do not strike the blades. This prevents over speeding of the turbine.

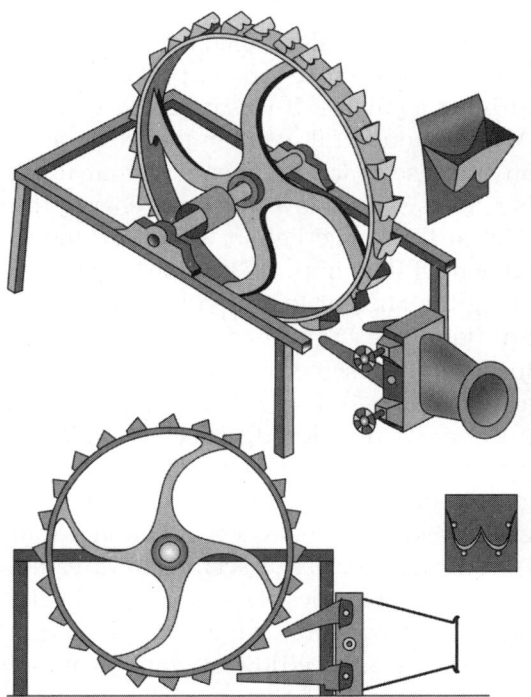

Fig. 5.9: Pelton wheel

5.16.2 Reaction Turbine

Reaction turbine: The pressure of liquid changes while it flows through the rotor of the machine. The change in fluid velocity and reduction in its pressure causes a reaction on the turbine blades; this is where the name reaction turbine may have been derived. Francis and Kaplan turbines fall in the category of reaction turbines.

Examples of reaction turbines are:

1. Francis turbine
2. Kaplan turbine

In reaction turbine, low head water with low kinetic energy and high potential energy glides over the turbine blades and are mainly for low and medium head plants. In reaction turbine the water enters the runner partly with pressure energy and partly with velocity head. Most water turbines in use are reaction turbines and are used in low (<30 m/98 ft) and medium (30–300 m/98–984 ft) head applications. In reaction turbine pressure drop occurs in both fixed and moving blades. In this turbine the runner blades changed with respect to guide vane opening. As the sudden decrease of load takes place, the guide vane limit decreases according to that runner blade closes.

5.16.2.1 Francis Turbine

Francis turbines are very versatile. These are reaction turbines, i.e. during energy transfer from water to the runner there is a drop in static pressure as well as a drop in velocity head. Francis turbine water flow is radial into the turbine and exits the turbine axially. Water pressure decreases as it passes through the turbine imparting reaction on the turbine blades making the turbine rotate. Francis turbine is the first hydraulic turbine with radial inflow. It was designed by the American scientist James Francis. Francis turbine is a reaction turbine. Reaction turbines have some primary features which differentiate them from impulse turbines. The major part of pressure drop occurs in the turbine itself, unlike the impulse turbine where complete pressure drop takes place up to the entry point and the turbine passage is completely filled by the water flow during the operation.

5.16.2.2 Design of Francis Turbine

Francis turbine has a circular plate fixed to the rotating shaft perpendicular to its surface and passing through its center. This circular plate has curved channels on it, the plate with channels is collectively called as runner. (Fig. 5.10). The runner is encircled by a ring of stationary channels called as guide vanes. Guide vanes are housed in a spiral casing called as volute. The exit of the Francis turbine is at the center of the runner plate. There is a draft tube attached to the central exit of the runner. The design parameters such as, radius of the runner, curvature of channel, angle of vanes and the size of the turbine as a whole depend on the available head and type of application altogether.

5.16.2.3 Working of Francis Turbine

Francis turbines are generally installed with their axis vertical (Fig. 5.11). Water with high head (pressure) enters the turbine through the spiral casing surrounding the guide vanes. The water looses a part of its pressure in the volute (spiral casing) to maintain its speed. Then water passes through guide vanes where

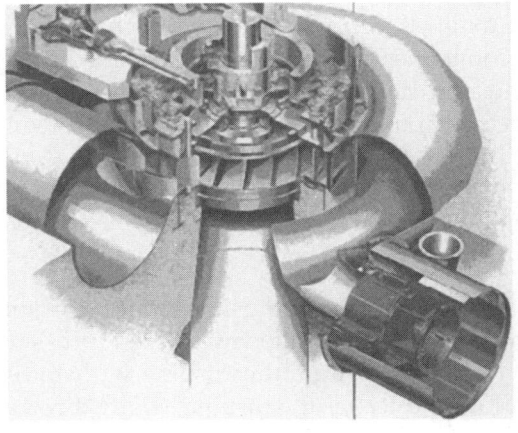

Fig. 5.10: Design of Francis turbine

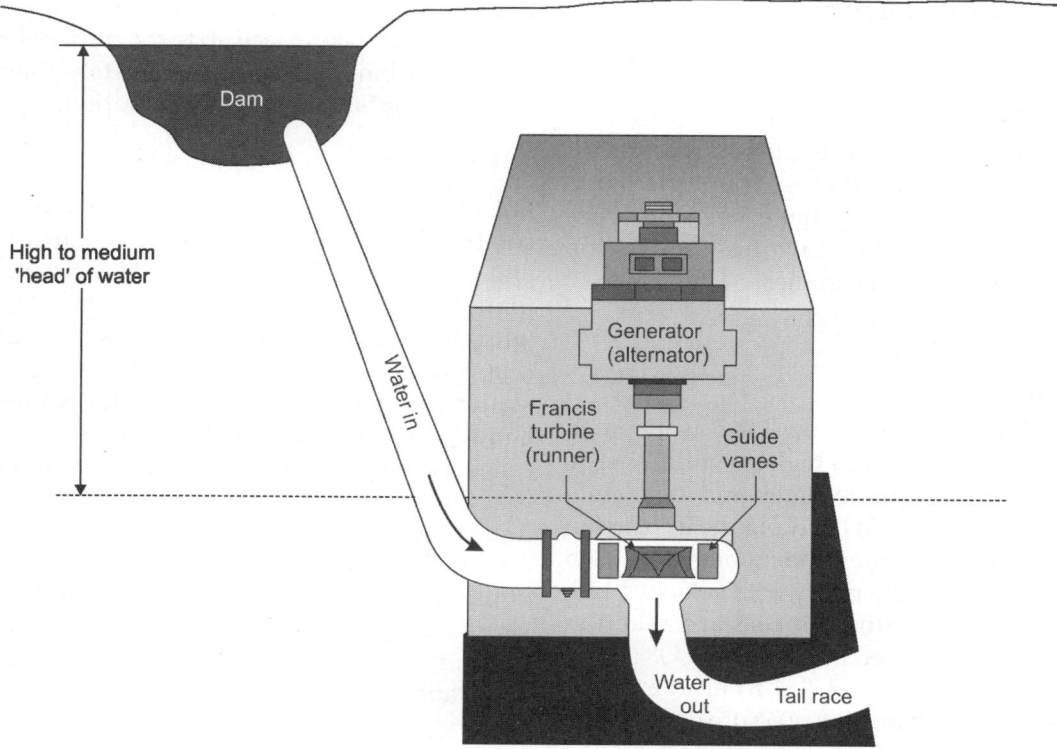

High to medium
'head' of water

Fig. 5.11: Francis turbine

it is directed to strike the blades on the runner at optimum angles. As the water flows through the runner its pressure and angular momentum reduces. This reduction imparts reaction on the runner and power is transferred to the turbine shaft. If the turbine is operating at the design conditions the water leaves the runner in axial direction. Water exits the turbine through the draft tube, which acts as a diffuser and reduces the exit velocity of the flow to recover maximum energy from the flowing water.

5.16.2.4 Power Generation Using Francis Turbine

For power generation using Francis turbine, the turbine is supplied with high pressure water which enters the turbine with radial inflow and leaves the turbine axially through the draft tube. The energy from water flow is transferred to the shaft of the turbine in form of torque and rotation. The turbine shaft is coupled with dynamos or alternators for power generation. For quality power generation speed of turbine should be maintained constant, despite the changing loads. To maintain the runner speed constant even in reduced load condition the water flow rate is reduced by changing the guide vanes angle.

5.16.2.5 Kaplan Turbine

Kaplan turbine, is also a reaction turbine, but with a different design than Francis. Kaplan turbine is designed for low water head applications. Kaplan turbine has propeller like blades but works just reverse. Instead of displacing the water axially using shaft power and creating axial thrust, the axial force of water acts on the blades of Kaplan turbine and generates shaft power.

The Propeller turbine is a reaction turbine used for low heads (4 m–80 m) and high specific speeds (300–1000). The propeller runner may be considered as a development of a Francis type in which the number of blades is greatly reduced and the lower band omitted. It is an axial flow device providing large flow of area utilizing a large volume flow of water with low flow velocity.

Most of the turbines developed earlier were suitable for large heads of water. With increasing demand of power, need was felt to harness power from sources of low head water, such as, rivers flowing at low heights. For such low head applications Viktor Kaplan designed a turbine similar to the propellers of ships. Its working is just reverse to that of propellers. The Kaplan turbine is also called as Propeller turbine. The Kaplan turbine is a propeller type water turbine which has adjustable blades. It was developed in 1913 by the Austrian Professor Vikter Kaplan, who combined automatically adjusted propeller blades with automatically adjusted wicket gates to achieve efficiency over a wide range of flow and water level. This turbine has attained popularity and rapid progress has been made in recent years in the design and construction of this turbine.

The Kaplan turbine was an evolution of the Francis turbine. Its invention allowed efficient power production in low head applications that was not possible with Francis turbines. The head ranges from 10–70 meters and the output from 5 to 120 MW. Runner diameters are between 2 and 8 meters. The range of the turbine is from 79 to 429 rpm. Kaplan turbines are now widely used throughout the world in high-flow and low head power production.

5.16.2.6 Design of Kaplan Turbine

To generate substantial amount of power from small heads of water using Kaplan turbine it is necessary to have large flow rates through the turbine. Kaplan turbine is designed to accommodate the required large flow rates. Except the alignment of the blades the construction of the Kaplan turbine is very

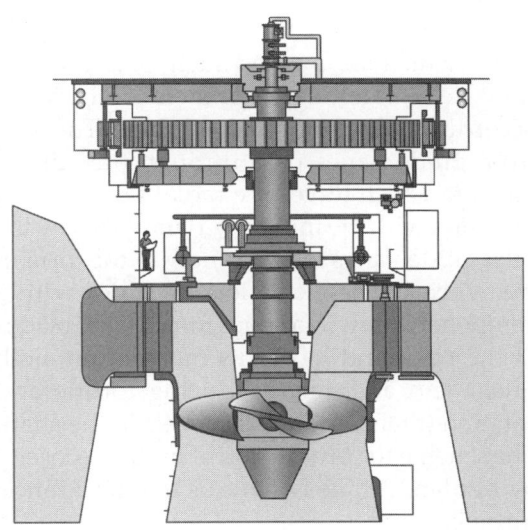

Fig. 5.12: Kaplan turbine

much similar to that of the Francis turbine. The overall path of flow of water through the Kaplan turbine is from radial at the entrance to axial at the exit. Similar to the Francis turbine, Kaplan turbine also has a ring of fixed guide vanes at the inlet to the turbine. Unlike the Francis turbine which has guide vanes at the periphery of the turbine rotor (called as runner in the case of Francis turbine), there is a passage between the guide vanes and the rotor of the Kaplan turbine. The shape of the passage is such that the flow which enters the passage in the radial direction is forced to flow in axial direction. The rotor of the Kaplan turbine is similar to the propeller of a ship. The rotor blades are attached to the central shaft of the turbine. The blades are connected to the shaft with moveable joints, such that the blades can be swiveled according to the flow rate and water head available. The blades of the Kaplan turbine, are not planer as any other axial flow turbine; instead they are designed with twist along the length so as to allow swirling flow at entry and axial flow at ·exit (Fig. 5.12).

5.16.2.7 Working of the Kaplan Turbine

The working head of water is low, so large flow rates are allowed in the Kaplan turbine.

The water enters the turbine through the guide vanes which are aligned such as to give the flow a suitable degree of swirl determined according to the rotor of the turbine. The flow from guide vanes pass through the curved passage which forces the radial flow to axial direction with the initial swirl imparted by the inlet guide vanes which is now in the form of free vortex. The axial flow of water with a component of swirl applies force on the blades of the rotor and looses its momentum, both linear and angular, producing torque and rotation (their product is power) in the shaft. The scheme for production of hydroelectricity by Kaplan turbine is same as that for Francis turbine.

5.17 COMPARISON OF TURBINES

The characteristic features of common types of turbines are summarized in Table 5.4.

5.18 CAVITATION

Cavitation is formation of vapour bubbles in the liquid flowing throught any hydraulic turbine. Cavitation occurs when the static pressure of the liquid falls below its vapour pressure. It is most likely to occur near the fast moving blades of the turbines and in the exit region of the turbines.

When the velocity of a fluid increases, it's pressure falls. In any turbine, if the pressure drops below the vapour pressure at that temperature some of the liquid flashes into vapour. The bubbles formed during vaporization are carried by the water stream to higher pressure zones, where the bubbles condense into liquid forming a county and vacuum. The surrounding liquid rushes towards the county giving rise to a very high local pressure which may be as high as 7000 atm. The formation of such a county and high pressure occurs repeatedly hundreds of times in a second. This phenomenon is known as cautation, which causes pitting on the metallic surface of runner blades and draft tube, accompanied by considerable vibration and noise. Cavitation should be minimized or avoided by selecting proper material like stainless steel or alloy steel, by adequate polishing of the surface, by selecting a runner of low specific speed or by keeping the runner under water.

Table 5.4: Comparison of common turbines

	Pelton wheel	Francis turbine	Kaplan/propeller turbine
1. Flow	Tangential, single stage, impulse	Inward radial flow, single stage reaction	Axial flow, single stage reaction
2. Maximum capacity	250 MW	720 MW	225 MW
3. Number of jets/kind of blades	1 to 6 maximum, 2 for horizontal and 6 for vertical shaft	Fixed blades	Propeller tubrbines have fixed blades, while Kaplan turbines have adjustable blades
4. Head	100–1750 m	30–550 m	1.3–77.5 m
5. RPM	75–1000	93.8–1000	72–600
6. Hydraulic efficiency	Single jet 85–90%	90–94%	85–93%
7. Specific speed	6–60	50–400	280–1100
8. Regulation mechanism	Spear nozzle and deflector plate	Guide vanes	Blade stagger

Causes of Cavitation

The liquid enters hydraulic turbines at high pressure, this pressure is a combination of static and dynamic components. Dynamic pressure of the liquid is generated by the flow velocity and the other components, static pressure is the actual fluid pressure which the fluid applies and which is acted upon it. Static pressure governs the process of vapour bubble formation or boiling. Thus, cavitation can occur near the fast moving blades of the turbine where local dynamic head increases due to action of blades which causes static pressure to full. Cavitation also occurs at the exit of the turbine as the liquid has lost major part of its pressure heads and any increase in dynamic head will lead to fall in static pressure causing cavitation.

Methods to Avoid Cavitation

To avoid cavitation while operating hydraulic turbines parameters should be set such that at any point of flow static pressure may not fall below the vapour pressure of the liquid. These parameters to control cavitation are pressure head, flow rate and exit pressure of the liquid. The control parameters for cavitation free operation of hydraulic turbines can be obtained by conducting tests on model of the turbine under consideration. The parameters beyond which cavitation starts and turbine efficiency falls significantly should be avoided while operation of hydraulic turbines. Flow separation at the exit of the turbine in the draft tube causes librations which can damage the draft tube. To dampen the libration and stabilize, the air flow is injected in the draft tube. To totally avoid the flow separation and cavitation in the draft tube, it is submerged below the level of the water in tail race. The cavitation effect can be reduced by selecting materials which can resist better the cavitation effect. The cast steel is better than cast iron and stainless steel or alloy steel is still better than cast steel. The pitting effect of cavitation on cast steel can be repaired mere economically by ordinary welding. It has been observed that the welded parts are mere resistant to cavitation than ordinary ones. The cavitation effect is less on polished surfaces than ordinary one.

5.19 SELECTION OF TURBINES

The major problem confronting the engineering is to select the type of turbine which will give maximum economy. The hydraulic prime-mover is always selected to match the specific conditions under which it has to operate and attain maximum possible efficiency. The hydraulic turbine is selected according to the specific conditions under which it has to operate and attain the maximum possible efficiency. The choice depends on the head available, power to be developed and the speed at which it has to run. The following factors basically govern the selection of suitable type of turbine.

a. **Operating head:** The present practice is to use Kaplan and Propeller type of turbines for heads upto 50 m. For head from 50 to 400 m , Francis turbines are used. For heads greater than 400 m, impulse or Pelton turbines are used.

b. **Specific speed:** It is better to choose turbines of high specific speeds. High speed turbines mean small sizes of turbines, generators, powerhouse, etc. and are therefore more economical. In all the modern power plants, it is a common practice to select a high specific speed runner because it is more economical as the size of the turbo-generator as well as that of powerhouse will be smaller.

c. **Size of turbine:** It is better to go in for as large a size of turbine as possible since this results in economy of size of the power-house, the number of penstocks, the gene-rator, etc. Bigger size means less number of runners. However, the number of runners should not be less than two so that at least one unit is always available for service in the case of a plant breakdown.

d. **Height of installation:** It is better to install the turbines as high above the tail water level as possible. This saves the cost of excavation for the draft tube. Care should be taken to ensure that cavitation does not occur.

5.20 ENVIRONMENTAL ASPECTS CONCERNED WITH HYDROPOWER

Hydro force plants are environment friendly and good with the Earth (dissimilar to fossil fuel plants which discharge CO_2, SOx, NOx, particulate matter and nuclear force plants connected with radioactive wastes. The real ecological issues to be investigated before arranging and authorizing a hydroplant are however, related with the supply volume and tail race water stream and their effect on the population, farming, geology, topography, backwoods, amphibian life, and so on. The effect is connected with the catchment region. The licenses from natural powers for extensive hydro undertakings ought to be taken ahead of time.

6

Magnetohydrodynamic Energy Conversion

6.1 INTRODUCTION

We all are cognizant of power generation by utilizing hydel, thermal and atomic resources. In all the systems, the potential energy or thermal energy is initially changed into mechanical energy and afterwards the mechanical energy is changed over into electrical energy. The change of potential energy into mechanical energy is respectably high (70% to 80%), yet transformation of thermal energy into mechanical energy is significantly poor (40% to 45%). In expansion to this concept, the mechanical segments needed for changing over heat energy into mechanical energy are vast in number and impressively costly. This requires immense capital cost and additionally upkeep require likewise. The researchers are of speculation to eliminate the mechanical system and change thermal energy into immediate electrical energy throughout the previous 50-years and more, unfortunately, no framework is yet advanced in vast capacity (MW) to compete with conventional systems. In expansion to this the effectiveness of such transformation remained respectably poor (less than 10%) therefore, these force producing generating systems are not improved on extensive scale.

6.2 THERMODYNAMIC ENERGY CONVERSION

The power generation process, regularly, is characterized by the movement of primary or secondary energy, from thermal to mechanical and after that to power. At the present state

of advancement, the greater parts of the power plants are dependent upon methodologies reputed to be accepted as conventional methods. The establishment of electric power plants, through conventional forms or commercial primary energy, concern just the hydroelectric and thermal power stations, where the thermal power stations are distinctive for utilization of essential source (typically fossil fuel energies, for example, natural gas, oil, coal, wood and biomass, metropolitan or industrial solid wastes, and so on, or atomic powers and geothermal energy). In hydroelectric power generation, mechanical vigor, in diverse structures (dynamic, potential and pressure) from streaming liquid, is changed over into electric power with a water turbine and an alternator. In the thermal power plants the thermal energy is changed over into mechanical energy and from this machine the mechanical energy into power. The dominant parts of thermal power plants are powered by fuels, generally fossil or atomic. Except a few cases, for example, power plants that utilize thermal energy accessible as a part of nature (principally sun oriented and geothermal), the type of energy at the base of each of course of action is the chemical potential energy of the fuel. The potential vigor of the fossil fuel is changed over into high temperature heat energy through a chemical exothermic response (combustion), characterized by generation of thermal energy proportionate, in absolute value, to the enthalpy variation for the same reaction. On account of atomic energy there is a fission reaction. The high temperature is then transmitted to versatile working fluid evolving in suitable machines (generally gas turbine or responding motor) preparing mechanical work. All things considered, it has changed over thermal energy to mechanical (thermodynamic change) energy. It ought to be noted that in any transformation process one can not fully completely change over the energy starting with one structure then into the next, each of the steps being described by a conversion efficiency, a coefficient that considers the portion of the energy at first accessible, which is changed over in the desired form. Fossil fuels (coal, fuel oil and natural gas) still meet our energy interest and are getting exhausted. Additionally, there are not kidding contamination perils like greenhouse impact and worldwide global warming which happen because of fossil fuel smoldering. The utilization of atomic power excessively has its issues and atomic combination is yet to be acknowledged in practice. We have hence been compelled to search for nonconventional energy generation systems in order to decrease fossil fuel utilization by expanding the transformation effectiveness of fuel to power. A significant efficiency might be attained by changing over "high temperature" (inner energy) specifically to power by taking out the connection procedure of transforming mechanical vigor through steam (Rankin cycle). Real coordinate vigor transformation units are magneto hydrodynamic, thermo-electric generators and power devices. In this section we will describe the accompanying unconventional energy change systems which have respectable impact on the energy situation of what's to come in future.

6.3 DIRECT ENERGY CONVERSION ROUTES

The possibilities of improving significantly the conventional energy conversion processes are mainly related to technological progress. In any case they have little edge and thus the scientists have turned to the improvement of different frameworks, the non-conventional. In the conventional conversion systems significant losses of energy take place in the transition, from thermal to mechanical energy (thermodynamic conversion). Exploration is centering its exertions on transformation forms that don't utilize this step. The absence of moving mechanical parts may permit the accomplishment of working temperatures much higher than those run of the mill of routine methods, coming about in this manner, at any rate possibly, a higher transformation productivity. These processes are known as direct conversion, as primary and secondary energy is changed over straight forwardly into electricity without the requirement to pass through a phase of mechanical energy. The direct energy conversion routines that these days are considered in terms of industrial application are:

- Photovoltaic generation systems (photovoltaic solar cells)
- Electrochemical energy conversion (fuel cells)
- Magnetohydrodynamic generation (MHD)
- Electrogasdynamic generation (EGD)
- Thermoeletric power generation

In the first two processes the changes are from the primary to the secondary energy form and takes place avoiding the conversion of the intermediate thermal energy. This (Fig. 6.1), shows the energy conversion stages in the direct generation of electric energy.

The outline of a energy converter is regularly managed by the sort of energy to be changed over, despite the fact that it is the obligation of the designer to search out new and more proficient methods for changing the essential wellsprings of the energy into power. There are numerous explanations behind the utilization of new and immediate transformation plans. These could be catagorizes into three imperative territories: productivity,

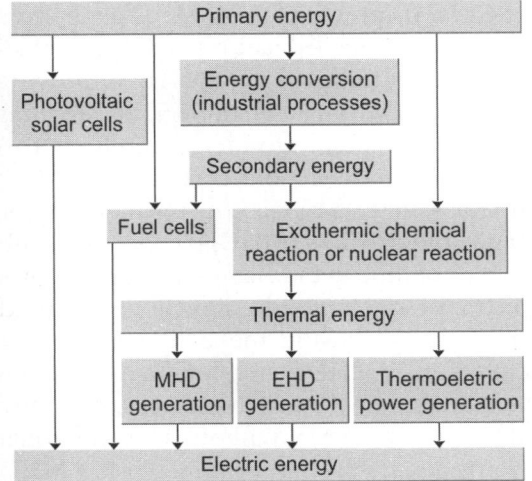

Fig. 6.1: Direct energy conversion stages

dependability, and the utilization of new sources of energy. It is trusted that when a technique happens specifically, instead of passing through numerous steps, it is liable to be more proficient. This will accelerate less consumption of the essential energy store and a more level venture for every (introduced) unit power. Efficiencies are, then again, still low at this phase of improvement of most immediate energy transformation plans. With respect to dependability, there are spots where energy change supplies must run for a long time without breaking down and without support. These are circumstances where a definitive dependability is needed. At last, the likelihood of utilizing new sources of energy appears upgraded by the improvement of the new direct energy converters. There are numerous ways whereby the immediate energy change of thermal to electrical energy might be acquired. In this section the fundamental one, magnetohydrodynamic force era, is specified in detail to give the general background of interest in direct conversion. MHD power era is a stylishly straightforward strategy. Magnetohydrodynamic (magneto-liquid flow or hydro-magnetic) is the scholastic order which studies the elements of electrically directing liquids. Illustrations of

such fluids incorporate plasmas, fluid metals, and salt water. The generator utilized as a part of this procedure is called magnetohydro-dynamic generator (MHD). It takes after the rocket motor encompassed by tremendous magnet. It has no moving parts and the real conductors are displaced by ionized gas (plasma). Hence it has a quite high effective-ness. The MHD power era engineering (MHD) is the handling of electrical force using a high temperature leading plasma traveling through an intensive magnetic field. The transforma-tion in MHD was first depicted by **Michael Faraday** in 1893. However the genuine use of this idea remained unbelievable. The primary known endeavor to improve a MHD generator was made at vesting house research lab (USA) around 1936. The efficiencies of all present day thermal power creating framework lies between 35% to 40% as they need to reject substantial amounts of high temperature to the Earth. In all other tried and true power plant, first the thermal energy of the gas is directly changed over into electrical energy. Henceforth, it is reputed to be run energy transformation framework. The MHD force plants are grouped into open and closed cycle depending upon the transforming of the working fluid. With the present examination and advancement programmes, the MHD power generation may assume an imperative part in the power industry in future to help the present emergency of force. The MHD methodology might be utilized for business power era as well as for such a variety of different provisions. MHD guarantees a sensational change in the expense of creating power from coal, helpful to the development of the national economy. The valuable environ-mental point of view of MHD are presumably of equivalent or considerably of more signi-ficant importance. The MHD energy change methodology can help enormously to the result of the serious air and thermal conta-mination resulting in issues confronted by all steam-electric force plants while it simul-taneously guarantees better use of natural resources. It can accordingly be asserted that the advancement of MHD for electric utility power generation is an goal of national significance. The high temperature of MHD procedure makes it conceivable to false advantage of the highest flame temperature which could be transformed by ignition from fossil energizes.

It is a non-conventional source of energy which is based upon Faraday's law of electromagnetic induction, which explains that energy is generated due to the movement of an electric conductor inside a magnetic field (Fig. 6.2).

In the world, 80% energy produced is hydal, while remaining 20% is produced from nuclear, thermal, solar, geothermal energy and from magnetohydrodynamic (MHD) generators. Magnetohydrodynamic is a field of science dealing with the interaction between magnetic field, flow of hot electrically ionized fluid (gaseous or liquid) and generation of electrical field – MHD generators change thermal energy to electrical energy directly (without rotating machines). The magnetohydrodynamic involves: strong magnetic fields (5 to 6 T), flowing ionized fluid at high temperatures (2500 K) containing thermal energy and enthalpy. The energy fluid is either heat ionized (seeded) gas of hot ionized liquid metal and generation of DC electrical energy from thermal energy.

The connection between the magnetic field and the ionized working liquid in movement is utilized for era of electrical energy directly from thermal energy and enthalpy. The plant for processing electrical vigor by MHD is called MHD power plant of MHD generators (Fig. 6.3). In propelled nations MHD generators are broadly utilized yet as a part of improving nations like India it is still under joint exertions of BARC (Bhabha atomic research centre) BHEL, associate cement corporation (ACC) and Russian technologies. As the name infers, magnetohydro flow (MHD) is concerned with the flow of fluid in the vicinity of magnetic and electric field. The liquid may be gas at lifted temperature or fluid metal like sodium or potassium. A MHD generator is a mechanism for changing heat energy of a fuel specifically into electrical energy without a conventional electric generator. In this framework, a MHD convertor system is a high temperature motor, in which heat energy consumed at a higher temperature is incompletely changed over into advantageous (electrical) work and the remnant is rejected at a lower level temperature. Like all high temperature motor, the thermal productivity of a MHD convertor (i.e. the extent of the hotness consumed is changed over into functional work) is expanded by supplying the heat at the highest practical temperature and dismissing it at the lowest practical temperature. MHD generation looks the most promising of the immediate change methods for the bigger scale handling of electric force. It is watched that monetary and physical variables will accelerate configuration yields of the request of 1000 MW. Infact MHD is truly of investment just for focal force era, its possibilities for a drive unit are remote. We all are attentive to power era utilizing hydel, thermal and atomic resources in all the

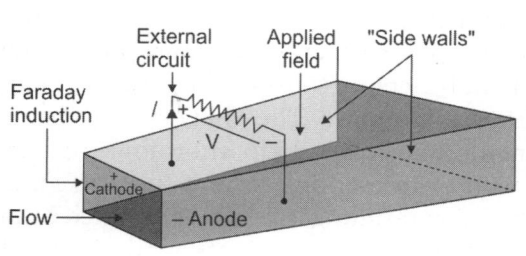

Fig. 6.2: Magnetic field

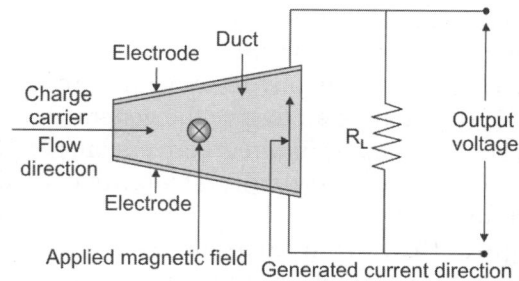

Fig. 6.3: MHD channel

systems, the potential energy or thermal energy is initially changed over into mechanical energy and after that the mechanical energy is changed over into electrical vigor. The transformation of potential energy into mechanical energy is impressively high (70% to 80%) yet the change of thermal energy into mechanical energy is significantly poor (40% to 45%). In expansion to this the mechanical segments needed for changing heat energy into mechanical energy are expansive in number and respectably costly. This requires colossal capital cost and upkeep require likewise. The researchers are in constant search to take out the mechanical system and convert thermal energy into immediate electrical energy throughout the previous 50-years and more. Unfortunately, no system is yet improved in substantial capacity (MW) to capital with conventional systems. In expansion to this the proficiency of such transformation remained respectably poor (less than 10%) therefore, these power creating systems are not advanced on vast scale.

6.4 PRINCIPLES OF MHD POWER GENERATION

Of all the direct energy conversion methods exploitable, the MHD power generation seems to be the most promising for a utility system. The maximum limiting temperature for turbine blades being 750°–800°C, the MHD generator is capable of tapping the vast potential offered by modern furnaces, which can reach temperatures of more than 2500 K, and upto 3000 K with preheating of air.

The standard principle of MHD generation is basically that identified by Faraday. The point when an electric transmitter moves over an magnetic field, a voltage is actuated in it which produces an electric current. The conductor may be a robust, fluid or gas. This is the guideline of the expected generator likewise where the conductors comprise of copper strips. In MHD generator, the strong conductors are displaced by a vapourous conductor, an ionized gas. In the event that

such a gas is passed at a high speed through an effective magnetic field, a current is created and can be removed by putting cathodes in a suitable position in the steam. Since a quite high temperature is obliged to ionize a gas (thermal ionization) which can't be induced by the materials accessible, the hot gas is seeded with an alkali metal, for example, cesium or potassium (K_2CO_3 or KOH) having a low ionization potential before the gas enters the MHD channel. A satisfactory electrical conductivity of the order of 10 mho/m can hence be acknowledged to a degree at lower temperature in the extent of 2200°–2700°C, (Fig. 6.4).

6.4.1 Faraday's Law of Electromagnetic Induction

When an electric conductor moves across a magnetic field, an emf is induced in it, which produces an electric current. MHD power generation process is governed by Faraday's law of electromagnetic induction, (i.e. when the conductor moves through a magnetic field, it generates an electric field perpendicular to the magnetic field and direction of conductor). The flow of the conducting plasma through a magnetic field at high velocity causes a voltage to be generated across the electrodes, perpendicular to both the plasma flow and the magnetic field according to Fleming's right hand rule (Fig. 6.5).

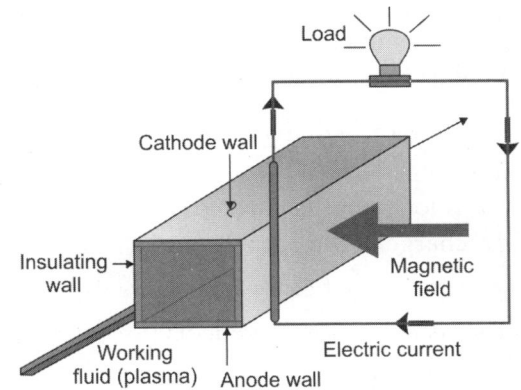

Fig. 6.4: A simple perspective of the MHD generator

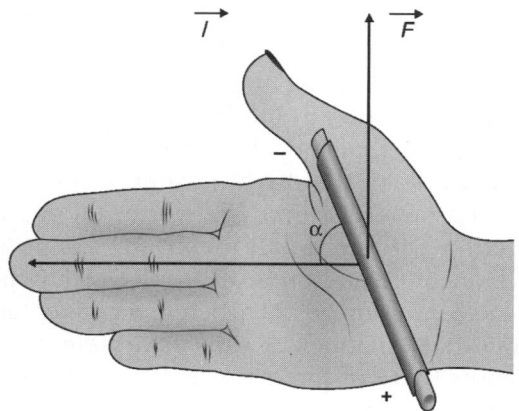

Fig. 6.5: Fleming's right hand rule

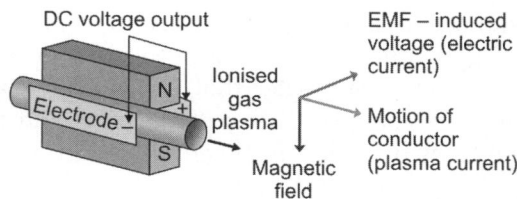

Fig. 6.6: Magnetohydrodynamic power generation (principle)

The principle can be explained as follows: An electric conductor moving through a magnetic field experiences a retarding force as well as an induced electric field and current. This effect is a result of Faraday's law of electromagnetic induction. The induced emf is given by

$$E_{ind} = \vec{\mu} \times \vec{\beta}$$

where, μ is the velocity of the conductor and B is the magnetic field intensity as shown, the induced current density is

$$J_{ind} = \sigma \vec{E}_{ind}$$

where σ is electric conductivity. The retarding force on the conductor is the Lorentz force. The Lorentz Force Law describes the effects of a charged particle moving in a constant magnetic field. The simplest form of this law is given by the vector equation.

Lorentz force on the charged particle (vector),

$$\vec{F} = q(v \times \vec{B})$$

where,

v = velocity of the particle (vector)

q = charge of the particle (scalar)

$\vec{B}$ = magnetic field (vector)

The vector $\vec{F}$ is perpendicular to both v and $\vec{B}$ according to the right hand rule.

From the energy perspective, the development of the energy through a displacement (mechanical work) is changed over into electrical work (current stream against potential difference) by the method of the electromagnetic induction standard. This is a work energy change and is not restricted by the Carnot rule, obviously, electromagnetic induction structures the foundation of operation for expected electric generators (transform it). The electromagnetic induction standard need not be constrained to solid channels, the development of a leading liquid through an magnetic field might additionally be utilized for electric energy change when a liquid is utilized, the energy transformation strategy is known as the magnetohydrodynamic (MHD), energy change. The direct transformation of motor (or movement) energy into electrical energy by the flow of an electrically leading liquid, generally a gas or a gas-fluid mix, through a stationary magnetic field. Assuming that the flow direction is at right angles to the magnetic field heading, an electromotive power (or electric voltage) is induced in the direction, and right angles to both flow and field directions (Fig. 6.6).

A schematic of MHD generator is demonstrated in Fig 6.7. The conducting flow liquid is constrained between the plates with a kinetic energy and weight differential sufficient to defeat the magnetic induction force find. The closure perspective drawing delineates the development of the flow channel. An ionized gas is utilized as the conducting liquid. Ionization is prepared either by thermal means, i.e. by a raised temperature or by seeding with substance like cesium or potassium vapours which ionize generally at low temperature. The molecules of the seed components separate from electrons. The vicinity of the adversely charged electrons makes the carrier gas an electrical transmitter. The other path is to join a fluid metal into a flowing carrier gas since the metal is a good electrical conductor, the gas metal mixture might be utilized as the working liquid within a MHD generator (refer Fig. 6.7).

6.4.2 Thermal Efficiency

Clearly the overall MHD cycle is thermal power cycle and as such is limited by the correct efficiency. The cycle thermal efficiency may be written as

$$n_{th} = \frac{\text{Work output}}{\text{Heat output}}$$

$$= \frac{(h_{20} - h_{30}) - (h_{10} - h_{40})}{(h_{20} - h_{10})}$$

where the indicated enthalpies are stagnation values which take into account the KE of the flow. The stagnation enthalpy is defined as

$$h_0 = \frac{h + u^2}{2gc}$$

where, u is the flow velocity. In practical MHD converters, high velocity ionized gases are usually employed for the conversion process, so that the KE of the flow represents a substantial portion of the total energy. The ideal process 2–3 is shown as entropic, which would assume perfect conversion of the flow pressure and kinetic energy into electric energy.

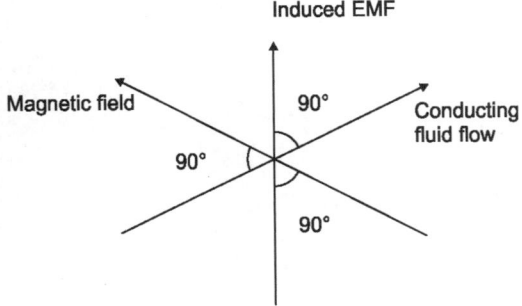

Fig. 6.7: Principle of magnetohydrodynamic conversion

Several variables can reduce the efficiency of the converter. These are:

i. Dissipation of energy in the internal resistance of the ionized,

ii. A space charge boundary at the electrode surface.

iii. Heat transfer through the electrode and insulator walls.

iv. Various losses connected with fluid friction etc.

v. Hall effect losses resulting in current induction in the direction of the flow.

Such losses could be significant when the seeded burning gases are utilized as the working liquid. The *Hall effect* causes a voltage slope in a bearing perpendicular to the connected magnetic field and current stream. The gas, encounters a breaking force because of electromagnetic interaction, which could be contrasted with the retarding energy processed by the turbine razor sharp edges in the conventional system. As specified above, the gas must be ionized in place that it may direct sufficiently for fruitful operation. In practice the level of ionization required is quite little (0.1%), so the gas is still made basically of neutral particles. It is these particles that carry recently all the KE of the stream, and obviously, they are unaffected by electromagnetic forces. The exact mechanism of the breaking power has not yet been demonstrated, yet it is felt that the connected attractive field shows itself through the force that it pushes on the electrons in the gas. This power is then coupled with the impartial particles by the electron-ion Coulomb powers and the ion-neutral collisions. The retarding energy in this way be an intricate capacity of collsion cross-area and attractive field thickness. In a functional MHD convertor (or generator), the energy of movement of the conduction liquid is determined from high temperature acquired by blazing a fossil fuel. Consequently, a MHD generator is a gadget for changing heat energy specifically into electrical energy without an accepted electric generator. Fig. 6.8, represents the comparison between a turbo generator and a MHD generator.

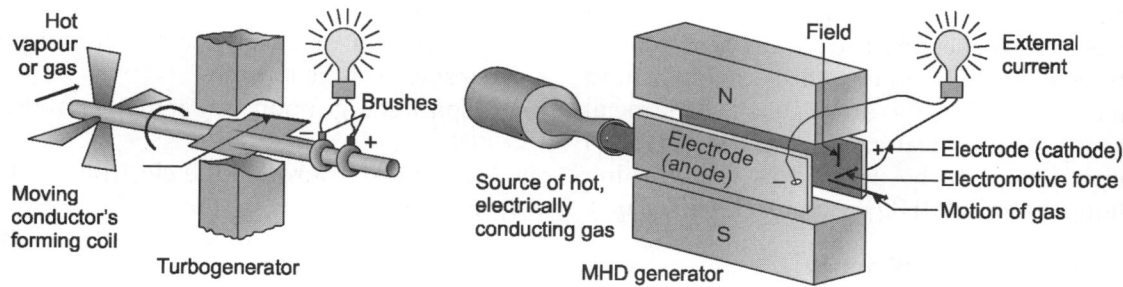

Fig. 6.8: Comparison between a turbo and MHD generator

6.5 CONSTRUCTION

Its development is exceptionally basic. MHD generator looks like the rocket motor encompassed by the tremendous magnet. It has no moving parts and the actual conductors are displaced by ionized gas (plasma). The magnets utilized could be electromagnets or superconducting magnets. Superconducting magnets are utilized within the bigger MHD generators to eliminate the misfortunes. As demonstrated in Fig. 6.8, the anodes are set parallel and inverse to one another. It is made to work at quite high temperature, without moving parts. Since the plasma temperature is normally over 2000 °C, the conduit holding the plasma must be developed from non-directing materials equipped for withstanding this high temperature. The anodes must obviously be directing and heat resistant. Due to the high temperatures, the non-conducting walls of the channel must be developed from an exceedingly heat resistant substance, for example, zirconium dioxide to retard oxidation. It might be acknowledged as liquid dynamo like mechanical dynamo. The key segment is superconducting magnets (Fig. 6.9).

There are two types of MHD power generation:

i. Open cycle MHD.
ii. Closed cycle MHD

6.5.1 Magnetohydrodynamic Power Generation

The magnetohydrodynamic force generator is a gadget that creates electric power by the method of the connection of a moving liquid (generally an ionized gas or plasma) and an magnetic field. Like all direct conversion processes the MHD generators can also convert thermal energy directly into electricity without moving parts. Along these lines the static energy converters, with no moving mechanical part, can enhance the dynamic change, working at temperature more higher than conventional processes. The ordinary arrangement of MHD generator is demonstrated in (Fig 6.9). The underlying rule of MHD force generation is stylishly simple. Commonly, an electrically directing gas is generated at high pressure by burning of a fossil fuel. The gas is then directed through a magnetic field, resulting due to the Hall effect. The MHD system constitutes a heat engine, involving an expansion of the gas from high to low pressure in a manner similar to that employed in a conventional gas turbogenerator (Fig. 6.8). In the turbogenerator, the gas interacts with blade surfaces to drive the

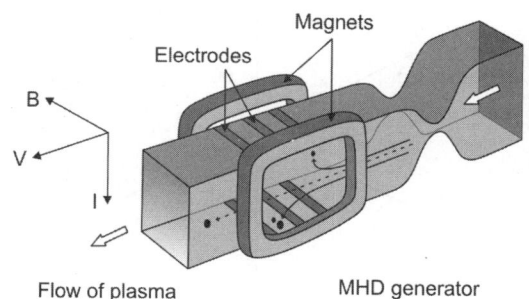

Fig. 6.9: Construction of a MHD generator

turbine and the joined electric generator. In the MHD system, the kinetic energy of the gas is changed directly to electric energy as it is allowed to expand. It is known that if we have a current flowing in a conductor immersed in a magnetic field, the same conductor will generate a Lorentz force that is perpendicular to the direction of the magnetic field and to the current.

In an MHD converter the electrical conductor is replaced by a plasma current at high speed and with high temperature to be partially ionized . So, the current flow is not only made of electrically neutral molecules but also with a mix of positive ions and electrons. When a high velocity gas flows into convergent-divergent duct and passes through the magnetic field, an emf is induced, mutual perpendicular to the magnetic field direction and to the direction of the gas flow. Electrodes in opposite side walls of the MHD flow channel provide an interface to an external circuit. Electrons pass from the fluid at one wall to an electrode, to an external load, to the electrode on the opposite wall, and then back to the fluid, completing a circuit. Thus the MHD channel flow is a direct current source that can be applied directly to an external load or can be linked with a power conditioning converter to produce alternating current. The electric energy produced is proportional to the reduction of kinetic energy and enthalpy of the fluid current. MHD effects can be produced with electrons in metallic liquids such as mercury and sodium or in hot gases containing ions and free electrons. In both cases, the electrons are highly mobile and move readily among the atoms and ions while local net charge neutrality is maintained. Any small volume of the fluid contains the same total positive charges to the negative charges in the ions because any charge imbalance would produce large electrostatic forces to restore the balance. Most theoretical and experimental work and power plant development and application studies have focussed on high-temperature ionized gas as the working fluid. Unfortunately, most common gases do not ionize significantly at temperatures obtainable with fossil fuel chemical reactions. This makes it necessary to seed the hot gasses with small amounts of ionizable materials such as alkali metals. Materials such as cesium and potassium have ionization potentials low enough that they ionize at temperatures obtainable with combustion reaction in air. Recovery and reuse of seed materials from the MHD channel exhaust are usually considered necessary from both economic and pollution standpoints. Engage in MHD power generation was initially fortified by the perception that the communication of a plasma with an magnetic field could happen at much higher temperatures than were conceivable in a rotating mechanical turbine. The restricting performance from the perspective of effectiveness of a heat motor is restricted by the Carnot cycle. A framework utilizing a MHD generator offers the potential of an extreme effectiveness in the reach of 60% to 65%. This is much superior to the 35% to 40% proficiency that could be accomplished in a cutting edge ordinary thermal power station.

The force yield of a MHD generator for every cubic metre of its channel volume is proprotional to the result of the gas conductivity, the square of the gas speed, and the square of the quality of the magnetic field through which the gas passes. For MHD generators to work intensely with great performance and sensible physical sizes, the electrical conductivity of the plasma must be in a temperature extend above in the vicinity of 1800 K. Separated from the MHD power generator, different apparatus are important to structure by and large the MHD framework. It is important to burn the fuel and the oxidizer, to include the seed, and to make plans for sending out the produced electrical force. The fuel is normally fossil and the oxidizer is air, for evident investment explanations. For extensive systems, a few insurances ought to be taken to breaking point the measure of misfortunes. The air may be advanced with additional oxygen, and preheating of the approaching oxidizer gets importance to permit thermal ionization. In practice various issues must be acknowledged in the usage of a MHD generator: generator productivity, economics, and toxic items. These issues are influenced by the decision of one of the three MHD generator outlines. These are the Faraday generator, the Hall generator, and the circle generator

6.5.2 Faraday Generator

A straightforward Faraday generator might comprise of a wedge-formed pipe or container of some non-conductive material. The point when an electrically conductive liquid moves through the tube, in the vicinity of a noteworthy perpendicular magnetic field, a charge is prompted in the field, which could be drawn off as electrical power by putting the terminals on the sides at 90° edges to the magnetic field. The main practical problem of a Faraday generator is that differential voltage and current in the liquid short through the electrodes on the sides of the channel. The most compelling waste is from the Hall impact current (Fig. 6.10).

6.5.3 Hall generator

The most common answer is to overcome the problems of Farady's generator is the Hall effect to create a current that flows with the fluid. The normal scheme is to place arrays of short, vertical electrodes on the sides of the duct. The first and last electrodes in the duct supply the load. Another electrode is shorted to an electrode on the opposite side of the duct. Losses are less than that of a Faraday generator,

and voltage is higher because there is less shorting of the final induced current. However, this design has problems because the speed of the material flow requires the middle electrodes to be offset to catch the Faraday current. As the load varies, the fluid flow speed varies, misaligning the Faraday current with its intended electrodes, and making the generator efficiency very sensitive to its load.

6.5.4 Disk Generator

The third, right now most effective response is the hall impact disk generator. This configuration right now holds the proficiency and energy density records for MHD generation. A disk generator has liquid flowing between the focal point of a plate, and a conduit wrapped around the edge. The magnetic excitation field is made by a couple of round Helmholtz loops above and underneath the disk. The Faraday current flow is a perfect dead short around the fringe of the plate. The Hall effect currents flows between ring electrodes close to the core and ring terminals close to the outskirts. Another significant advantage of this design is that the magnet is more efficient. First, it has simple parallel field lines. Second, because the fluid is processed in a disk, the magnet can be closer to the fluid, and magnetic field strengths increase as the 7th power of distance. Finally, the generator is compact for its power, so the magnet is also smaller. The resulting magnet uses a much smaller percentage of the generated power (Fig. 6.11).

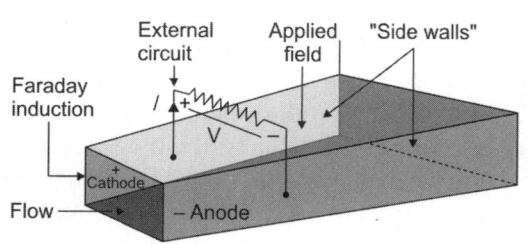

Fig. 6.10: Faraday's generator

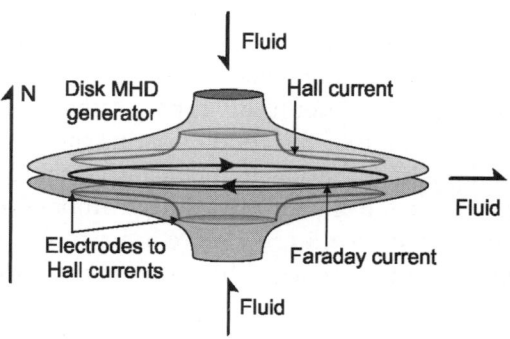

Fig. 6.11: Disk generator

6.6 POWERCYCLE FOR MHD GENERATOR

MHD generator replaced the gas turbine used in conventional cycle is shown in Fig. 6.12. A compressor is used to elevate the pressure and then heat is added to increase the gas temperature which is sufficient to ionize the gas. Then the gas flow is accelerated by passsing through the nozzle before entering MHD generator. The gas passing through the MHD generator is deaccelerated and electrical energy is generated. It is obvious that the MHD cycle is thermal power cycle and the thermal efficiency is given by

$$\eta = \frac{\text{work output}}{\text{heat input}}$$

$$= \frac{(h_3 - h_4) - (h_2 - h_1)}{(h_3 - h_2)}$$

where indicated enthalpies are stagnation values which takes into account the KE of the flowing gas The stagnation enthalpy of the following gas is given by

$$h_o = h + u_{22}$$

where, u is the velocity of the flowing gas. In actual MHD-generator, the gas velocity is sufficiently high so that the KE of the flowing gas represents substantial portion of the total energy Fig. 6.12.

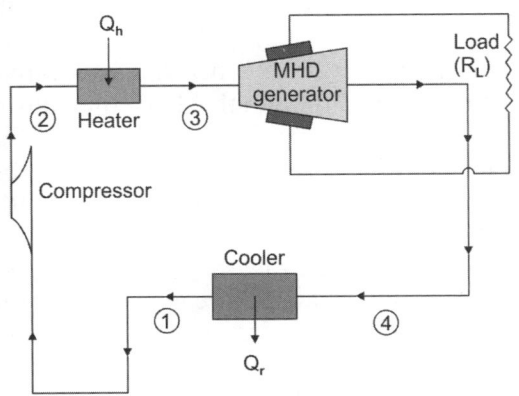

Fig. 6.12: Power cycle for MHD power generation

6.7 MHD SYSTEM

The MHD power generation is in advanced stage today and closer to commercial utilization. Significant progress has been made in development of all critical components and subsystem technologies. Coal burning MHD combined steam power plant promises significant economic and environmental advantages compared to other coal burning power generation technologies. It will not be long before the technological problem of MHD systems will be overcome and MHD system would transform itself from non-conventional to conventional energy sources.

Magnetohydrodynamic conversion systems can operate in either open or closed cycles. In an open cycle system, the working fluid is used on the once through basis. The working fluid after generating electrical energy is discharged to the atmosphere through a stack. In the closed cycle system the working fluid is continuously re-circulated and the discharged working fluid is reheated and returned to the converter. In an open cycle system the working fluid is converter 'air'. In closed cycle systems helium or argon is used as the working fluid. In open-cycle-systems, the hot combustion gases offer seeding, can be used directly as the working fluid, in closed cycle systems, however, heat is transferred from the combustion gases to the working fluid by means of a heat exchanger. A higher working temperature and a better thermal efficiency are thus possible in open cycles, provided suitable construction materials are available. The MHD power era is in progressed stage today and closer to business usage. Huge advancement has been made being developed of all discriminating parts and subframework innovations. Coal blazing MHD joined steam force plant guarantees critical budgetary and ecological points of interest contrasted with other coal smoldering force era innovations. It won't be much sooner than the innovative issue of MHD frameworks will be overcome and MHD framework might convert itself from non-ordinary to expected vigor sources.

Thus the MHD systems can be classified broadly as follows:
1. Open cycle systems
2. Closed cycle systems.
This may be further subclassified as:
   i. Seeded inert gas systems and
   ii. Liquid metal systems.

6.7.1 Description of a Typical Open Cycle MHD Plant

Average open cycle plant is depicted in this segment with the assistance of Fig. 6.13. In the open cycle, the working fluid, i.e. the seeded combustion gas is discharged from the MHD generator to the ambient. In the closed system heat is transformed from the combustion gas to the working fluid by means of heat exchanger. In this system, fuel utilized may be oil through an oil tank or gasified coal through a coal gasification plant. The fuel (coal, oil or characteristic gas) is blazed in the combustor or ignition chamber. The hot gases from combustor is then seeded with a little measure of an ionized soluble base metal (cesium or potassium), to build the electrical conductivity of the gas. The seed material, for the most part of potassium carbonate, is infused into the ignition chamber, the potassium is then ionized by the hot burning gases at temperatures of approximately (2300° to 2700°C). To achieve such high temperatures the compacted air used to burn the coal (or other fuel) in the ignition chamber, must be preheated to atleast 1100°C. An alternative is to utilize compressed oxygen alone for burning of the fuel, practically no preheating is then needed. The hot pressurized working liquid leaving the combustor moves through a convergent-divergent nozzle. In passing through the nozzle the arbitrary movement vigor of the particles in the hot gas is generally changed into directed, mass-motion energy, the gas rises up out of the nozzle and enters the MHD generator unit at a high speed. The hot gas expands through the rocket like generator encompassed by powerful magnet. Throughout the movement of the gas the positive and negative particles move to the electrodes and constitute an electric current. The magnetic field direction, which is at right angle to the liquid stream, might be perpendicular to the plane of paper. Various oppositely located electrode pairs are embedded in the channel to direct the electric current created to an external load. An MHD generator, unlike a conventional generator generates current, this could be changed over into commonly used alternating current by an inverter.

By the application of Hall effect the magnetic field acts on the MHD-generated (Faraday current) current and produce a voltage in the flow direction of the working fluid rather than at right angle to it. The resulting current in an external load is then called the Hall current. As the working fluid travels along the MHD generator and its energy is converted into electricity, its temperature falls. When the gas temperature reaches about 199°C, the extent of ionization of potassium is insufficient to maintain an adequate electrical conductivity. The large residual heat available from the hot discharge working gas can then be utilized in several ways. At this stage some 25% to 35% of the heat energy in the working fluid should have been converted into electrical energy. The still hot gas leaving the air preheater would be used in waste heat boiler to produce steam for operating a turbine generator. In this way another 25% to 35% of the initial heat should be recovered as electrical energy in a combined cycle system (Fig. 6.13).

The main components in the MHD generator are:
1. **Channel:** The hot ionized gases move through the (channel). The duct has a convergent-divergent shape such that the gases achieve supersonic speed in the divergent portion. Cross segment of the channel is either rectangular or roundabout.
2. **Combustor:** The fuel, hot compressed air and seeding material are conceded in the combustor. The fuel is lighted by auxiliary burners to get hot gases at high temperatures (2500°–3000°K) and high pressure. Seeding material is potassium or cesium,

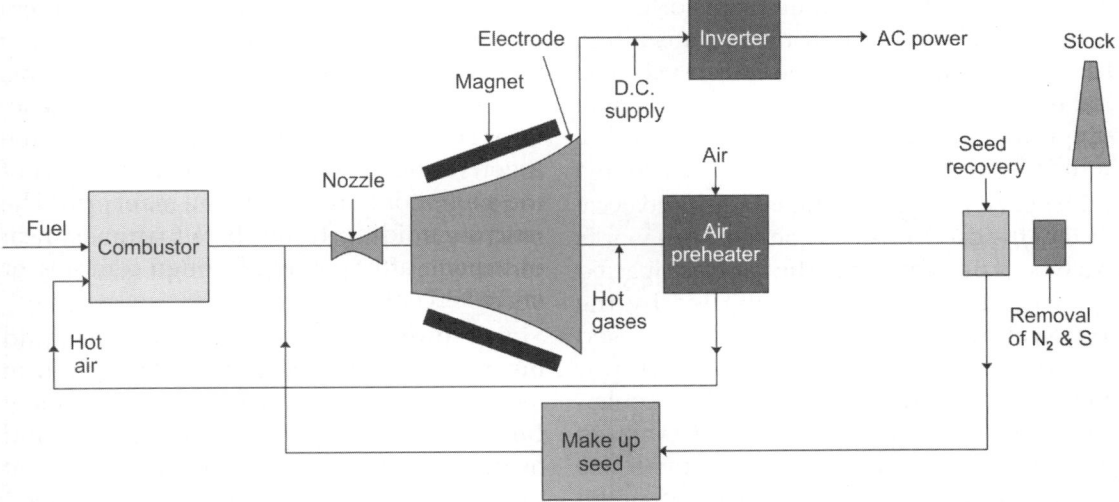

Fig. 6.13: Open cycle MHD plant

which is blended with the hot gases to acquire ionized gas, reputed to be working liquid. The working liquid is quickened to very nearly supersonic speeds in the different bit of the pipe (channel).

3. **Electromagnetic field coils:** The magnetic field coils are put such that the magnetic fields is in course perpendicular to the hub of the channel. The field coils are energized by auxillary DC supply got from a rectifier set. The field coils are by and large super-conducting DC. Auxiliary DC power supply is given by the MHD generator set via rectifier.

4. **Air heater:** The duct is connected to the air heater. The hot air is used for driving the air turbine-compressor unit. The compressor delivers high pressure air to the combustor via the air heater.

5. **Seed recovery, flue gas cleaning, heat recovery:** The exhaust gases from the duct are passed through the stack into atmosphere in case of "open cycle".

6. **Electrode system:** Two sets of segmented electrodes (+ and –) are placed in the direction of the electrical field E which is perpendicular to B and HG. The power generated by the MHD is of DC form.

7. **The invertor:** DC to 3-phase, AC conversion is done by a thyristor inverter. For effective practical acknowledgment a MHD framework must have the accompanying characteristics:

1. Air super heating course of action to heat the gas to around 2500°C, (the channel temperatures of MHD is about 2500°C), with the goal that the electrical conductivity of the gas is expanded.

2. The combustion chamber must have low heat losses.

3. Arrangement to add a low ionization potential seed material to the gas to increases its conductivity.

4. A water cooler but electrically insulating expanding duct with long life electrodes.

5. Seed recovery apparatus–necessary for both environmental and economic reasons.

6.7.2 Closed-Cycle Systems

Two general sorts of closed cycle MHD generators are, no doubt examined. In first type, electrical conductivity is administered in the working liquid by ionization of a seed material, as in open cycle systems and in the other type, a fluid metal gives the conductivity. The carrier is generally a synthetic inactive gas, in spite of the fact that a fluid transporter has been utilized with a fluid metal transmitter. The working liquid is flowed in a closed circle and is warmed by the ignition gases utilizing a high temperature exchanger. The working liquid is helium or argon with cesium seedling.

i. **Seeded inert gas system:** In a closed cycle system the carrier gas (argon/helium) works in a type of Brayton cycle. The gas is compacted and high temperature is supplied by the source, at basically steady weight, the packed gas then extends in the MHD generator, hotness is evacuated from the gas by a cooler, this is the heat rejection phase of the cycle. At last the gas is recompressed and returned for heating. A closed cycle MHD system is shown in Fig. 6.16. The complete system has three different yet interlocking circles. On the left is the outside warming circle. Coal is gasified and the gas (having a high top worth of around the range of 5.35 MJ/kg and at a temperature of about 525°C) is blazed in a combustor to give heat. In the primary heat exchanger, this heat is transferred to a carrier gas argon/helium of the MHD cycle. The combustion products after passing through the air preheater (to recover a part of the heat of combustion products) and purifier (to remove harmful emissions) are discharged to atmosphere. Because the combustion system is separate from the working fluid, so also are the ash and flue gases. Hence, the problem of extracting the seed material from flyash does not arise. The flue gases are used to preheat the incoming combustion air and then treated for flyash and sulphur dioxide removal, if necessary prior to discharge through a stack to the atmosphere. The hot argon gas is seeded with cesium and resulting working fluid is passed through the MHD generator at high speeds. The DC power out of MHD generator is converted to AC by the inverter and is then fed into the grid. The loop shown on the right hand side in figure is the steam loop for further recovery of the heat of working fluid and converting this heat into electrical energy in the diffuser the working fluid is slowed down

to a low subsonic speed. Then hot fluid enters a secondary heat exchanger, which serves as a waste heat boiler to generate steam. This steam is partly utilized to drive a turbine generator and for driving a turbine which runs the argon (or helium) compressor. The output of the generator is also fed to the main grid. The working fluid is returned back to primary heat exchanger after passing through compressor and intercooler.

ii. **Liquid metal system:** When a liquid metal provides the electrical conductivity, an inert gas (e.g. argon or helium) is a convenient carrier. The carrier gas is pressurized and heated by passage through a primary heat exchanger within the combustion chamber. The hot gas is then incorporated into the liquid metal, usually hot sodium, to form the working fluid. The latter consists of gas bubbles uniformly dispersed in an approximately equal volume of liquid sodium. The working fluid is introduced into the MHD generator through a nozzle in the usual ways and the carrier gas then provide the required high directed velocity of the electrical conductor (i.e. the liquid metal). After passage through the generator, the liquid metal is separated from the carrier gas. Part of the heat remaining in the gas is transferred to water in a secondary heat exchanger to produce steam for operating a turbine generator. Finally the carrier gas is cooled, compressed and returned to the combustion chamber for reheating and mixing with the recovered liquid metal. The working fluid temperature is usually around 800°C, as the boiling point of the sodium, even under moderate pressure is below 900°C. A liquid metal MHD cycle is given in (Fig. 6.14), in which liquid potassium after being heated in breeder reactor is passed through a nozzle to increase its velocity. The vapour formed due to nozzle action to increase its velocity are separated in the separator and condensed and then pumped back to the reactor as shown. Then the liquid metal with high velocity attained is passed through MHD generator to produce DC power. The liquid potassium

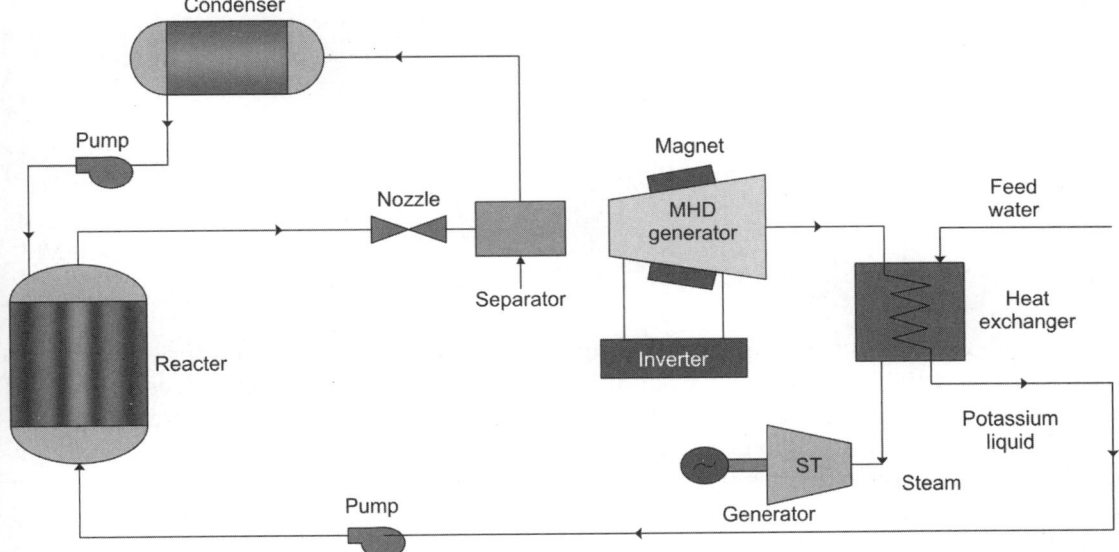

Fig. 6.14: Closed cycle MHD generator

coming out of MHD generator is passed through the conventional steam plant, where in the heat exchanger the heat of liquid potassium is utilized to generate steam to run steam turbine generator.

6.8 DIFFERENCE BETWEEN OPEN CYCLE AND CLOSED CYCLE SYSTEMS

Open Cycle System (OC)

1. Working fluid after generating electrical energy is discharged to the atmosphere through a stack.

2. Operation of MHD generator is done directly on combustion products.

3. Temperature requirement: 2300° to 2700°C.

4. More developed.

Closed Cycle System (CC)

1. Working fluid is recycled to the heat sources and thus is used again.

2. Helium or argon (with cesium seeding) is used as the working fluid.

3. Temperature requirement is about 530°C.

4. Less developed.

Working fluid is potassium seed combustion in open cycle MHD generator. Temperature in OC-MHD is about 2500°C. DC super-conducting magnets of 4~6 Tesla are used. Here exhaust gases are left out to atmosphere and the capacity of these plants are about 100 MW. In closed cycle MHD system, the working fluid is cesium seeded helium. Temperature of CC-MHD plants is very less compared to OC-MHD plants. It is about 1400°C. DC Superconducting magnets of 4~6 Tesla are used. Here exhaust gases are again recycled and the capacities of these plants are more than 200 MW.

6.9 TECHNICAL PARTICULARS OF CONCEPTUAL MHD STEAM POWER PLANTS

Table 6.1, gives the technical data regarding some conceptual designs of MHD plants.

6.10 GAS CONDUCTIVITY

The gas conductivity (O) is one of the parameters in the statement for the force produced. It is this parameter that has hindered the methodology of the MHD generation. To achieve sensible conductivity in a gas is a troublesome undertaking. In this context a

Table 6.1: Technical particulars of MHD plants for conceptual design

Plant rating feature	1400 MW, advanced energy conversion systems	600 MW conceptual study of potential early commercial MHD power plants (CSPEC)	500 MW parametric study of potential early commercial MHD power plants	85 MW (ETF)
				Parametric analysis of closed cycle MHD power plants

reasonable conductivity refers to a value between 10 and 100 mmhos/m, such conductivity is very small in comparison with those of materials which are negligible. For instance, most metals have conductivities of the order of 10^7 mhos/m and even transistor materials have conductivities exceeding 100 mhos/m. The gas has to be made sufficiently conducting, so that it may be used as a working fluid. In other words, it has to be ionized—a methodology in which electrons are evacuated from an atom. Conduction is due to free electrons and positive ions which are more under the effect of a magnetic field. There are several types of ionization, viz.

i. Thermal ionization;

ii. Magnetically induced ionization;

iii. Radio-frequency wave induced ionization;

iv. Radioactivity;

v. Photo ionization

vi. Electron-beam ionization and

vii. Flames.

Thermal ionization is the most significant method of ionizing a plasma. Ionization is obtained by imparting enough thermal energy to the gas. This method of ionization is directly dependent upon temperature. As a temperature of the gas is raised, the kinetic energy of its constituents expands. When the temperature has reached a sufficiently high value than an electron can ionize, a molecule collides thus produces more electrons, giving rise to what is called thermal ionization. In magnetically induced ionization, a DC voltage is connected across a gas at reduced pressure to create an electric field which will supply energy to electrons. Radio frequency waves can be used to produce a low amount of ionization in a gas. The working gas, although ionized is still composed mainly of neutral atoms, which carry nearly all the kinetic energy gas stream. The magnetic field connected pushes a power on the free electrons in the working medium with the goal that they move in tight rings in the attractive field and are viably trapped. These electrons thus follow up on the positive particles because of the coulomb electrostatic magnetic strengths. Ion neutral collisions then provide the last link in the chain which provides an effective coupling between the magnetic field and the neutral atoms of the gas (Fig. 6.15).

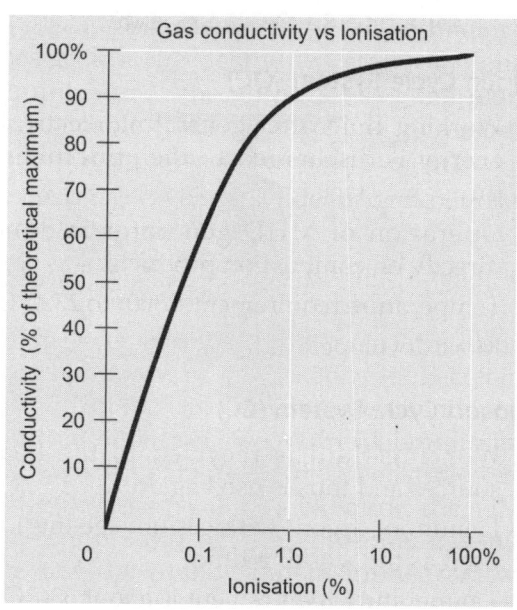

Fig. 6.15: Gas conductivity

6.11 MATERIALS FOR MHD GENERATORS

Owing to the high temperature of plasma (2700°C), refracting materials are required in several parts of generators like electrodes, channel or duct wall. Due to stringent environmental conditions, selection of materials is very limited. Following are the important factors to be considered for selection of materials:

i. Thermal shock resistance

ii. Electrical conductivity

iii. Corrosion resistance

iv. Erosion resistance

v. Oxidation reduction resistance

vi. Melting point of materials

vii. Density

Requirement of materials in MHD generators:

1. **Electrodes:** For high conductivity, high temperature is required. They should be electrically conductive and structurally stable at high temperature.

2. **Duct liner:** Cooler than electrodes.

   Must be electrical insulator.

   Must be thermal insulator.

3. **Magnet:** Stronger than permanent magnets or ferromagnetic substances can be used

   Few materials have high meeting points:

   a. *For insulators*—Oxides, nitrides, zirconates etc.

   b. *For conductors*—Refractory metals, nitrides besides and carbides.

6.11.1 Electrode Materials

Electrodes with higher electrical conductivity contribute towards better charge transfer across the plasma interface. Most of the initial channel wall designs involved water cooled metallic electrode due to the oxidizing condition. Important aspects of any practical electrode system for MHD generator are high reliability and long life of operation (10,000) and efficient in transferring current densities of 1–5 amp/cm^2. There are several potential ceramic materials for the electrode system that operate near 1500°–2000°C. The carbides (SiC, ZrC, MbC), besides (ZrB$_2$, TiB$_6$, LaB$_6$), silicides (MO Si$_2$, Ws$_1$), chromites (Sr, Mg doped), oxides and their solid solutions (ZrO$_2$ – CeO$_2$, CeO$_2$) and graphite as electrically conducting refractory materials. Most promising of these, however, stabilized zirconia.

6.11.2 Materials for Channels Duct Wall

Construction of channel through which to pass the precise hot, some of the time exceptionally corrosive gas is a second significant issue. Materials to withstand temperatures of the order of 3000°C are hard to find. Search of suitable ceramics as coatings of ceramics is continuing. Steam cooled walls which are made of insulated laminations of very thin nickel or stainless steel have been tried. Apart from high temperature, the channel divider need to withstand high temperature angles and also withstand the burdens experienced being used and additionally the warm stun in the event of incidental close down of the gas supply the duct material has to be electrically insulated the operating temperatures.

6.13 MAGNETIC FIELD

As the yield force conveyed by a MHD generator is corresponding to B$_2$ (attractive flux density), hence the magnet should be as large as possible. The magnet being one of the most expensive items of an MHD generator it is imperative that the production of large magnetic fields is optimized, constant research in this direction has been in progress. In this connection various types of magnets are identified which are:

1. Permanent magnets

2. Water cooled electromagnets

3. Cryogenically cooled electromagnets

4. Super conduction magnets.

The first two sorts of magnets are self explanetary. The cryogenically cooled magnet employs refrigerator principle for cooling. In a super conducting magnet there is no Joule on alloy when they are in a magnetic fields and below a certain temperature usually a few degree Kelvin, their resistance decreases the zero. Every material has a discriminating field quality, if the field is expanded, the super conductivity is pulverized.

6.14 WORKING PROCEDURE

It is the generation of electric power utilizing the high temperature conducting plasma (stream of high temp working fluid) moving through an intense magnetic field. It converts the heat energy of fuel (thermal energy) directly into electrical energy. The fuel is burnt in the presence of compressed air in combustion chamber. During combustion seeding materials are added to increase the ionization and this ionized gas (plasma) is made to expand through a nozzle into the generator. Magnetic field, a current is generated and it can be extracted by placing electrodes in a suitable stream. This generated EMF is DC ionization of GAS: various strategies for ionizing the gas are accessible, all of which depend on upon granting sufficient energy to the gas. The ionization can be produced by thermal or nuclear means. Materials, for example, potassium carbonate or cesium are frequently included little measures, ordinarily something like 1% of the sum mass stream to build the ionization and enhance the conductivity,

especially burning of gas plasma. 90% conductivity might be attained with an equitably low level of ionization of just in the vicinity of 1%. Figure 6.16, illustrates the MHD electricity generation procedure.

6.15 ADVANTAGES OF MHD SYSTEMS

1. Conversion efficiency of about 50%
2. Less fuel consumption
3. Large amount of pollution free power generated
4. Ability to reach full power level as soon as started
5. Plant size is considerably smaller than conventional fossil fuel plants
6. Less overall generation cost
7. No moving parts, so more reliable

In MHD the thermal pollution of water is eliminated. (Clean energy system) use of MHD plant operating in conjunction with a gas turbine power plant might not require to reject any heat to cooling water. These are less complicated than the conventional generators, having simple technology. There are no moving parts in generator which reduces the energy loss. These plants have the potential to raise the conversion efficiency upto 55–60%, since conductivity of plasma is very high (can be treated as infinity).

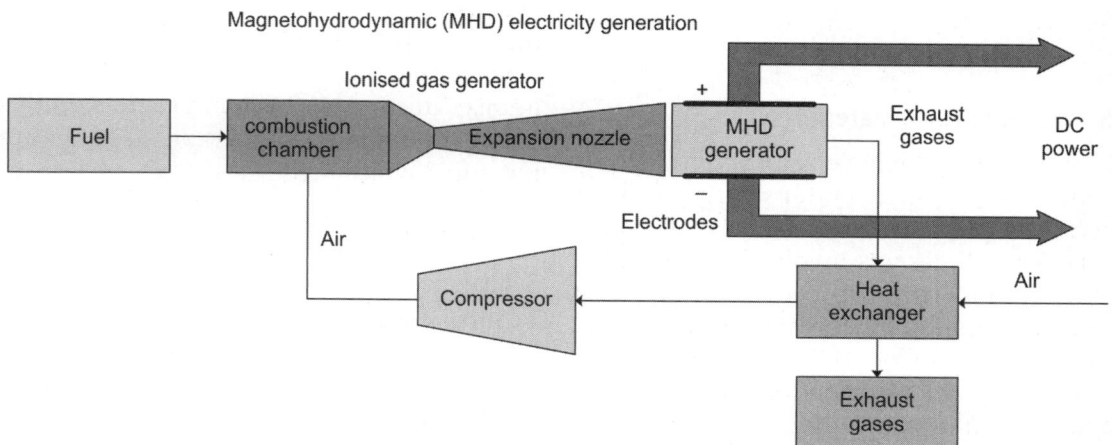

Fig. 6.16: MHD electricity generation

6.16 DISADVANTAGES OF MHD SYSTEM

1. Suffers from reverse flow (short circuits) of electrons through the conducting fluids around the ends of the magnetic field.
2. Needs very large magnets and this is a major expense.
3. High friction and heat transfer losses.
4. High operating temperature.
5. Coal used as fuel poses problem of molten ash which may short circuit the electrodes. Hence, oil or natural gas are much better fuels for MHDs. Restriction on use of fuel makes the operation more expensive.

The construction of superconducting magnets for small MHD plants of more than 1 kW capacity is only on the drawing board. Difficulties may arise from the exposure of metal surface to the intense heat of the generator and form the corrosion of metals and electrodes. Construction of generator is uneconomical due to its high cost. Construction of heat resistant and non conducting ducts of generator and large superconducting magnets is difficult. MHD without superconducting magnet is less efficient when compared with combined gas cycle turbine.

6.17 CONCLUSION

Improvement in corrosion science and superconducting magnets can make rapid commercialization possible. Saving billions of dollars towards fuel prospects of much better fuel utilization. It can therefore be claimed that the development of MHD for electric utility power generation is an objective of national significance. The practical efficiency of this type of power generation will not be less than 60%. Hence, it will be more significant in upcoming decades.

7

Geothermal Energy

7.1 INTRODUCTION

With the increase of fossil fuel prices and the increased pollution they produce, there has been an increase in research and investment in renewable energy. With increased energy demand, man has been forced to use energy sources that in the past were unexplored. This has led to the development of new methods of research and exploration and improvements through new techniques. This chapter deals about geothermal energy, a renewable energy whose utilization has grown in recent years due to investment and research in new technologies and due to the necessity of reducing atmospheric pollutants. The word geothermal comes from the Greek words "geo" means "earth" and "thermal" means "heat". Geothermal energy is heat from within the Earth. The thermal energy contained in the interior of the Earth is called the geothermal energy (Fig. 7.1). This energy is present as heat (thermal energy) in the earth's crust, the more readily accessible heat in the uppermost (10 km) or so, of the crust constitutes a potentially useful and almost inexhaustible source of energy. This heat is apparent from the increase in temperature of the earth with increasing depth below the surface. The centre of the earth is around 6000°C is hot enough to melt rock. Even a few kilometres down, the temperature can be over 250°C, if the earth's crust is thin. In general, the temperature rises one degree celsius for every 30–50 metres you go down, but this does vary depending on location. In volcanic areas, molten rock can be

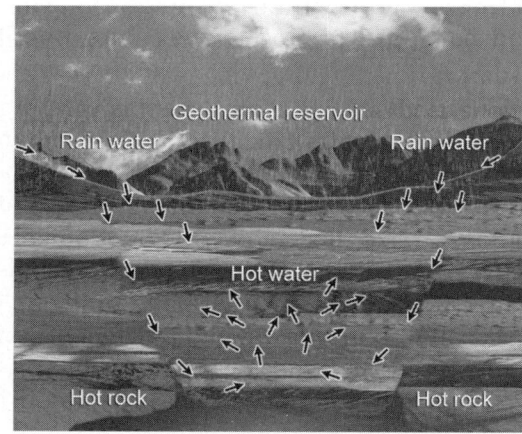

Fig. 7.1: Geothermal reservoir

very close to the surface. Geothermal resources are typically underground reservoirs of hot water or steams created by heat from the Earth, but also include subsurface areas of dry hot rock. Geothermal energy is a clean, renewable resource that provides energy be around the world and is essentially limitless. Heat flows constantly from the earth's interior and will continue to radiate for billions of years to come, ensuring an inexhaustible supply of energy. Utility-scale geothermal power production adds reliability to the power system. Geothermal power can be produced as a baseload renewable energy resource, meaning it operates 24 hours a day, 7 days a week regardless of changing weather, providing a uniquely reliable and continuous source of clean energy. As a baseload power source, geothermal energy is well suited as a substitute for coal in our utility system or, geothermal power can be flexible to support the needs of intermittent renewable energy resources such as wind and solar. Because geothermal energy can also be ramped up or down and depending on need it can be used to supplement the integrity of the power grid, enhancing the efficiency of the entire system while providing clean, reliable power. It is also capable of achieving high capacity factors—a measure of actual output over a period of time–usually at or above 90%, which is on par with, or higher than, other baseload power sources such as coal-fired or nuclear power plants, and much greater than intermittent sources.

Geothermal energy is a kind of thermal energy that determines the temperature of matter. The geothermal energy of the Earth's crust originates from the original formation of the planet (20%) and from radioactive decay of minerals (80%). It occurs due to the difference in temperature between the core of the planet and its surface, drives a continuous conduction of thermal energy in the form of heat from the core to the surface. The total amount of energy in the outer 10 km of the earth's crust exceeds greatly that obtainable by the combustion of coal, oil and natural gas. There is a ample scope to develop geothermal power in India, but still development in geothermal field is in initial stage. There are about 340 known thermal areas in India, each represented by hot/warm spring. This energy is also one of the renewable energy sources, which are defined as those resources that draw on the natural energy flows of the Earth. Renewable energy sources are so named because they recur, are seemingly inexhaustible and are free for the taking. Water is pumped down, an "injection well", filters through the cracks in the rocks in the hot region, and comes back up the "recovery well" under pressure. It "flashes" into steam when it reaches the surface. Hot rocks underground, heat water to produce steam. We drill holes down to the hot region, steam comes up, is purified and used to drive turbines, which

drive electric generators. There may be natural "groundwater" in the hot rocks anyway, or we may need to drill more holes and pump water down to them. The steam may be used to drive a turbogenerator, or passed through a heat exchanger to heat water to warm houses. A town in Iceland is heated this way. The steam must be purified before it is used to drive a turbine, or the turbine blades will get "furred up" like your kettle and be ruined.

Evidences of the geothermal energy found deep inside the earth surface is apparent only in a few countries and a few locations in the world in the form of, hot water springs, the geysers, fumaroles and volcanic eruptions. Historically the, first application of geothermal energy were for space heating, cooking, and for medicinal purposes. The earliest record of space heating dates back to 1300 in Iceland. It has been analysed that about 40% to 50% of total heat required by the society is a heat below 200°C. 30% heat is required at about 150°C. 20% heat is required at about 100°C. Thus there is a vast scope to use geothermal energy for low temperature applications. There is a ample scope to develop geothermal power in India, but still development in geothermal field is in initial stage. There are about 340 known thermal areas in India, each represented by hot/warm spring. Till now only one pilot plant is in operation in puga valley, in Jammu and Kashmir, having 20 MW capacity. Another plant is at Parvati valley, Himachal Pradesh is under construction. A 7.5 tonne capacity cold storage pilot plant based on geothermal energy was installed at Manikaran, Himachal Pradesh. The geothermal energy is enormous and will last for several millions of years and is therefore called renewable.

Figure 7.2 shows a typical geothermal field. The hot magma near the surface (A) solidifies into igneous rock (B). Igneous is a latin word, enormous meaning "of fire" specially formed by volcanic action or heat. The heat of the magma is conducted upwards to this igneous rock. Ground water that finds its way down

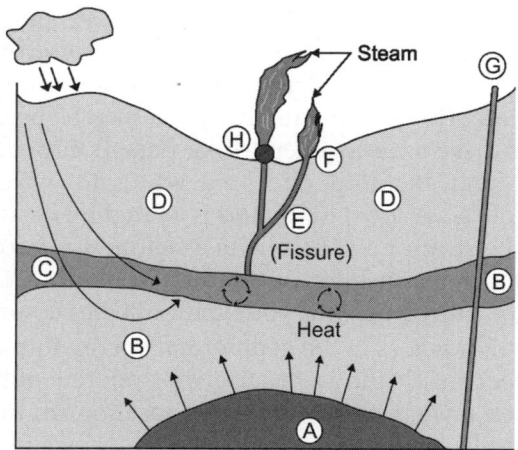

Fig. 7.2: A typical geothermal field

to this rock through fissures in it, will be heated by the heat of the rock or by mixing with hot gases and steam emanating from the magma. The heated water will then rise convectively upward and into a porous and permeable reservoir (C) above the igneous rock. The reservoir is capped by a layer of impermeable solid rock, however, has fissures (E) that act as vents of the gaint underground boiler. The vents show up at the surface as geysers fumarols (F) (steam is continuously vented through fissures in the ground , these vents are called fumarols) or hot spring (G). A well (H) taps steam from the fissures for use in a geothermal power plant. Geothermal steam is of two types, steam originating from the magma itself called magmetic steam and that from ground water heated by the magma called meteoritic steam.

The earth's *core* lies almost 4,000 miles beneath the earth's surface. The double-layered core is made up of very hot *molten* iron surrounding a solid iron center. Estimates of the temperature of the core range from 5,000 to 11,000 °F. Heat is continuously produced within the Earth by the slow decay of radio-active particles that is natural in all rocks. Earth is made up of 3 layers namely, crust, mantle and metallic core. The crust, mainly composed of granites is split into 6 main plates

which move away from each other at the rate of 2 to 20 cm/yr. Relative motion of the plates occasionally gives rise to areas where internal heat of the Earth reaches the surface and where most volcanoes and most of the world's geothermal fields are found. Water is the most important carrier of the energy in geothermal wells as it has a high heat capacity and latent heat of vapourization. Impermeable rocks cover permeable layers which contain reservoir of hot water thereby preventing heat loss and maintaining water under pressure. Geothermal fields such as geysers require a combination of 3 geological factors—a natural underground source of water, an impermeable layer that traps the water and permits formation of steam and a mass of hot rock near the natural water system. Geothermal energy is derived from the hot interior of the Earth. Crustal rock up to 70 km thick insulates it from earth's surface. Due to this insulation temperature at the surface is relatively cool but increases rapidly with growing depth inside the rock. A small portion of the rock is permeable so that over a long period of time it acquires a store of natural ground water which gets heated. In places where the heated water finds its way to the surface naturally, the geothermal energy can be extracted with relative ease. In other places it may be mined using deep bore holes. Surrounding the earth's core is the *mantle*, thought to be partly rock and partly magma. The mantle is about 1,800 miles thick. The outermost layer of the Earth, the insulating crust, is not one continuous sheet of rock, like the shell of an egg, but is broken into pieces called *plates*. These slabs of continents and ocean floor drift apart and push against each other at the rate of about one inch per year in a process called *continental drift*. *Magma* (molten rock) may come quite close to the surface where the crust has been thinned, faulted, or fractured by plate tectonics. When this near-surface heat is transferred to water, a usable form of geothermal energy is created.

7.2 HISTORY OF GEOTHERMAL ENERGY

Early humans probably used geothermal water that occurred in natural pools and hot springs for cooking, bathing and to keep warm. We have archeological evidence that the Indians of the Americas occupied sites around these geothermal resources for over 10,000 years to recuperate from battle and take refuge. Many of their oral legends describe these places and other volcanic phenomena. Recorded history shows uses by Romans, Japanese, Turks, Icelanders, Central Europeans and the Maori of New Zealand for bathing, cooking and space heating. Baths in the Roman empire, the middle kingdom of the Chinese, and the Turkish baths of the Ottomans were some of the early uses of balneology; where, body health, hygiene and discussions were the social custom of the day. This custom has been extended to geothermal spas in Japan, Germany, Iceland, and countries of the former Austro-Hungarian empire, the Americas and New Zealand. Early industrial applications include chemical extraction from the natural manifestations of steam, pools and mineral deposits in the Larderello region of Italy, with boric acid being extracted commercially starting in the early 1800s. At Chaudes-Aigues in the heart of France, the world's first geothermal district heating system was started in the 14th century and is still going strong. The oldest geothermal district heating project in the United States is on Warm Springs Avenue in Boise, Idaho, going on line in 1892 and continues to provide space heating for upto 450 homes. Many ancient peoples, including the Romans, Chinese, and Native Americans, used hot mineral springs for bathing, cooking, and heating. Geothermal resources have been used for centuries for balneological purposes. Water from hot springs is now used world-wide in spas, for heating buildings, and for agricultural and industrial uses. Many people believe hot mineral springs have natural healing powers. Using geothermal energy to produce electricity is a relatively new industry. It was initiated by a group of Italians who built an electric generator at Lardarello in 1904. Their generator was powered by the natural steam erupting from the Earth. The first attempt to develop geothermal power in the United States came in 1922 at The Geysers steam field in northern California. The project failed because the pipes and turbines of the day could not stand up to the abrasion and corrosion of the particles and impurities that were in the steam. Later, a small but successful hydrothermal plant opened at the Geysers in 1960. Today 28 plants are operating there. Electricity is now produced from geothermal energy in 21 countries, including the United States.

The spread of technology to other parts of the world was slow during the first part of the twentieth century. Electricity was first generated at Wairakei (New Zealand) in 1958, from a high temperature liquid geothermal resource. In the US, the electricity production from geothermal resources only commenced in 1960, at the Geysers in California. Since that time, the development of steam and liquid-dominated resources for power production has begun in many countries worldwide, with a total installed capacity of 11180.2 MW using 551 units. With the use of geothermal heat pumps, direct applications of geothermal fluids are found all parts of the world. The use of geothermal heat pumps grew so rapidly that it is difficult to present accurate data related to the number of installations. Another point is that an elevated number of geothermal plants is installed by private entities, and there are countries like Portugal where these installations are not registered. ICA data base reports an installed capacity of 32679.3 MWt and an annual use of 196185.1 TJ/Year. This represents 67% of the direct utilization of geothermal fluids.

The former USSR produced power from the first true binary power plant, 680 kWe using 81°C water at Paratunka on the Kamchatka peninsula–the lowest temperature, at that time. Iceland first produced power at Namafjall in northern Iceland, from a 3 MWe non-condensing turbine.

7.3 MODE OF UTILIZATION

Geothermal hot water and steam can naturally reach the earth's surface in the form of hot springs, geysers, mud pots, or steam vents. Geothermal reservoirs of hot water and steam are also found at various depths below the Earth's surface. Geothermal resources can be accessed by wells and used to provide heat directly. The heat can also be captured and used to commercially generate electricity.

7.4 FORMATION OF GEOTHERMAL RESERVOIR

A geothermal system requires heat, water, and permeable host rock (called a reservoir). Heat from the earth's interior flows continuously outward. In some places, heat flow causes the partial melting of crustal rock creating magma (molten rock), which rises to the earth's surface. Magma that reaches the earth's surface and erupts from volcanoes is called lava. Magma remaining below the earth's surface heats nearby rock and water, sometimes to levels as hot as 700°F (371°C). As a result, hot water and steam become trapped in the permeable and porous rocks underlying impermeable rock layers, forming a geothermal reservoir.

7.5 TRANSMISSION MODE

Electricity generated from geothermal energy is sent to users through a transmission system consisting of electric transmission lines, towers, substations, and other components. The integration of geothermal energy into a transmission system requires careful planning to balance the mix of geothermal energy with other sources of energy generation.

7.6 GEOTHERMAL SPRINGS FOR POWER PLANTS

The most common current way of capturing the energy from geothermal sources is to tap into naturally occurring "hydrothermal convection" systems where cooler water seeps into earth's crust, is heated up, and then rises

to the surface (Fig. 7.3). When heated water is forced to the surface, it is a relatively simple matter to capture that steam and use it to drive electric generators. Geothermal power plants drill their own holes into the rock to more effectively capture the steam. The largest geo-thermal system now in operation is a steam-driven plant in an area called the geysers, north of San Francisco, California. Despite the name, there are actually no geysers there, and the heat that is used for energy is all steam, not hot water. Although the area was known for its hot springs as far back as the mid-1800s, the first well for power production was drilled in 1924. Deeper wells were drilled in the 1950s, but real development didn't occur until the 1970s and 1980s. By 1990, 26 power plants had been built, for a capacity of more than 2,000 MW. Geothermal power is considered to be renewable because any projected heat extraction is small compared to the Earth's heat content. The earth has an internal heat content of 1031 joules (3.10^{15} TW/hr). About 20% of this residual heat is from planetary accretion, and the remainder is attributed to higher radioactive decay rates that existed in the past.

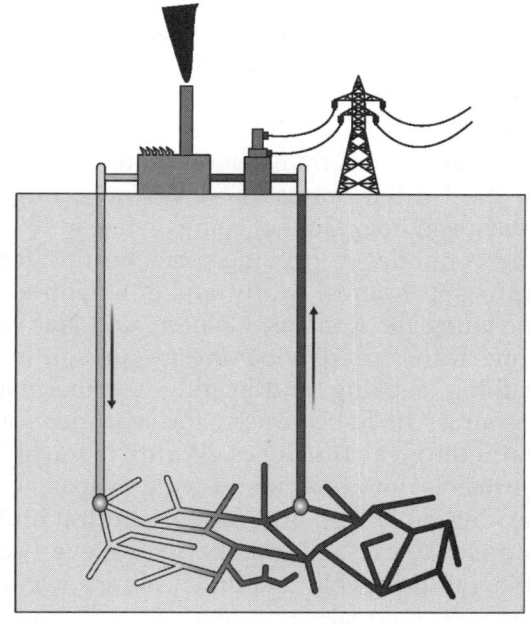

Fig. 7.3: Geothermal springs

7.7 THE GEOTHERMAL RESOURCE

Below the earth's crust, there is a layer of hot and molten rock called magma. Heat is continuously produced there, mostly from the decay of naturally radioactive materials such as uranium and potassium. The amount of heat within 10,000 meters (about 33,000 feet) of earth's surface contains 50,000 times more energy than all the oil and natural gas resources in the world. The areas with the highest underground temperatures are in regions with active or geologically young volcanoes. These "hot spots" occur at plate boundaries or at places where the crust is thin enough to let the heat through.

Earthquakes and magma movement break up the rock covering, allowing water to circulate.

Seismically active hotspots are not the only places where geothermal energy can be found. There is a steady supply of milder heat-useful for direct heating purposes-at depths of anywhere from 10 to a few hundred feet below the surface virtually in any location on the Earth. In addition, there is a vast amount of heat energy available from dry rock formations very deep below the surface (4–10 km). Using a set of emerging technologies known as enhanced geothermal systems (EGS), we may be able to capture this heat for electricity production on a much larger scale than conventional technologies allow.

If these resources can be tapped, they offer enormous potential for electricity production. The geothermal energy association estimates that 132 projects now under development around the country could provide up to 6,400 MW of new capacity. As EGS technologies improve and become competitive, even more of the largely untapped geothermal resource could be developed. With the combination of both the size of the resource base and its consistency, geothermal can play an indispensable role in a cleaner, more sustainable power system. Geothermal sources are therefore of three basic kinds: *(1) hydrothermal (2) geopressured and*

(3) petrothermal. The total amount of energy in the outer 10 km of the earth's crust exceeds greatly that obtainable by the combustion of coal, oil and natural gas. At present, however, only the relatively small proportion of the geothermal energy in wet reservoirs (a geothermal reservoir is defined as s region where there is a concentration of extractable heat), may be regarded as economically useful.

7.8 NATURE OF GEOTHERMAL FIELDS

Geothermal energy originates from the characteristic era of hotness essential because of the decay of the regularly happening radioactive isotopes of uranium, thorium and potassium in the earth. On account of the inside high temperature era, the earth's surface hotness stream midpoints 82 MW/m^2 which sums to a sum heat misfortune of something like 42 million megawatts. The evaluated aggregate warm vigor above mean surface temperature to a profundity of 10 km is 1.3 × 1027 J, equal to smoldering of 3.0 × 1017 barrels of oil. Since the worldwide vigor utilizations for numerous types of vigor, is equal to utilization of about 100 million barrels of oil every day, the earth's vigor to a profundity of 10 kilometers could hypothetically supply all of humanity's vigor needs for six million years. On normal, the temperature of the earth builds something like 30°C/km above the mean surface encompassing temperature. In this manner, accepting a conductive slope, the temperature of the earth at 10 km might be over 300°C. In any case, most geothermal investigation and utilization happens where the angle is higher, and consequently where boring is shallower and less costly. These shallow depth geothermal resources occur due to:

1. Intrusion of molten rock (magma) from depth, bringing up great quantities of heat;
2. High surface heat flow, due to a thin crust and high temperature gradient;
3. Ascent of groundwater that has circulated to depths of several kilometers and been heated due to the normal temperature gradient;

4. Thermal blanketing or insulation of deep rocks like shale, whose thermal conductivity is low;

5. Anomalous heating of shallow rock by decay of radioactive elements, perhaps augmented by thermal blanketing . These geothermal resources range from the mean annual ambient temperature of around 20°C to over 300°C. In general, resources above 150°C are used for electric power generation, although power has recently been generated at Chena hot springs resort in Alaska using a 74°C geothermal resource (Lund, 2006). Resources below 150°C are usually used in direct-use projects for heating and cooling. Ambient temperatures in the 5° to 30°C range can be used with geothermal (ground-source) heat pumps to provide both heating and cooling.

It is convenient to classify earth's surface into three broad groups according to their temperature gradient.

1. Non-thermal areas having a temperature gradient of 10–40°C per km depth.

2. Semi-thermal areas having a temperature gradient of 70°C per km depth.

3. Hyper-thermal areas where the temperature gradients are many times greater than the non-thermal areas.

7.9 TYPES OF PLANTS AND RESERVOIRS

In order to have a plant handling power at a given area, it is important to have at profundity a repository with high temp water or steam, a hotness source which warms the water or steam, and a method of convey water or steam to the surface where the plant is placed. Geothermal stores might be found in nature in districts with aquifers filling pores or blames and splits, or could be handled by man, in areas framed by dry rocks having high temperatures (HDR) (Table 7.1). In these cases, water must be sent from the surface to the repository and, once warmed by the rock, return again to the surface to be utilized. This technique is here and there utilized as a part

Table 7.1 Geothermal resource types

Resource type	Temperature range (°C)
Convective hydro-thermal resources	
Vapor dominated	240°
Hot-water dominated	20 to 350°+
Other hydrothermal resources	
Sedimentary basin	20 to 150°
Geopressured	90 to 200°
Radiogenic	30 to 150°
Hot rock resources	
Solidified (hot dry rock)	90 to 650°
Part still molten (magma)	>600°

of ordinary supplies when the water supply is less than the measure of water or steam withdrawn from the repository. There are two principle sorts of high-temperature geothermal stores: vapour dominated and liquid dominated (both fluid and vapour). A plant for creation and operation of a steam repository holds steam with weight near steam-static, and temperature close immersions. Store conditions in the geysers were at first those of immersed steam. In a steam dominated supply, the liquid distribution is regulated by the steam of stream climbing and the water moving down. Assuming that the mass fluxes of water moving down and steam climbing are harshly comparative, and the vertical weight inclination is close steam-static, the relative porousness to water must be low. The streaming steam involves a large portion of the crack space, and water possesses the remaining pore space. The mass of water in the repository is much more terrific than the mass of steam. Because of the reduction of steam handling, it is important to infuse water through wells into the supply. Water infusion likewise has the profit of diminishing H_2S in the liquid. In high temperature geothermal sources, a decrease in weight brought about by abuse may start bubbling to some extent or the greater part of the reservoir. The progressions created by misuse will incorporate changes in the steam/water proportion, and additionally weight and temperature changes. The store could be recognized consistently blended, holding steam and water, or a vapour-dominated zone overlying a fluid ruled zone. The first approach ignores gravity, the second assumes gravity which is dominant. Both accept steam and water in warm balance. Vapour-dominated geothermal fields are spotted in districts of later volcanism, close to the outskirts of tectonic plates. So as to structure a hotness supply, the bizarre magmatic interruption may well experience permeable and penetrable water filled rock strata. Inside the reservoir, convection currents and flows of boiling water or steam are situated up and the temperature contrast between the top and base of the store is not critical. With a specific end goal to avoid the break of hot supply liquids through convection, there is an impermeable top rock or a top rock with low porousness, overlying the store. Dry-steam power plants were the first sort of geothermal force plant to attain business status. Since the geofluid comprised exclusively of steam, it was not difficult to snare a mechanical mechanism to exploit the accessible vigor. Dry-steam plants have a tendency to be less complex and less costly than the flash steam plants, on the grounds there is no geothermal salt water to battle with. There was 63 units of this sort, in operation in 2012. When the geothermal wells transform a mixture of steam and fluid, the single-blaze plant is a moderately straightforward approach to changeover the geothermal vigor into power. A regular 30 MW single-flash power plant needs 5–6 creation wells and 2–3 infusion wells. The twofold flash plant is a change of the single flash outline. It can produce 15%–25% more power yield for the same geothermal liquid conditions. The plant is more intricate, more expensive and requires more support yet the additional force yield defends the utilization of such plants. The single-blaze units (150) produce 42% of the geothermal vigor. In some geothermal stores, water temperature is insufficient to transform steam so parallel plants are utilized. This kind of units work with low enthalpy repositories. They are utilized as a part of hot dry rock stores and in little power plants. A type of geothermal environment whose boiling point is very nearly (totally) fixed from trade with encompassing rocks is known as a geopressure system. Such system are found in bowls in which extremely fast loading with silt happens, bringing about higher than typical weight of the aqueous water. The initially recognized field of this type was in the Gulf of Mexico, at a profundity between 6 and 8 km, with pore weights of up to 130 Mpa and temperatures in the extent 150°–180°C. Supplies of this type are not yet investigated. One wellspring of water at temperature above the mean surface is from aquifers that are deep to the point that their temperature is raised by the ordinary geothermal inclination. The component of warming the water is by conduction. The liquid stream in the aquifer must be abate enough for the water to be warmed by conductive high temperature stream. The point when warm conductivity of the rock is low, for example, in sedimentary regions, the geothermal inclination expands and heated water might be found in areas with ordinary hotness stream. In these cases water could be utilized within requisitions like local warming.

A lot of warm springs could be found along major fault system and fractures. Such channels furnish the methods for flow of meteoric water to profundities where it is warmed to neighborhood temperatures and after that came back to the surface. These are a manifestation of convective framework. The main thrust for the course is the thickness distinction between the cool diving water and the hot rising water. The production of electricity by geothermal fluids can be present in any type of reservoir. Binary plants, which operate with lower temperatures, are the most widely used. Hot-dry-rock reservoirs are operated, or are being prepared to operate, using binary plants, while older reservoirs, located in specific regions, continue to be explored (Table 7.2). The development of geothermal heat pumps and the possibility of their being assembled anywhere means that geothermal energy can be seen as a viable method for heating and cooling of buildings. Direct uses of geothermal fluids have also increased in recent years.

7.10 GEOTHERMAL ENERGY SOURCES

The temperature of Earth increases with the depth rather non-uniformly with average increase of 30°C per 1000 m (geothermal gradient). It is therefore, generally necessary to drill 10 km deep production wells to obtain geothermal fluids at significant temperature of the order of 3000°C. Five general categories of geothermal resources have been identified:
1. Hydrogeothermal energy resources
2. Geopressure resources
3. Petrothermal
4. Magma resources
5. Volcanoes

Among these geothermal resources, hydrothermal systems are considered the best resources for geothermal energy exploitation at present. Hot-dry-rock is also being considered.

7.10.1 Hydrothermal-Geothermal Resources

These are reserves at moderate depths containing steam/hot water under pressure at temperature up to about 350 °C. These systems are further subdivided, depending upon

Table 7.2 Units installed capacity and produced energy

Type	Units	Installed Capacity (MW)	Energy (GWh/Year)
Dry steam	63	2862.0	15750.6
Single flash	150	4715.1	28358.0
Double flash	64	2182.7	14426.2
Binary	246	1267.3	6308.1
Back pressure	26	146.6	2393.4
Hybrid	2	6.5	10.0

whether steam or hot water is the dominant product. These resources occur where the earth's heat is carried upward by convective circulation of naturally occurring hot water or steam. Hydrothermal resources arise when water has access to high temperature rocks, this accounts for the description as 'hydrothermal'.

Boiling hot water and steam stores are spotted in the breaks inside the hot aquiferous porous rocks. The water from sprinkle, lakes, sea and so forth over some tens or many kilometer surface region permeates into the Earth through privileged and the deformities in the penetrable rocks to the profundities of 2 to 10 km. The water gets warmed and climbs through deformities in the robust impermeable shakes and gets gathered in the breaks inside the penetrable rocks. The upper impermeable rock furnishes covering blanket to the boiling hot water stores. Heated water or steam frequently escapes through gaps in the rock, in this manner structuring hot springs, fountains fumarols, and etc. are constructed. Keeping in mind the end goal to use the aqueous vigor, wells are bored either to catch a crevice or, all the more usually, into holding the water, (i.e. aqueous store). Most aqueous wells go in profundity from something like 600 to 2100 m, despite the fact that there are some shallower and deeper creation wells. Two principle sorts of aqueous assets are utilized to create power:
• Dry-steam (vapour-dominated) reservoirs, and
• Hot-water (liquid-dominated) reservoirs.

i. **Vapour-dominated/dry steam** repositories are uncommon, however, remarkably proficient at transforming power. They have the most reduced expense and minimum number of genuine issues. Be that as it may constitute just a couple of percent of aqueous assets and a much more diminutive extent of the receptive geothermal vigor assets. The geyser in California is the biggest and best known dry steam store. Here, steam is acquired by boring wells from 7,000 to 10,000 feet profundity.

In a dry steam supply, the common steam is funneled specifically from a geothermal well to power a turbine generator. The used steam (dense water) might be utilized as a part of the plant's cooling framework and infused go into the repository to look after water and weight levels. Geothermal vigor as dry steam is changed over into mechanical vigor by the turbine. The mechanical energy is changed over to electrical energy by the generator. The electrical force is transmitted by electrical transmission framework and nourished to 30 AC supply arrange. Steam geothermal force plants are grid associated and worked constantly as base burden force plants. Dry-steam force plants were the first kind of geothermal power plant to accomplish business status. Since the geofluid comprised exclusively of steam, it was not difficult to snare a mechanical unit to exploit the accessible vigor. Dry-steam plants tend to be simpler and less expensive than the flash-steam plants, because there is no geothermal brine to contend with. There were 63 units of this type, in operation in 2012.

Vapour-dominated geothermal fields are located in regions of recent volcanism, near the borders of tectonic plates. In order to form a heat reservoir, the anomalous magmatic intrusion should encounter porous and permeable water filled rock strata. Within the reservoir, convection currents of hot water and/or steam are set up and the temperature difference between the top and bottom of the reservoir is not significant. In order to prevent the escape of hot reservoir fluids through convection, there is an impermeable cap rock or a cap rock with low permeability, overlying the reservoir. The geothermal fluid for such plants is dry steam at temperature between 180 °C to 240 °C with low content of particulate impurities and dissolved solid impurities. Mass flow rate per well is around 10 kg/sec with power rating per well at a rate of 6 MWe. Figure 7.4 gives a typical schematic diagram of a vapour dominated geothermal power plant. Dry-steam from the geothermal reservoir flows upwards through the production well and is admitted in the centrifugal separator. The temperature and pressure of the steam at

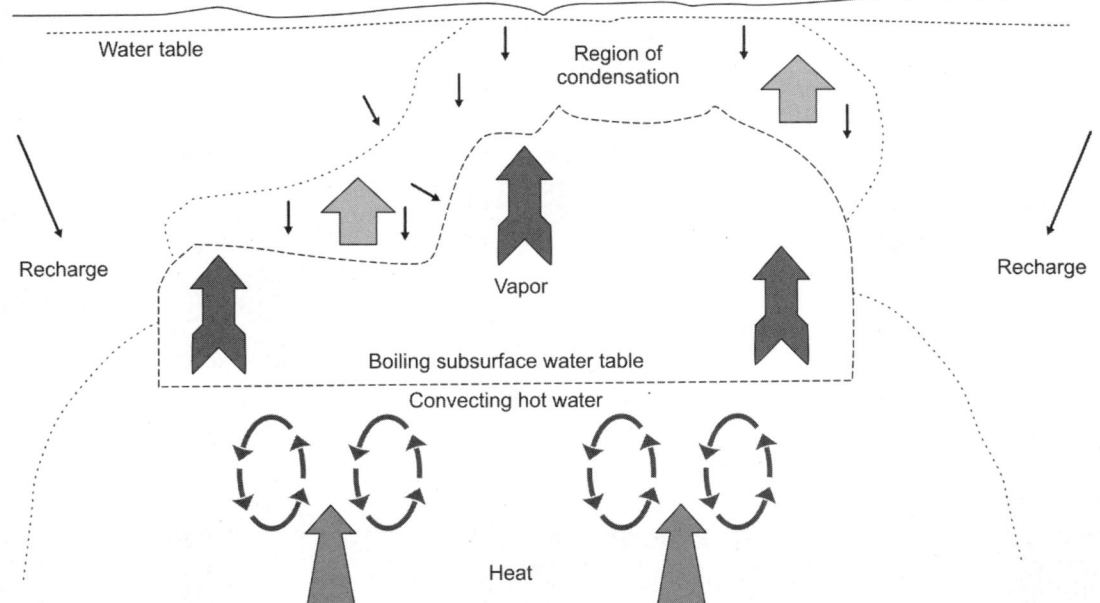

Fig. 7.4: Vapor dominated geothermal system. (White et al., 1971)

the bottom of the well are 280 °C and 35 bar. As steam flows towards the ground surface, it expands and cools. Temperature and pressure at the well head is about 250 °C and 8 bar. The centrifugal separator removes particulate matter from the steam. The steam is admitted into the steam turbine. The steam expands in the turbine buckets producing rotary kinetic energy. The low pressure steam at the exhaust of the turbine is condensed in condenser the condensated is reinjected into the earth via the reinjection wall. Cooling water for condensing the steam is circulated through the cooling tower by means of cooling water pump. The synchronous generator generates power at high voltage of 50 Hz AC (Fig. 7.5). The turbine and generator form one unit. Complete power plants has several units. Geothermal energy in the form of dry steam is converted into mechanical energy by the turbine. The mechanical energy is converted to electrical energy by the generator. The electrical power is transmitted by electrical transmission system and fed to 30 AC supply network.

Steam geothermal power plants are grid connected and operated continuously as base load power plants. These reservoirs are few in number, with the geysers in northern California, rderello in Italy and Matsukawa in Japan being ones, where the steam is exploited to produce electric energy.

ii. **Liquid-dominated/hot-water geothermal** sources are the most widely recognized power source. In a fluid ruled repository, the heated water has not vapourized into steam on the grounds that the supply is soaked with water and is underweight. To produce power, the boiling hot water is piped from geothermal wells to one or more separators where the weight is brought down and the water flashes into steam. The steam then moves a turbine generator to handle power. The steam is cooled, dense and either utilized as a part of the plant's cooling framework or infused into the geothermal supply. These frameworks are prepared by ground water circling to profundity and climbing from lightness in penetrable stores that are with uniform

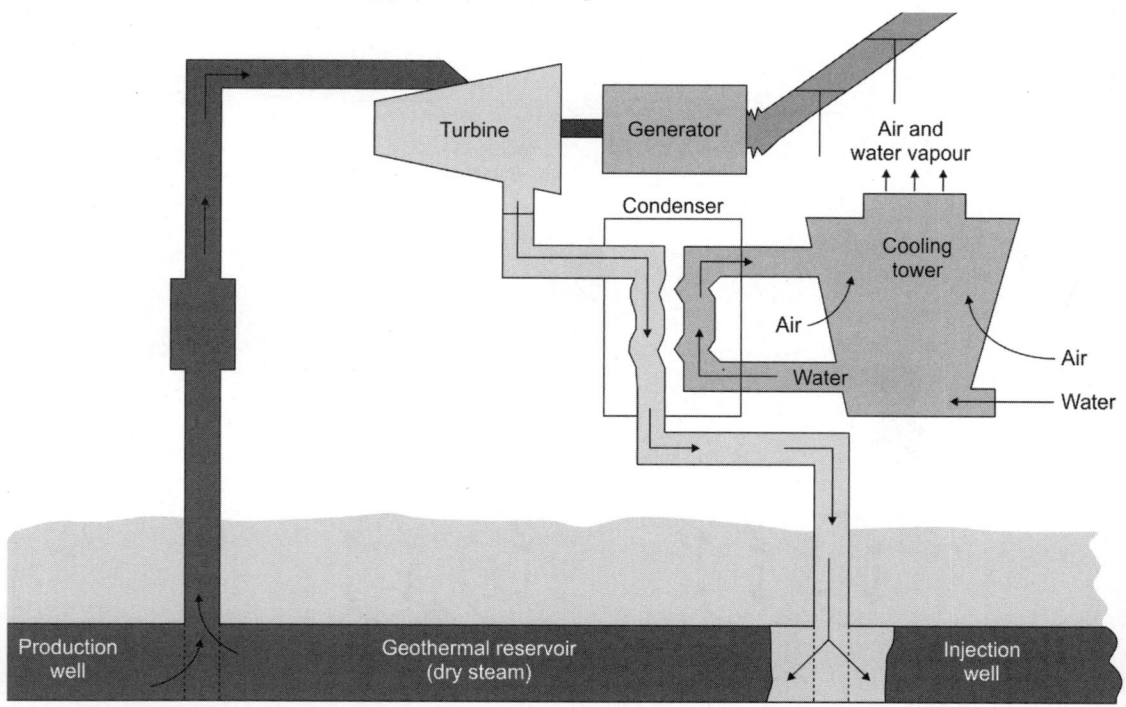

Fig. 7.5: Steam plant using vapour or dry steam dominated geothermal resource

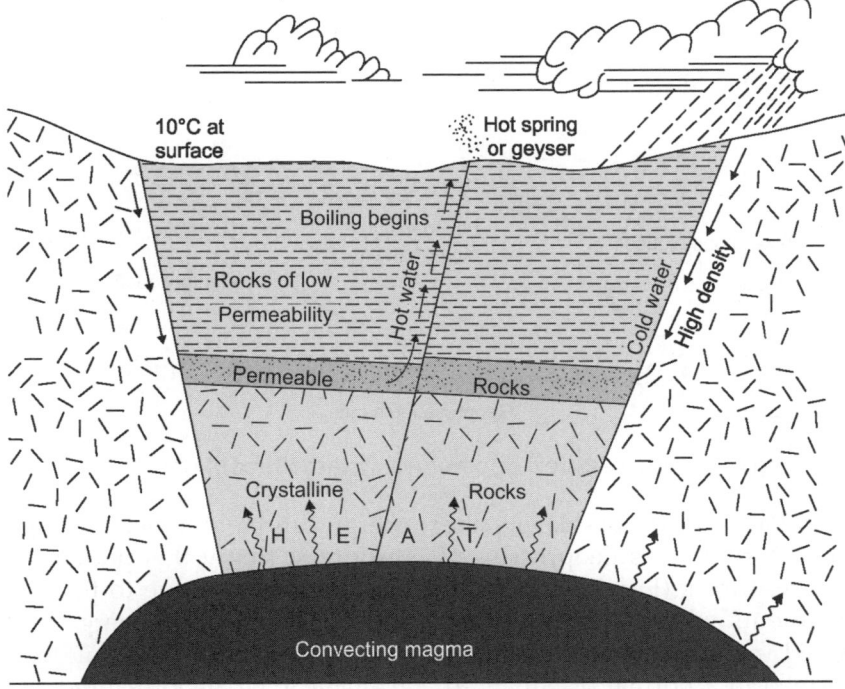

Fig. 7. 6: Liquid dominated geothermal resource

temperature over extensive volumes. There is commonly an upflow zone, at the middle of every convection cell, an outpouring zone or crest of warmed water moving along the side far from the core of the framework, and a downflow zone, where recharge is occurring. Surface indications incorporate hot springs, fumaroles, geysers, travertine stores, artificially modified rocks, or at times, no surface signs (Fig. 7.6).

iii. **A binary cycle** power plant is used when the water in a hot water reservoir is not hot enough to flash into steam (Fig. 7.7). Instead, the lower-temperature hot water is used to heat a fluid that expands when warmed. The turbine is powered from the expanded, pressurized fluid. Afterwards, the fluid is cooled and recycled to be heated over and over again.

7.10.2 Geopressured Resources

The geopressured hot water reservoirs were apparently formed by accumulation of geo-

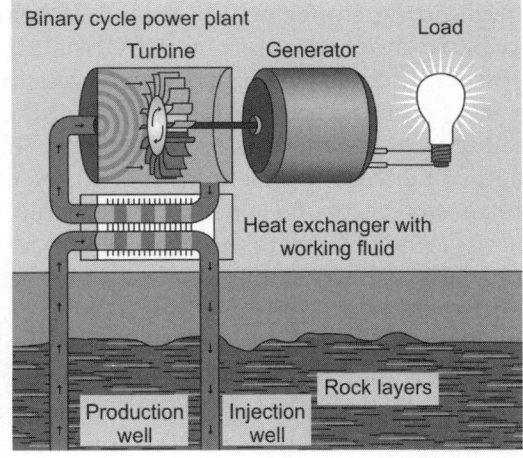

Fig. 7.7: Binary power plant

thermal heat stored over several million years in water trapped in a porous sedimentary medium by the overlying impervious layers. These are hydro-geothermal resources at greater depths of 3 to 10 km. The water is stored in underground cavities. Because of abnormally high pressure of the water, up to

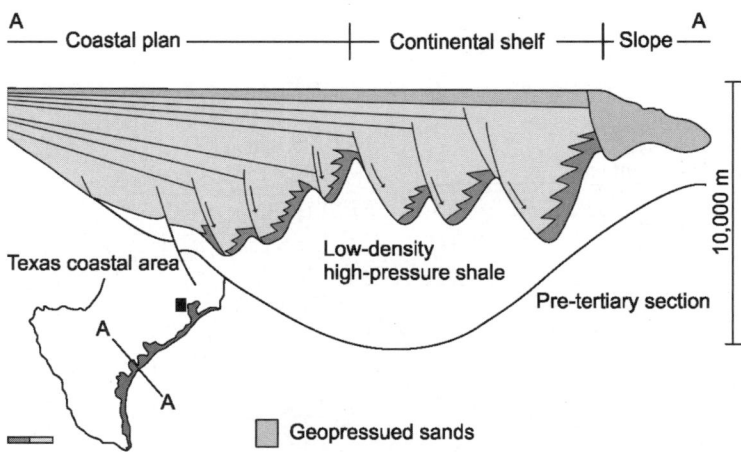

Fig. 7.8: Geopressured geothermal system (Bebout et al., 1978)

1350 atm (137 MPa) in the deepest layers, the reservoirs are referred to as geopressured. Geopressured resources (Fig. 7.8) occur in basin environments where deeply buried fluids contained in permeable sedimentary rocks are warmed in a normal or enhanced geothermal gradient by their great burial depth. The fluids are tightly confined by surrounding impermeable rock and bear pressure much greater than hydrostatic pressure. Thermal waters under high pressure in sand aquifers are the target for drilling, mainly as they contain dissolved methane. The source of energy available from this type of resource consists of: (1) heat; (2) mechanical energy; and, (3) methane. The Texas and Louisiana Gulf Coast in the United States has been tested for the geothermal energy, however, due to the great depths of several kilometers, they have not proved economic. Geopressure deposits are located at different levels and at different places. The quantities of hot water vary. Production wells may have active life of six to fifteen years. A geopressured geothermal resource consists of deeply buried water that contains dissolved methane. It is found in large, deep aquifers under high pressure. The water and methane are trapped in sedimentary formations at a depth of about 3 to 6 km. The temperature of the water is in the range of 90° to 200°C. Three forms of energy can be obtained from geopressured resources: firstly thermal energy, secondly hydraulic energy due to the high pressures and thirdly chemical energy by burning the dissolved methane gas. While technologies are available to tap geopressured resources, they are not currently economically competitive.

7.10.3 Hot-Dry-Rock (HDR) Resources

Hot dry rock assets, (Fig. 7.9) are characterized as hotness archived in rocks inside something like 10 km of the surface from which vigor can't be financially removed by characteristic high temp water or steam. These hot rocks have few pore space, or cracks, and in this way, hold little water and practically zero interconnected penetrability and necessities to be cracked to expand its hotness exchange surface keeping in mind the end goal to concentrate the hotness, test undertakings have misleadingly broken the rock by water driven weight, accompanied by circulating cold water down one well to concentrate the high temperature from the rocks and after that handling from a second well in a closed system. There, it is utilized within a force plant to prepare power. Hot-dry-rock (HDR) assets are found in regions where the stream of high temperature from the inside of the Earth to the surface is higher than normal yet there is no water since no aquifers or cracks are

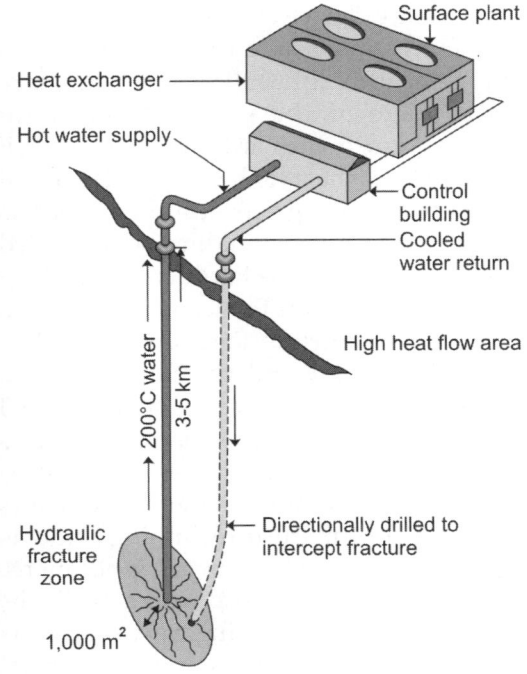

Fig. 7.9: Hot-dry-rock exploitation

available. Water is pumped into hot rock through an injection wall, becomes super-heated as it flows through open joints in the hot rock reservoir. The boiling water is then carried up to the surface through an alternate borehole and used to produce power. The water could be reused again and again. This resource is quite vast and is by and large more receptive than hydrothermal resources.

There are two strategies to tap this geo-thermal energy, one conceivable strategy is to explode a high hazardous at the bottom of an overall bored into the rock. This may be an atomic hazard. Water might be infused into this decently, coursed through the holes, so structured to concentrate heat from the rock. The water or water-steam mixture is with-drawn through an alternate well. An alternate technique is to utilize pressure driven breaking to transform the high temperature exchange surface and porousness needed to concentrate vigor at a high rate from hot dry rock. Water driven cracking, which is per-formed by pumping water at high pressure into the rocks, is usually utilized as a part of oil and gas fields to enhance the stream. High temperature can then be concentrated from the hot shake by coursing water through the split. It is accepted that HDR frameworks offer more adaptability in operation and outline than other geothermal frameworks. Case in point, the architect can have a decision of water stream rates and temperatures by penetrating to different profundities, and the specialist can change pumping weight and thus stream rates to suit burden condition.

7.10.4 Molten Rock or Magma Resources

Magma, the biggest geothermal asset, is liquid rock found at profundities of 3–10 km and deeper, and in this way not effectively approachable. It has a temperature extent of 700°–1,200 °C. Engineering is not accessible to use magma as a vigor source in a few cases, particularly in the region of generally later volcanic action liquid or partially liquid rock (i.e. magma) happens at moderate profundi-ties (e.g. not exactly 5 km). The precise high temperature above 650°C and the expansive volume make a significant geothermal asset. Be that as it may, extraction of the high tempe-rature from the liquid rock will be challenging and may not be achievable for quite a while

7.11 PRODUCTION WELL

Are located at a distance of about 0.5 km. Life of a steam power production well is 6 to 15 years. Thereafter new well should be used. Steam production from a well declines by about 20% in about 6 years.

7.11.1 Reinjection Wells

The reason for reinjection of condensate into the Earth are:

   i. To ensure supply of fresh water from pro-duction well for several years. The Earth acts as a steam generator.

ii. To prevent environmental pollution. The condensate contains certain pollutants for, e.g. the steam at the geyser contains hydrogen sulphide up to 200 ppm. If condensate is let out on ground level, it causes environmental hazard.

7.12 APPLICATIONS OF GEOTHERMAL ENERGY

Geothermal energy can be used in two ways as a source of direct heat or for electricity generation.

Direct Use

Hydrothermal resources of low to moderate temperature (20°–150°C) are used to provide direct heating for a range of applications in the residential, commercial and industrial sectors. Direct use of geothermal resources is primarily for direct heating and cooling. The main utilization categories are:

1. Swimming, bathing and balneology;
2. Space heating and cooling including district energy systems;
3. Agricultural applications such as greenhouse and soil heating;
4. Aquaculture application such as pond and raceway water heating;
5. Industrial applications such as mineral extraction, food and grain drying; and
6. Geothermal (ground-source) heat pumps (GHP), used for both heating and cooling. About twice as much geothermal energy is used for direct heat as for electricity generation. Direct use geothermal systems usually consist of a production facility (e.g. a well) to bring the heated water to the surface, a mechanical system (e.g. piping, heat exchanger, pump, controls) to move the heat energy to places where it is required, and a disposal system (e.g. injection well or storage pond) to receive the cooled fluid. Heat exchangers are usually needed when the geothermal fluid contains salt and other dissolved solids. Heat pumps are often used to move the heat energy from place to place. Geothermal heat

pumps are devices which operate on the same principle as the refrigerator but can move heat in either direction. They take advantage of the relatively constant temperature of the Earth's interior, using it as a source of heat for heating and as a heat sink for cooling. In summer, heat is extracted from the building being cooled and dumped into the Earth. In winter, heat is removed from the Earth and pumped into the building. Such systems are used widely in Switzerland and the Scandinavian countries. Through the use of geothermal heat pumps, marginal geothermal resources with temperatures as low as 20°C can be used. The direct use of geothermal resources is a proven, mature technology and is commercially viable for many applications. The use of this resource, where available, can result in a net saving in energy costs for consumers in homes and commercial operations.

Electricity Generation

High temperature geothermal resources can be used for electricity production. There is currently 8 GW of installed geothermal electricity generating capacity worldwide. There are a number of energy conversion technologies which use the geothermal resource. These include dry steam, flash steam and binary cycle systems. Geothermal electricity can be used to supply base load power, as well as for peak load demand as required. Where the resource is in good supply, geothermal electricity can be competitive with conventional energy sources. Geothermal power is generated by using steam or a secondary hydrocarbon vapour to turn a turbine-generator set to produce electrons. A vapour dominated (dry steam) resource can be used directly, whereas a hot water resource needs to be flashed by reducing the pressure to produce steam. In case of low temperature resource, generally below 150°C, the use of a secondary low boiling point fluid (hydrocarbon) is required to generate the vapour, in a binary or organic

Rankine cycle plant. Usually a wet or dry cooling tower is used to condense the vapour after it leaves the turbine to maximize the temperature drop between the incoming and outgoing vapour and thus increase the efficiency of the operation. The worldwide installed capacity has the following distribution: 29% dry steam, 37% single flash, 25% double flash, 8% binary/combined cycle/hybrid, and 1% backpressure. Factors affecting economic growth of geothermal energy based power generation are the well head temperature, well flow rate and the cost of wells. Optimum plant size is in 50 to 100 MW range. Scope of cost reduction in large plants through economics of scale is limited and hence geothermal power plants are more appropriate for supplying power to public buildings and commercial establishments. Total cost of geothermal plant can be divided in 3 categories:

1. Power plant capital costs
2. Operating cost
3. Energy supply cost

Power plant costs mainly cover the initial capital required to build it with operating cost forming a small proportion. As the proportion of operating costs to total costs is low, uncertainties in operating costs would not have a significant impact on power costs. Geothermal power generation is thus capital intensive with 75% of generating costs as fixed costs pertaining to capital investment.

7.13 BENEFITS OF GEOTHERMAL ENERGY

1. Using modern emission controls, geothermal energy is one of the least polluting sources of energy.
2. Geothermal power stations are very reliable with a high availability and capacity factor and are designed to run 24 hours a day, and operation is independent of weather or fuel delivery.
3. This technology is modular in design and highly flexible and the output of a geothermal plant can be expanded as required, avoiding the need for a high initial capital outlay.
4. Geothermal resources are used for local energy supply and can reduce the economic pressures of importing fuels and can provide local technical infrastructure and employment.
5. Geothermal energy has an inherent energy storage capability and there is no need for other technology to store the energy as, for example, batteries are needed with solar cells to store the electrical energy.
6. Geothermal power stations do not need large area for its extraction.

7.14 GEOTHERMAL ENERGY AND THE ENVIRONMENT

Geothermal energy is a renewable energy source that does little damage to the environment. Geothermal steam and hot water do contain naturally occurring traces of hydrogen sulphide (a gas that smells like rotten eggs) and other gases and chemicals that can be harmful in high concentrations. Geothermal power plants use "scrubber" systems to clean the air of hydrogen sulphide and the other gases. Sometimes the gases are converted into marketable products, such as liquid fertilizer. Newer geothermal power plants can even inject these gases back into the geothermal wells. Geothermal power plants do not burn fuels to generate electricity as do fossil fuel plant and release less than 1%–4% of the amount of carbon dioxide (CO_2) emitted by coal plants. Geothermal power plants, on the other hand, emit only about 1%–3% of the sulphur compounds that coal and oil-fired power plants do. Well-designed binary cycle power plants have no emissions at all. These power plants are compatible with many environments. They have been built in deserts, in the middle of crops, and in mountain forests. Geothermal development is often allowed on federal lands because it does not significantly harm the environment. Before permission is granted, however, studies must be made to determine what effect a plant may have on the environment.

7.15 ENVIRONMENTAL CONSIDERATIONS

Geothermal resources are considered renewable and "green", however, there are several environmental impacts that must be considered during utilization that are usually mitigated. These are emission of harmful gases, noise pollution, water use and quality, land use, and impact on natural phenomena, wildlife and vegetation.

1. **Emissions:** These are usually associated with steam power plant cooling towers that produce water vapour emission (steam), not smoke. The potential gases that can be released, depending upon the reservoir type are carbon dioxide, sulphur dioxide, nitrous oxides, hydrogen sulphide along with particulate matter geothermal fluids (steam or hot water) usually contain gases such as carbon dioxide (CO_2), hydrogen sulphide (H_2S), ammonia (NH_3) and methane CH_4). These gases, if released, can not only add to greenhouse warming but can also be toxic and smelly.

2. **Geothermal fluids:** These fluids also usually contain dissolved chemicals, which commonly include sodium and potassium chlorides, arsenic or mercury. These fluids would be a source of pollution if discharged into the environment. Modern emission control techniques and re-injection of contaminated fluids back into the ground is needed to minimize the impacts of these pollutants.

3. **Water use:** Geothermal plants use about 20 liters of freshwater per MWh, while binary air-cooled plants use no fresh water, as compared to a coal plant that uses 1,370 liters per MWh. An oil plant uses about 15% less and nuclear power plant about 25% more than the coal plant. The only change in the fluid during use is to cool it, and usually the fluid is returned to the same aquifer, so it does not mix with the shallow groundwater. Geothermal energy resources will be depleted if used beyond their natural recharge rate

4. **Land use:** Geothermal power plants are designed to "blend-in" with the surrounding landscape, and can be located near recreational areas with minimum land and visual impacts. Subsidence and induced seismicity are two land use issues that must be considered while withdrawing fluids from the ground. These are usually mitigated by injecting the used fluid back into the same reservoir. In addition, utilizing geothermal resources eliminates the mining, processing and transporting required for electricity generation from fossil fuel and nuclear resources. The technique of injecting the geothermal fluid back into the ground can effectively remove this risk.

5. **Seismic activity:** Geothermal energy production has been associated with increased seismic activity specially those geothermal fields which are located in regions that are already prone to earthquakes.

6. **Noise:** The majority of the noise produced at a power plant or direct-use site is during construction (e.g. drilling of wells, and the escape of high pressure steam during testing) but this is not significant once the plants are operating. The noise from a power plant is not considered an issue of concern, as it is extremely low, unless you are next to or inside the plant. Most of the noise comes from cooling fans and the rotating turbines.

7. **Impact on natural phenomena, wildlife and vegetation:** Plants are usually prevented from being located near geysers, fumaroles and hot springs, as the extraction of fluids to run the turbines, might impact these thermal manifestations. Most plants are located in areas with no natural surface discharges. Any site considered for a geothermal power plant, must be reviewed and considered for the impact on wildlife and vegetation, and if significant, provide a mitigation plan. Direct use projects are usually small and thus have no significant impact on natural features.

In summary, the use of geothermal energy is reliable, providing base load power, is renewable and has minimum air emission and offsets the high air emissions of fossil fuel fired plants has minimum environmental impacts, is combustion free, and is a domestic fuel source.

7.16 CONCLUSIONS

Geothermal power can become a valuable source of energy, if properly harnessed. Continued energy shortages have created added interest in geothermal energy for both power generation and direct applications. Geothermal growth and development of electricity generation has increased significantly over the past 30 years approaching 15% annually in the early part of this period, and dropping to 3% annually in the last ten years due to an economic slowdown in the far east and the low price of competing fuels. The electricity generation from geothermal resources is currently performed in all types of reservoirs except those associated with resources under high pressure, found at great depths, in sedimentary basins. Now-a-days, plants of all types are in operation. Binary plants are widely used because they can work with low-enthalpy resources and are used in *hot-dry-hot* reservoirs. The use of these types of reservoirs has grown in recent years due to discoveries of new reservoirs located outside the borders of the tectonic plates. The technology is expensive and it is necessary that for real breakthrough the cost be reduced substantially. Finally, the largest growth will include the installation and use of geothermal heat pumps, as they can be used anywhere in the world, as shown by the large developments in Switzerland, Sweden, Austria, Germany.

8

Nuclear Energy

8.1 INTRODUCTION

Energy, 'the ability to do work', is essential for meeting basic human needs, extending life expectancy and providing comfort in living standards. Energy can be divided into two categories—primary and secondary. Primary energy is energy in the form of natural resources, such as wood, coal, oil, natural gas, natural uranium, wind, hydropower, and sunlight. Secondary energy is the more useable form of energy derived from the primary energy sources, such as electricity and petrol. Primary energy can be renewable or non-renewable: renewable energy sources include solar, wind and wave energy, biomass (wood or crops such as sugar), geothermal energy and hydropower. Non-renewable energy sources include the fossil fuels—coal, oil and natural gas, which together provide 80% of our energy today, plus uranium. The advantages and disadvantages of using nuclear power are given in Table 8.1.

When the term nuclear power or nuclear energy gets utilized typically the first things that strike a mind are bombs, destruction, war, and deformed people. Which in fact are four things that occur the least in a nuclear power field. Atomic force is substantially more than just bombs and destruction, bombs scarcely even consume a rate of the aggregate sum of atomic energy utilized within the planet today. Some radioactive materials (uranium, thorium) accessible in nature undergoes fission reaction in atomic reactor force plants to get gigantic measure of high temperature,

Table 8.1: Advantages and disadvantages of nuclear energy

Advantages	Disadvantages
Nuclear power costs about the same as coal, so it's not expensive to make	Although not much waste is produced, it is hazardous. It must be sealed up and buried for many years to allow the radioactivity to die away
Does not produce smoke or carbon dioxide, so it does not contribute to greenhouse effect	Nuclear power is reliable, but a lot of money has to be spent on safety–if it does go wrong, a nuclear accident can be a major disaster

steam and subsequently electrical energy. Mostly the greater part of the atomic energy utilized today comes quite close to handling of power. Atomic force plants are answerable for 16% of the sum of the planets power preparation and which truly may not sound like a great deal, yet when you think about the measure of electricity utilized as a part of the planet it truly places it into prospective, how essential atomic force is to every one of us. Nuclear power plays a vital role in our society largely due to the global warming trend, and the fact that nuclear power is much cleaner and more "environmentally friendly" than other forms of energy production, which is just one of many reasons nuclear power is a growing trend. Nuclear power is not just a very simple energy source to come by. It is produced through a very tedious and meticulous process that if controlled is very powerful and if uncontrolled is extremely dangerous, that is why it is a much regulated energy source, but without it where would we be?

Nuclear energy, the vigor archived in the core of an atom and discharged through fission, fusion combination, or radioactivity. In these techniques a little measure of mass is changed over to energy, consistent with the relationship $E = mc^2$, where E is energy , m is mass, and c is the speed of light. The most striking issues concerning atomic energy are the probability of a mishap or system failure at an atomic reactor or fuel plant, for example, those which happened at Three Mile Island (1979), Chernobyl (1986), and Fukushima (2011), and the potential danger to the proceeded presence of humanity postured by atomic weapons.

The Nucleus and its Constituents

An atom consists of a centrally located nucleus surrounded by electrons revolving in certain physically permitted orbitals. The nucleus itself is made up of neutrons and protons, collectively called nucleons. The number of protons (Z) is called the atomic number and the total number (A) of nucleons in a nucleus is called the atomic or (nuclear) mass number. The number of neutrons (A–Z) is represented as N. The basic properties of the atomic constituents are summarized in Table 8. 2

8.2   WHAT IS NUCLEAR ENERGY?

Nuclear energy is energy in the nucleus (core) of an atom. Atoms are tiny particles that make up every object in the universe. There is enormous energy in the bonds that hold atoms together. Nuclear energy can be used to make electricity. But first the energy must be released. It can be released from atoms in two ways: nuclear fusion and nuclear fission. In nuclear fusion, energy is released when atoms are combined or fused together to form a larger atom. This is how the sun produces energy. In nuclear fission, atoms are split apart to form smaller atoms, releasing energy. Nuclear power plants use nuclear fission to produce electricity.

8.3   NUCLEAR REACTIONS AND RADIATION

Traditional chemical reactions happen as an aftereffect of the association between valence electrons around a molecule's core. In 1896, Henri Becquerel extended the field of science to incorporate atomic progressions when he invented that uranium emits radiation. Not long after Becquerel's disclosure, Marie Sklodowska Curie started concentrating on radioactivity and did a great part of the pioneering breakthrough. Curie found that radiation was corresponding to the measure of radioactive component present, and she recommended that radiation was a property of atoms (instead of a chemical property of a compound). In 1902, Frederick Soddy proposed the hypothesis that 'radioactivity is the consequence of a characteristic change of an isotope of one component into an isotope of an alternate component'. Atomic responses include changes in particles in a particle's core and subsequently cause a change in the atom itself. All components heavier than bismuth (Bi) (and some lighter) show regular radioactivity and therefore can decay into lighter components. Dissimilar to normal chemical responses that structure atoms, atomic responses bring about the transmutation of one component into an alternate isotope or an alternate component inside and out.

Nuclear reactions involve changes in an atom's nucleus and thus cause a change in the atom itself. Unlike normal chemical reactions that form molecules, nuclear reactions result

Table 8.2 Properties of atomic constituents

Fundamental particle	Charge	Mass (u)
Proton	e	1.007276
Proton	0	1.008665
Electron	–e	0.000549

in the transmutation of one element into a different isotope or a different element altogether. The release of nuclear energy is associated with changes from less stable to more stable nuclei and produces far more energy for a given mass of fuel than any other source of energy. The most stable nuclei, those with the highest binding energies per nucleon holding their components together are in the middle range of atomic weights, with the maximum stability at weights near 60. Thus, fission, which produces two lighter fragments, occurs for very heavy nuclei, while fusion occurs for the lightest nuclei. Nuclear responses include changes in an atom's core and thus cause a change in the particle itself. Dissimilar to typical chemical reaction that form molecules, nuclear reactions result in the transmutation of one element into an alternate isotope or an alternate component inside and out. The release of nuclear energy is connected with progressions from less stable to additionally stable nucleus and transforms significantly more energy for a given mass of fuel than whatever viable source of energy available. There are two types of nuclear reactions. The first is the *radioactive decay of bonds* within the nucleus that emit radiation as it decays or transforms to a more stable state. The second is the *"billiard ball" type of reactions*, where the nucleus or a nuclear particle (like a proton) is slammed into by another nucleus or nuclear particle. In fission processes, a fissionable nucleus absorbs a neutron, becomes unstable, and splits into two nearly equal nuclei. In fusion processes, two nuclei combine to form a single, heavier nucleus. The most stable nuclei those with the most astounding tying energies for every nucleon holding their parts together-are in the center extent of nuclear weights, with the greatest solidness at weights close to 60. Consequently, splitting, which handles two lighter parts, happens for quite overwhelming nuclei, while combination happens for the lightest nuclei. There are three common types of radiation and nuclear changes:

a. **Alpha radiation (α)** is the emission of an alpha particle from an atom's nucleus. A α molecule holds 2 *protons* and 2 *neutrons* (and is similar to a He nucleus: $^4He_2$). The point when an atom transmits a α molecule, the particle's nuclear mass will diminish by 4 units (since 2 *protons* and 2 *neutrons* are lost) and the nuclear number (Z) will diminish by 2 units. The component is said to "transmute" into an alternate component that is 2 units of Z (smaller). A case of a α transmutation happens when uranium decay into the component thorium (Th) by emanating an alpha molecule as delineated in the accompanying mathematical statement:

$$\,^{238}_{92}U \rightarrow \,^{234}_{90}Th + \,^4_2He$$

b. **Beta radiation (β)** is the transmutation of a *neutron* into a *proton* and an *electron* (followed by the emission of the *electron* from the atom's nucleus: $^{\,0}_{-1}e$). When an atom emits a β particle, its *mass* will not change (since there is no change in the total number of nuclear particles), however the atomic number will increase by 1 (because the *neutron* transmutated into an additional *proton*). An example of this is the decay of the isotope of carbon named carbon-14 into the *element* nitrogen:

$$\,^{14}_{6}C \rightarrow \,^{14}_{7}N + \,^{\,0}_{-1}e$$

c. **Gamma radiation (γ)** involves the emission of electromagnetic *energy* (similar to *light energy*) from an atom's nucleus. No particles are emitted during gamma radiation, and thus gamma radiation does not itself cause the transmutation of *atoms*, however γ radiation is often emitted during, and simultaneous to, α or β radioactive decay. X-rays, emitted during the beta decay of cobalt-60, are a common example of gamma radiation.

d. **Half-life:** Radioactive decay proceeds according to a principal called the *half-life*. The *half-life* ($T_{1/2}$) is the amount of time necessary for ½ of the radioactive material to decay. For example, the radioactive *element* bismuth (^{210}Bi) can undergo alpha decay to form the *element* thallium (^{206}Tl) with a reaction *half-life* equal to 5 days. If we begin an experiment starting with 100 g of bismuth in a sealed lead container, after 5 days we will have 50 g of bismuth and 50 g of thallium in the jar. After another 5 days (10 from the starting point), ½ of the remaining bismuth will decay and we will be left with 25 g of bismuth and 75 g of thallium in the jar.

The fraction of *parent* material that remains after radioactive decay can be calculated by using the following equation:

$$\text{Fraction remaining} = \left(\frac{1}{2}\right)^n$$

where n = half-lives elapsed

The amount of a radioactive material that remains after a given number of half-lives is therefore:

Amount remaining

= original amount × fraction remaining

The decay reaction and $T_{\frac{1}{2}}$ of a substance are specific to the isotope of the element undergoing radioactive decay. For example, ^{210}Bi can undergo α decay to ^{206}Tl with a $T_{\frac{1}{2}}$ of 5 days. ^{215}Bi, by comparison, undergoes β decay to ^{215}Po with a $T_{\frac{1}{2}}$ of 7.6 minutes, and ^{208}Bi undergoes yet another mode of radioactive decay (called electron capture) with a $T_{\frac{1}{2}}$ of 3,68,000 years!

8.4 WHAT IS NUCLEAR DAMAGE?

Nuclear damage is explained as any injury or death, sickness or disease of a person, or damage to the environment including loss of property which arises from ionizing radiation associated with a nuclear installation, nuclear vessel or handling of radioactive materials.

8.5 WHAT IS NUCLEAR SAFETY?

Nuclear safety means the achievement of safe operating conditions, prevention of nuclear accidents or mitigation of nuclear accident consequences, resulting in the protection of workers, the public and the environment against the potential harmful effects of ionizing radiation or radioactive material.

8.6 A BRIEF HISTORY OF NUCLEAR ENERGY

The science of atomic radiation, atomic change and nuclear fission was developed from 1895 to 1945, much of it in the last six of those years. Over 1939-45, most development was focused on the atomic bomb. From 1945 attention was given to harness this energy in a controlled fashion for naval propulsion and for making electricity. Since 1956 the prime focus has been on the technological evolution of reliable nuclear power plants. It was also infamously used to create nuclear bombs to destroy the cities of Hiroshima and Nagasaki in Japan during the Second World War.

8.7 NUCLEAR RENAISSANCE

In the new century several factors have combined to revive the prospects of nuclear power. First is realization of the scale of projected increased electricity demand worldwide, but particularly in rapidly-developing countries. Secondly is awareness of the importance of energy security, and thirdly is the need to limit carbon emissions due to concern about global warming. These factors coincide with the availability of a new generation of nuclear power reactors, and in 2004 the first of the late third-generation units was ordered for Finland – a 1600 MWe European PWR (EPR). A similar unit is planned for France as the first of a full fleet replacement there. In the USA the 2005 Energy Policy act provided incentives for establishing new-generation power reactors there. But plans in Europe and North America are overshadowed by those in China, India, Japan and South Korea. China alone plans a sixfold increase in nuclear power capacity by 2020, and has more than one hundred large units proposed and backed by credible political determination and popular support. A large portion of these are the latest western design, expedited by modular construction. The history of nuclear power thus starts with science in Europe, blossoms in UK and the USA with the latter's technological might, languishes for a few decades, then has a new growth spurt in East Asia.

8.8 NUCLEAR ENERGY APPLICATIONS

Although nuclear energy is mainly used for the production of electricity in nuclear power plants is not the only use of nuclear energy. Nuclear power has other applications in various fields:

1. **Industrial applications:** for analysis and process control.
2. **Medical applications:** in diagnosis and therapy of diseases.
3. **In food processing:** in the production of new species, conservation treatments of food, pest insects and vaccine preparation.
4. **Environmental applications:** The determination of significant amounts of pollutants into the environment.

Other applications such as carbon dating, which uses the properties of carbon-14 fixation to bone, wood or organic waste, determining their chronological age, and applications in geophysics and geochemistry, taking advantage of the existence of naturally occurring radioactive materials for fixing the dates of the deposits of fossil.

8.9 CASE STUDY: THE CHERNOBYL DISASTER

On April 26, 1986, a sudden surge of power during a reactor system's test destroyed unit 4 of the nuclear power station at Chernobyl, Ukraine, in the former Soviet Union. The accident and the fire that followed, released massive amount of radioactive material into the environment. Emergency crews responding to the accident used helicopters to pour sand and boron on the reactor debris. The sand was to stop the fire and additional release of radioactive material, and the boron was to prevent additional nuclear reactions. A few weeks after the accident, the crews completely covered the damaged unit in a temporary concrete structure, called the "sarcophagus," to limit further release of radioactive material. The Soviet government also cut down and buried about a square mile of pine forest near the plant to reduce radioactive contamination

at and near the site. Chernobyl's three other reactors were subsequently restarted but all eventually shut down for good, with the last reactor closing in 1999. The Soviet nuclear power authorities presented their initial report to an International Atomic Energy Agency meeting in Vienna, Austria, in August 1986. After the accident, officials closed off the area within 30 kilometers (18 miles) of the plant, except for persons with official business at the plant and those people evaluating and dealing with the consequences of the accident and operating the undamaged reactors. The Soviet (and later on, Russian) government evacuated about 1,15,000 people from the most heavily contaminated areas in 1986, and another 2,20,000 people in subsequent years. The Chernobyl accident's severe radiation effects killed 28 of the site's 600 workers in the first four months after the event. Another 106 workers received high doses to cure acute radiation sickness. Two workers died within hours of the reactor explosion from non-radiological causes. Another 2,00,000 cleanup workers in 1986 and 1987 received doses of between 1 and 100 rem. (The average annual radiation dose for a US citizen is about 0.6 rem). Chernobyl cleanup activities eventually required about 6,00,000 workers, although only a small fraction of these workers were exposed to elevated levels of radiation. Government agencies continue to monitor cleanup and recovery of workers' health.

8.10 TYPES OF NUCLEAR ENERGY

There are two basic types of nuclear energy-fission and fusion.

8.10.1 Nuclear Fission

Nuclear fission is a nuclear reaction in which a heavy nucleus (such as uranium) splits into two lighter nuclei and in doing so, releases a significant amount of energy as well as more neutrons. These neutrons then go on to split more nuclei and a chain reaction takes place. In such radioactive heavy nuclei, the balance between the solid atomic power alluring energy and the electrostatic repulsive force might be knocked out of balance, by the introduction of energy as a consumed neutron or photon, the nucleus oscillates in an attempt to recapture balance until the electrostatic force picks up more power than the shorter distance nuclear force, at which the core parts separated, discharging energy. Nuclear fission is a mature technology that has been in use for more than 50 years. The latest designs for nuclear power plants build on this experience to offer enhanced safety and performance, and are ready for wider deployment over the next few years. There is great potential for new developments in nuclear energy technology to enhance nuclei's role in a sustainable energy in future. Nevertheless, important barriers to a rapid expansion of nuclear energy remain. Governments need to set clear and consistent policies on nuclear power to encourage private sector investment. Picking up more terrific open acknowledgement will additionally should be key, and this will be aided by un-anticipated usage of arrangements for land transfer of radioactive waste, and proceeded safe and powerful operation of atomic plants.

Working: If a massive nucleus like uranium-235 breaks apart (fissions), then there will be a net yield of energy because the sum of the masses of the fragments will be less than the mass of the uranium nucleus. If the mass of the fragments is equal to or greater than that of iron at the peak of the binding energy curve, then the nuclear particles will be more tightly bound than they were in the uranium nucleus, and that decrease in mass comes off in the form of energy according to the Einstein equation. For components lighter than iron, fusion will yield energy. The fission of uranium-235 in reactors is triggered by the retention of a low energy neutron, regularly termed a "slow neutron" or a "thermal neutron". Other fissionable isotopes which could be prompted to fission by slow neutrons are plutonium-239, uranium-233, and thorium-232 (Fig. 8.1).

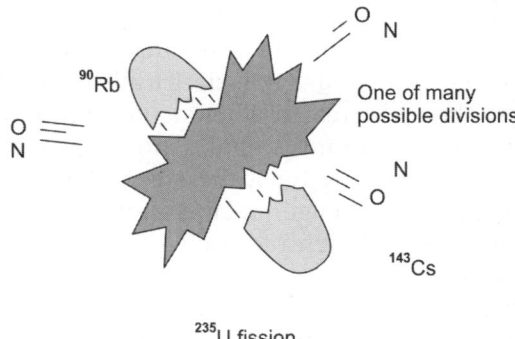

Fig. 8.1: Nuclear fission reaction

8.10.2 Nuclear Fusion

Fusion offers important advantages: no carbon emissions, no air pollution, unlimited fuel, and is intrinsically safe. While fusion technology is not at the deployment stage, the potential is substantial. The fusion reaction is about four million times more energetic than a chemical reaction such as the burning of coal, oil or gas. Fusion is a process where nuclei collide and join together to form a heavier atom, usually deuterium and tritium. When this happens a considerable amount of energy gets released at extremely high temperature: nearly 150 million degrees celsius. Nuclear fusion in stars and supernovae is the primary process by which new natural elements are created. It is this reaction that is harnessed in fusion power. It takes considerable energy to force nuclei to fuse, even those of the lightest element, hydrogen.

Combination offers essential preferences: no carbon emmissions, no air contamination, unlimited fuel, and is inherently protected. While fusion technology is not at the development stage, the potential is considerable. The fusion reaction is in the vicinity of four million times more energetic than a chemical reaction, for example, the blazing of coal, oil or gas. Fusion is a methodology where nuclei collide and join together to form a heavier atom, generally deuterium and tritium. The point when this happens a lot of energy gets discharged at greatly high temperature, about 150 million degrees celsius. At extreme temperatures, electrons are separated from nuclei and a gas becomes a plasma–a hot, electrically charged gas. The fuel (created when deuterium combines with tritium) is abundant; it gives off very little radioactivity, there is no need for underground storage, and there is no environmental risk of high radioactive fuel leakage in case of an accident, the plasma dissipates. Much more work is needed to achieve deployment of fusion technologies. It is accompanied by the release or absorption of energy depending on the masses of the nuclei involved. Iron and nickel nuclei have the largest binding energies per nucleon of all nuclei and therefore are the most stable. The fusion of two nuclei lighter than iron or nickel generally releases energy while the fusion of nuclei heavier than iron or nickel absorbs energy and vice-versa for the reverse process, nuclear fission. Nuclear fusion of light elements releases the energy that causes stars to shine and hydrogen bombs to explode. Nuclear fusion of heavy elements (absorbing energy) occurs in the extremely high-energy conditions of supernova explosions. Nuclear fusion in stars and supernovae is the essential process by which new natural elements are made. It is this reaction that is harnessed in fusion power. It consumes considerable energy to force nuclei to fuse , even those of the lightest element, hydrogen.

Working

If light nuclei are forced together, they will fuse with a yield of energy because the mass of the combination will be less than the sum of the masses of the individual nuclei. If the combined nuclear mass is less than that of iron at the peak of the binding energy curve, then the nuclear particles will be more tightly bound than they were in the lighter nuclei, and that decrease in mass comes off in the form of energy according to the Einstein relationship. For elements heavier than iron, fission will yield energy. For potential nuclear energy sources for the Earth, the deuterium-tritium fusion reaction contained by some kind of magnetic confinement seems the most likely path. However, for the fueling of the stars, other fusion reactions will dominate Fig. 8.2.

Abundant fuels

Deuterium is abundant as it can be extracted from all forms of water. If the entire world's electricity were to be provided by fusion power stations, present deuterium supplies from water would last for millions of years.

Tritium does not occur naturally and will be obtained from lithium within the machine. Therefore, once the reaction is established, even though it occurs between deuterium and tritium, the external fuels required are deuterium and lithium.

Lithium is the lightest metallic element and is plentiful in the earth's crust. If all the world's electricity were to be provided by fusion, known lithium reserves would last for at least one thousand years.

The energy gained from a fusion reaction is enormous. To illustrate, 10 grams of deuterium (which can be extracted from 500 litres of water) and 15 g of tritium (produced

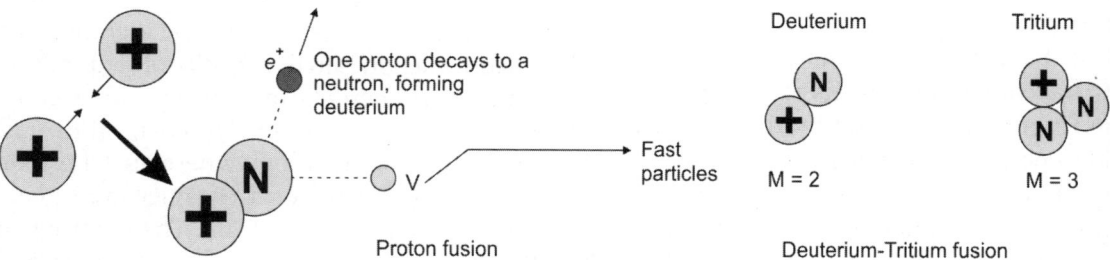

Fig. 8.2: Nuclear fusion reaction

from 30 g of lithium) reacting in a fusion power plant would produce enough energy for the lifetime electricity needs of an average person in an industrialized country.

Hydrogen Bomb: The Basics

A fission bomb, called the essential, produces a flood of radiations including a large number of neutrons. This radiation encroaches on the nuclear bit of the shell, regarded as the optional. The auxiliary comprises generally of lithium deuteride. The neutrons respond with the lithium in this synthetic compound, processing tritium and helium.

$$_3^6\text{Li} + _0^1 n \rightarrow _2^4\text{He} + _1^3\text{H}$$

The production of tritium from lithium deuteride is shown in Fig. 8.3.

This reaction produces the tritium on the spot, so there is no need to include tritium in the bomb itself. The extreme heat which exists in the bomb, the tritium fuses with the deuterium in the lithium deuteride.

The shock waves processed by the essential (A-shell) might spread excessively and gradually to allow get together of the nuclear fusion 112 nuclear stage before the bomb blew itself separated. This issue was settled by Edward Teller and Stanislaw Ulam. To do this,

they presented a high energy gamma beam absorbing material (styrofoam) to catch the energy of the radiation. As high vigor gamma radiation from the essential is retained, radial compression forces are pushed along the whole chamber at just about the same moment. This processes the compression of the lithium deuteride. Extra neutrons are likewise processed by different segments and reflected towards the lithium deuteride. With the compacted lithium deuteride center now bombared with neutrons, tritium is shaped and the combination methodology starts.

8.11 EXISTING NUCLEAR POWER PLANTS

Today, nuclear power plants are either under construction, or fully functional in all the developed nations of the world and in many of the developing countries. The biggest user of nuclear power in the world is the United States of America, followed by France, Russia, South Korea, China, Germany, Canada, Ukraine, United Kingdom and Sweden. In India, twenty nuclear power reactors produce 4,780.00 MW power. Some of the plants are situated in the states of Karnataka (Kaigapur), Gujarat (Kakrapar), Tamil Nadu (Kalpakkam), Uttar Pradesh (Narora), Rajasthan (Rawatbhata), Maharashtra (Tarapur) and Andhra Pradesh (Kudankulam).

8.12 NUCLEAR FUEL

Nuclear fuel is a material that can be 'burned' by nuclear fission or fusion to derive nuclear energy. Nuclear fuel can refer to the fuel itself, or to physical objects (for example bundles composed of fuel rods) composed of the fuel material, mixed with structural, neutron moderating, or neutron reflecting materials.

Most nuclear fuels contain heavy fissile elements that are capable of nuclear fission. When these fuels are struck by neutrons, they are in turn capable of emitting neutrons when they break apart. This makes possible a self-sustaining chain reaction that releases energy

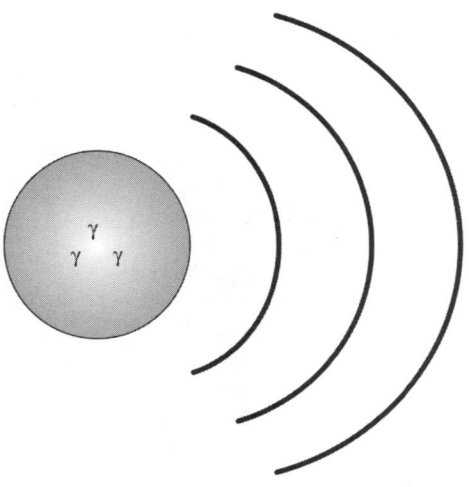

Fig. 8.3

with a controlled rate in a nuclear reactor or with a very rapid uncontrolled rate in a nuclear weapon.

The most common fissile nuclear fuels are uranium-235 (^{235}U) and plutonium-239 (^{239}Pu). The actions of mining, refining, purifying, using, and ultimately disposing of nuclear fuel together make up the nuclear fuel cycle.

Not all types of nuclear fuels create power from nuclear fission. Plutonium-238 and some other elements are used to produce small amounts of nuclear power by radioactive decay in radioisotope thermoelectric generators and other types of atomic batteries. Also, light nuclides such as tritium (^3H) can be used as fuel for nuclear fusion. Nuclear fuel has the highest energy density of all practical fuel sources.

8.12.1 Oxide fuel

For fission reactors, the fuel (typically based on uranium) is usually based on the metal oxide, the oxides are used rather than the metals themselves because the oxide melting point is much higher than that of the metal and because it cannot burn, being already in the oxidized state.

8.12.2 Nuclear Fuel Cycle

The atomic fuel cycle is the arrangement of mechanical courses of action which include the handling of power from uranium in atomic force reactors (Fig. 8.4). Uranium is generally a regular component that is found all around the globe. It is mined in various nations and must be prepared before it might be utilized as fuel for an atomic reactor. Fuel uprooted from a reactor, after it has arrived at the close of its handy life, might be reprocessed to process new fuel. The different exercises connected with the generation of power from nuclear energy are alluded to all in all as the atomic fuel cycle. The atomic fuel cycle begins with the mining of uranium and finishes with the transfer of atomic waste. With the reprocessing of utilized fuel as a choice for nuclear energy, the stage structure forms an accurate cycle. To prepare uranium for use in a nuclear reactor, it undergoes the steps of mining and milling, conversion, enrichment and fuel fabrication. These steps make up the 'front end' of the nuclear fuel cycle.

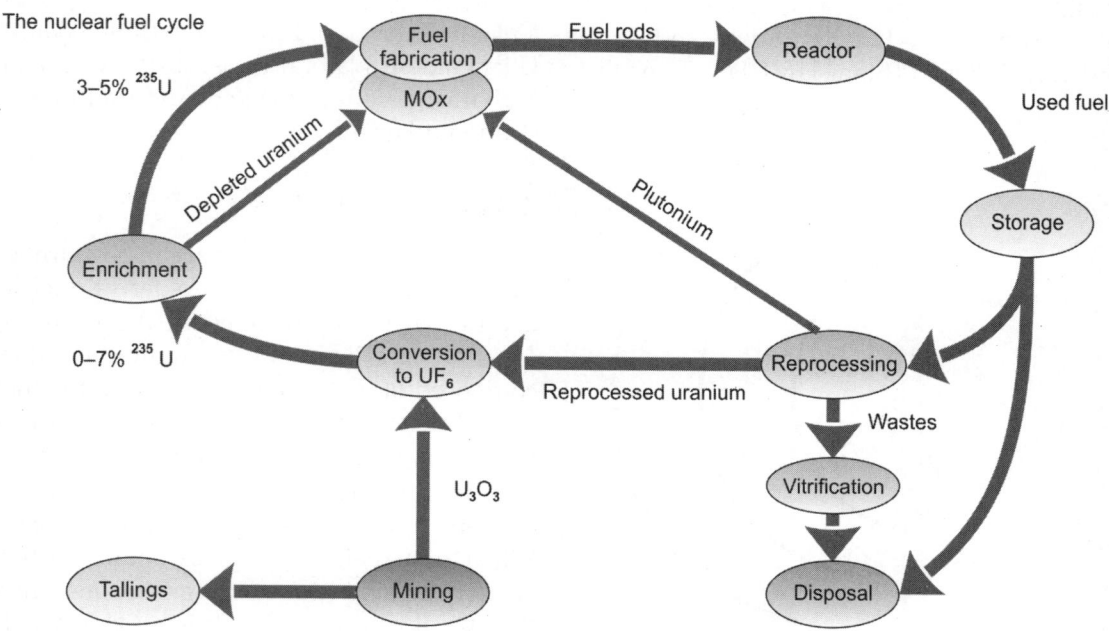

Fig. 8.4: Nuclear fuel cycle

After uranium has spent about three years in a reactor to produce electricity, the used fuel may undergo a further series of steps including temporary storage, reprocessing, and recycling before wastes are disposed. Collectively these steps are known as the 'back end' of the fuel cycle.

8.13 STORAGE AND TRANSPORTATION

Radioactive waste contains radioactive elements that send out higher levels of radiation than natural background radiation. Radioactive waste can be classified into three main categories: low, intermediate and high.

1. Low-level waste (LLW)—contains enough radioactive material to require action for the protection of people, but not so much that it requires shielding in handling or storage.

2. Intermediate-level waste (ILW)—requires shielding. If it has more than 4000 Bq/g of long-lived (over 30 year half-life) alpha emitters, it is categorized as "long-lived" and requires more sophisticated handling and disposal.

3. High-level waste (HLW)—sufficiently radioactive too, require both shielding and cooling, generates >2 kW/m^3 of heat and has a high level of long-lived α-emitting isotopes.

In most nations, atomic force era and different requisitions of radioactive materials began before arrangements for the transfer of the ensuing radioactive waste were generally advanced. As wastes emerged, it was most regularly archived in different types of built regulation on the surface and at destinations to which access was regulated. Exploration and advancement deal with waste transfer has indicated that, in guideline, numerous types of radioactive waste might be discarded in a way that furnishes insurance for the health and wellbeing of individuals and the Earth. For high level and long lived radioactive wastes, the consensus of the waste management experts internationally is that disposal in deep underground engineered facilities—geological disposal—is the best option that is currently available or likely to be available in the foreseeable future. This option is under investigation in most countries with significant amounts of such waste, and two countries have now made formal (government) decisions to go ahead with facilities for the disposal of high level waste.

Universal regulations for the transport of radioactive material have been published by the International Atomic Energy Agency (IAEA) since 1961. These regulations have been generally received into national regulations, and additionally into modal regulations, for example, the International Maritime Organisation's (IMO) Dangerous Goods Code. Regulatory control of shipments of radioactive material is free of the material's planned requisition.

- About twenty million consignments of all sizes containing radioactive materials are routinely transported worldwide annually on public roads, railways and ships.

- These use robust and secure containers. At sea, they are generally carried in purpose-built ships.

- Since 1971, there have been more than 2,0,000 shipments of used fuel and high-level wastes (over 80,000 tonnes) over many million kilometres.

- There have been accidents over the years, but never one in which a container with highly radioactive material has been breached, or has leaked.

8.14 ENERGY FROM NUCLEAR FISSION REACTIONS

A **nuclear chain reaction** occurs when one nuclear reaction causes an average of one or more nuclear reactions, thus leading to a self-propagating series of these reactions. The specific nuclear reaction may be the fission of heavy isotopes (e.g. ^{235}U). The nuclear chain reaction releases several million times more energy per reaction than any chemical reaction. Chain reactions naturally give rise to reaction rates that grow (or shrink) exponentially, whereas a nuclear power reactor needs to be able to hold the reaction rate reasonably constant. To maintain this control, the chain reaction criticality must have a slow enough time-scale to permit intervention by additional effects, (e.g. mechanical control rods or thermal expansion). Consequently, all nuclear power reactors (even fast-neutron reactors) rely on delayed neutrons for their criticality. An operating nuclear power reactor fluctuates between being slightly subcritical and slightly delayed-supercritical, but must always remain below prompt-critical.

It is impossible for an atomic power plant to experience an atomic chain reaction that brings about an outburst of force equivalent with an atomic weapon, but even low-powered explosions due to uncontrolled chain reactions, that would be considered "fizzles" in a bomb, may still cause considerable damage and meltdown in a reactor. For instance, the Chernobyl disaster included a runaway chain reaction however the effect was a low-controlled steam eruption from the moderately small release of high temperature, as contrasted and a bomb. Then again, the reactor complex was destroyed by the heat, and by customary blazing of the graphite presented to air. Such steam explosions would be typical of the very diffuse assembly of materials in a nuclear reactor, even under the worst conditions.

The energy from this reaction is generally used to heat water in thermal power plants for generation of power for commercial consumption.

8.14.1 Fast Breeder Reactors

Under appropriate operating conditions, the neutrons given off by fission reactions can "breed" more fuel from otherwise non-fissionable isotopes. The most common breeding reaction is that of plutonium-239 from non-fissionable uranium-238. The term "fast breeder" refers to the type of configurations which can actually produce more fissionable fuel than they use, such as the LMFBR (Fig. 8.5). This scenario is possible because the non-fissionable uranium-238 is 140 times more abundant than the fissionable U-235 and can be efficiently converted into Pu-239 by the neutrons from a fission chain reaction.

The reactor will use sodium as the coolant. It shall generate electrical power of 500 MW. It will make use of MOX fuel, a mixture of PuO_2 and UO. A fuel burnup of 100 GWd/t is expected. It will have an operational life of 40 years.

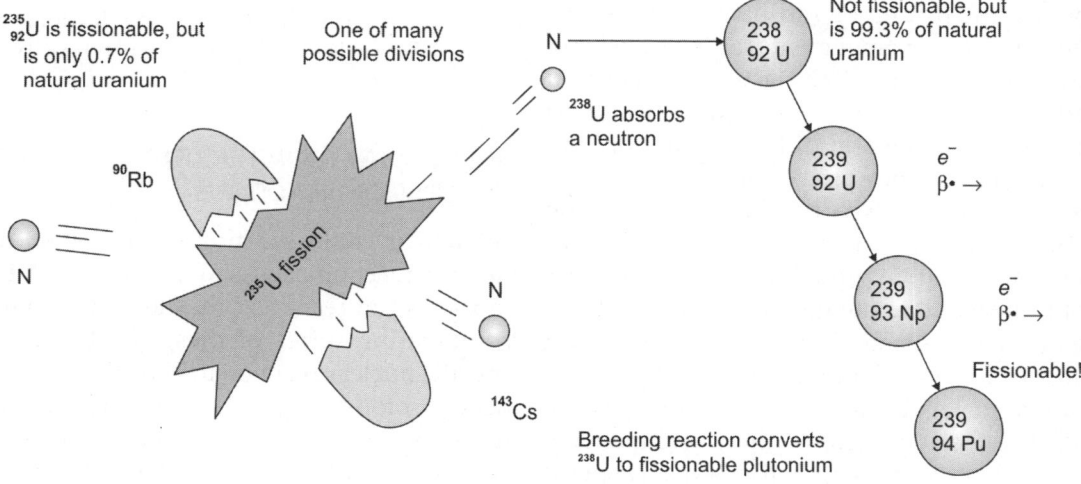

Fig. 8.5 Fast breeder reactor

Safety Considerations

The fact that PFBR will be cooled by liquid sodium creates additional safety requirements to isolate the coolant from the environment, since sodium explodes if it comes into contact with water and burns when in contact with air. Another hazard associated with the use of sodium as a coolant is the absorption of neutrons to generate the radioactive isotope ^{24}Na. There are two independent shutdown systems installed, designed to shut the reactor down effectively within a second. The reactor also has decay heat removal systems consisting of four independent circuits of 8 MWt capacity each.

8.14.2 Boiling Water Reactor (BWR)

BWRs' actually heat up the water. In both types, water is changed over to steam, and afterward reused, go into water by a part called the condenser, to be utilized again within the high temperature process as shown in Fig. 8.6. Since radioactive materials might be dangerous, atomic force plants have numerous safety systems to secure specialists, general society, and the Earth. These security systems include closing the reactor down rapidly and stopping the fission process, systems to chill the reactor off and divert heat from it, and restraints to hold the radioactivity and prevent it from getting away into the environment. Radioactive materials, if not used properly, can damage human cells or even cause cancer over long periods of time. BWRs' have been originally developed by GE. GE started its development in 1950s as light water reactor type nuclear power reactors, and the Dresden Unit-1 (2,00,000 kWe) commissioned in July 1960 is the first BWR nuclear power station. After that, the GE company has supplied many BWRs, Siemens (KWU, Germany), ABB-Atom (Switzerland/Sweden) and Toshiba and Hitachi (Japan) also supplied many BWRs. In the following, features and types of BWRs, mainly of conventional BWRs are explained. For BWRs, the steam void due to reactor coolant boiling has a negative-reactivity effect, which can suppress a power rise even if a positive reactivity is added. The reactor power can be controlled by two methods: reactor-coolant recirculation-flow control and control rod operation. A BWR nuclear power plant consists of the reactor coolant recirculation

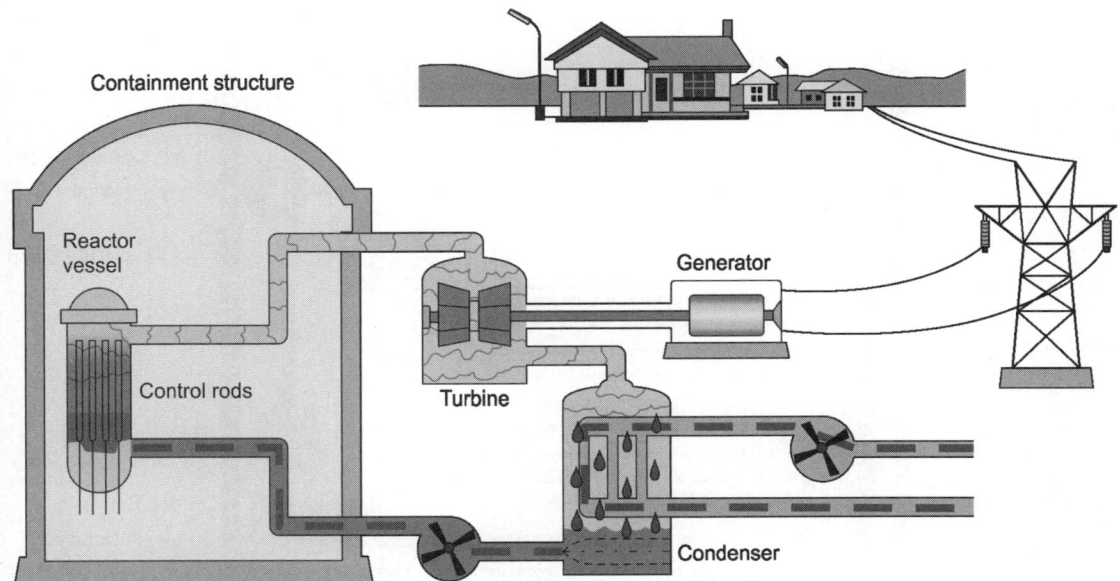

Fig. 8.6: Boiling water reactor (BWR)

system and main steam system that compose a nuclear reactor, engineered safety features that consist of the emergency core cooling system, reactor core isolation cooling system, containment cooling system and boric-acid injection system, turbine and generator equipment and other systems, such as the reactor coolant purification system, waste processing equipment, fuel handling equipment, other auxiliary equipment, etc. (Fig. 8.5).

8.14.3 Pressurized Heavy Water Reactor (PHWR)

A pressurized heavy water reactor (PHWR) is an atomic force reactor, usually utilizing unenriched characteristic uranium as its fuel, that uses heavy water (deuterium oxide-D_2O) as its coolant and mediator. The heavy water coolant is kept under pressure, allowing it to be heated to higher temperatures without boiling, much as in a typical pressurized water reactor (Fig. 8.7). While heavy water is essen-

tially more unreasonable than common light water, it yields extraordinarily upgraded neutron economy, permitting the reactor to work without fuel enhancement facilities (relieving the extra capital expense of the heavy water) and for the most part enhancing the capacity of the reactor to proficiently make utilization of substitute fuel cycles.

The pressurized heavy water reactor is a horizontal pressure tube reactor using natural uranium dioxide fuel with heavy water as moderator and coolant. The moderator is at low-pressure and temperature. The coolant is maintained in single phase by pressurization technique. Heat extracted by the coolant from the fuel is transferred to the secondary side light water to produce steam. The power output is enhanced by utilizing margin in the fuel linear heat rating and further flux flattening. Extraction of additional heat is achieved by allowing boiling of coolant near the channel exit. Thus, it is seen that the same

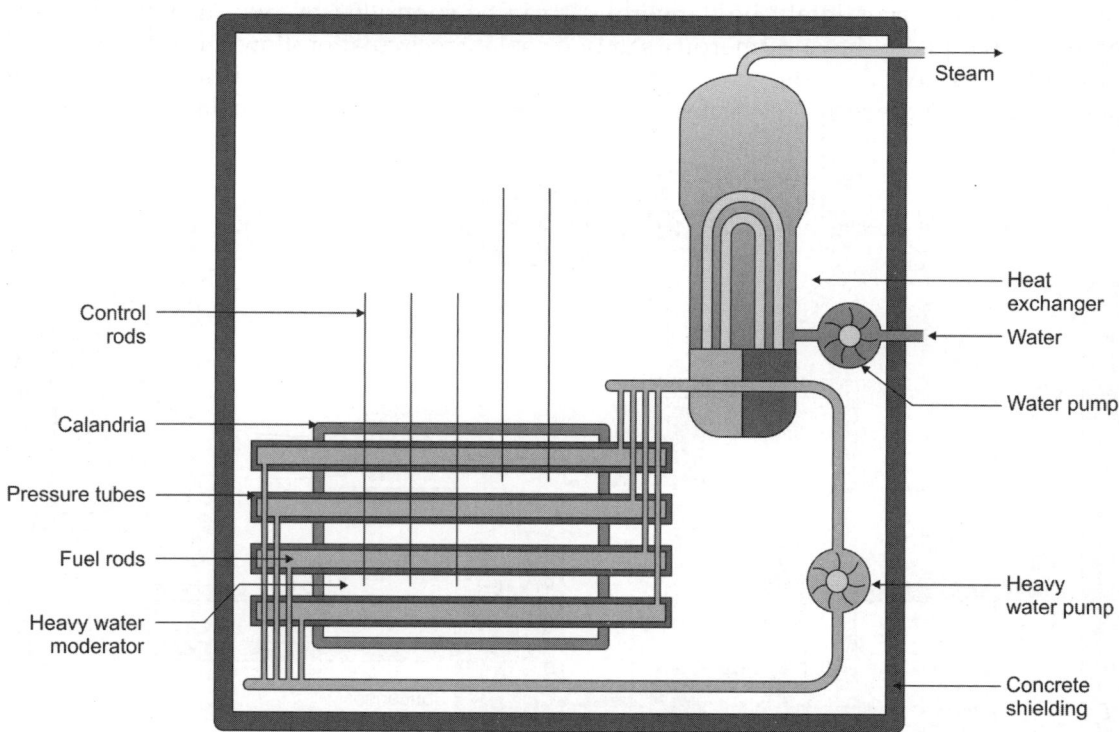

Fig. 8.7: Diagram of a typical pressurized heavy water reactor

reactor assembly and the same primary coolant loop are capable of delivering thermal energy equivalent to 700 MWe output resulting in a significant economic advantage. In a pressurized water reactor, the water is under enormous pressure so it can't boil. Instead, it's pumped to a heat exchanger outside the reactor. This heat exchanger transfers heat to water in another pipe, which boils to create the high-pressure steam that drives a generator.

PWRs' are used for marine propulsion in aircraft carriers, nuclear submarines and ice breakers. They are popular in several countries because they use less expensive natural (not enriched) uranium fuels and can be built and operated at competitive costs. The continuous refueling process used in PHWRs' has raised some proliferation concerns because it is difficult for international inspectors to monitor. PHWRs, like most reactors, can use fuel other than uranium. They run close to each other with small gaps and have several bends. Thus they represent a complex piping system. Under seismic loading, the adjacent feeder pipes may impact each other. In this paper a simplified procedure has been established to assess such impacting effects. The results of the proposed analysis include bending moment and impact force, which provide the stresses due to impacting effects. These results are plotted in nondimensional form so that they could be utilized for any set of feeder pipes. The procedure used to study the impacts includes seismic analysis of individual feeder pipes without impacting effects, selection of pipes for impact analysis, and estimating their maximum impact velocity. Based on the static and dynamic characteristics of the selected feeder pipes, the maximum bending moment, impact force, and stresses are obtained. The results of this study are useful for quick evaluation of the impacting effects in feeder pipes.

PWRs are utilized for marine propulsion as a part of plane carrying warships, atomic submarines and ice breakers. They are popular in some nations since they utilize less costly characteristic (not enhanced) uranium powers and might be constructed and operated at competitive costs. The constant refueling methodology utilized within PHWRs has raised some expansion concerns on the grounds that it is challenging for worldwide assessors to screen. Most reactors for instance PHWRs, can utilize fuel other than uranium. The core of a pressurized heavy water reactor (PHWR) consists of a large number of fuel channels. These fuel channels are connected to the feeder pipes through which the heavy water flows and transports heat from the reactor core to the steam generators. The feeder pipes are several hundreds in number. They run near one another with small gaps and have a few twists. Accordingly they represent to a complex system. Under seismic stacking, the contiguous feeder funnels may affect one another.

8.14.4 Gas Cooled Reactors (GCR)

The gas-graphite reactors operate using graphite as moderator and some gas (mostly CO_2, lately helium) as coolant. This belongs to the oldest reactor types. The first GCR was the Calder Hall power plant reactor, which was built in 1955 in England. This type is called MAGNOX after the special magnesium alloy (Magnox), of which the fuel cladding was made. The fuel is natural uranium. These reactors account for 1.1% of the total NPP power of the world and are not built any more. The AGR (advanced gas cooled reactor) is a development from MAGNOX: the cladding is not Magnox and the fuel is slightly enriched. The moderator is also graphite and the coolant is CO_2. Contribution to total world capacity is 2.5% and this type is not manufactured any longer. Figure 8.8, represents the arrangement of a GCR.

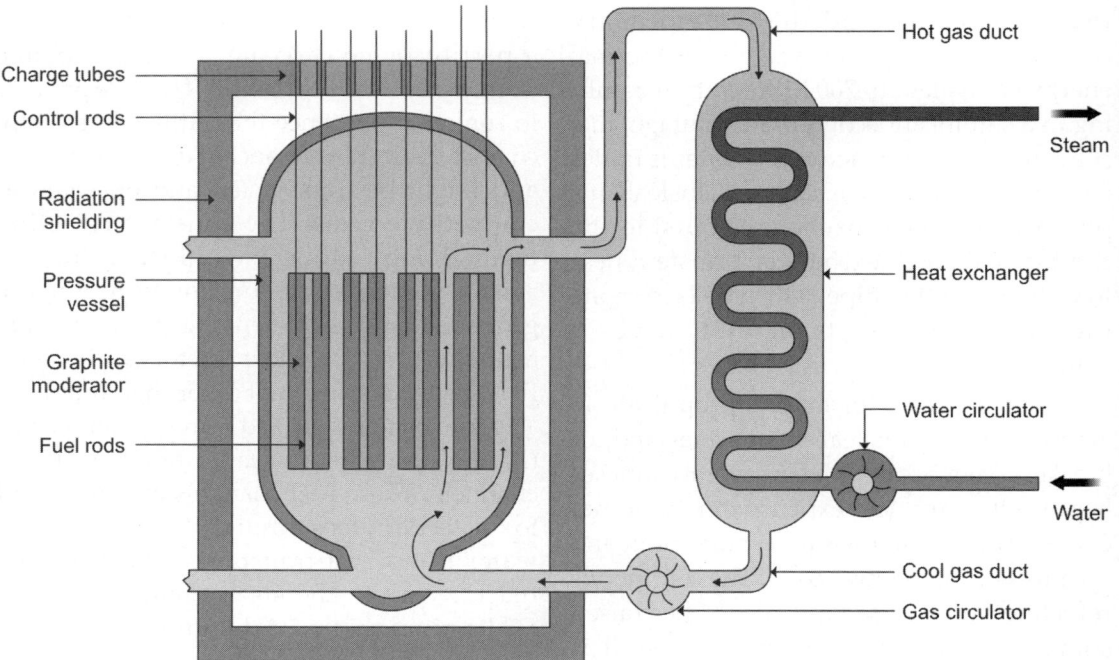

Charge tubes
Control rods
Radiation shielding
Pressure vessel
Graphite moderator
Fuel rods

Hot gas duct
Steam
Heat exchanger
Water circulator
Water
Cool gas duct
Gas circulator

Fig. 8.8: Gas cooled reactors (GCR)

8.15 NUCLEAR WASTE DISPOSAL

The time frame in question when dealing with radioactive waste ranges from 10,000 to 1,000,000 years, according to studies based on the effect of estimated radiation doses. Researchers suggest that forecasts of health detriment for such periods should be examined critically.

Above-ground Disposal

Dry cask storage typically involves taking waste from a spent fuel pool and sealing it (along with an inert gas) in a steel cylinder, which is placed in a concrete cylinder which acts as a radiation shield. It is a relatively inexpensive method which can be done at a central facility or adjacent to the source reactor. The waste can be easily retrieved for reprocessing.

Geologic Disposal

The basic concept is to locate a large, stable geologic trait and use mining technology to excavate a tunnel, or large-bore tunnel boring machines (similar to those used to drill the Channel Tunnel from England to France) to drill a shaft of 500 meters (1,600 ft) to 1,000 meters (3,300 ft) below the surface where rooms or vaults can be excavated for disposal of high-level radioactive waste. The goal is to permanently isolate nuclear waste from the human environment.

Transmutation

There have been proposals for reactors that consume nuclear waste and transmute it to other, less-harmful nuclear waste. In particular, the integral fast reactor was a proposed nuclear reactor with a nuclear fuel cycle that produced no transuranic waste and in fact, could consume transuranic waste. It proceeded as far as large-scale tests, but was then cancelled by the US government. Another approach, considered safer but requiring more development, is to dedicate subcritical reactors to the transmutation of the left-over transuranic elements. An isotope that is found in nuclear waste and that represents a concern in terms of proliferation is Pu-239. The estimated world total of plutonium in the year 2000 was of 1,645 MT, of which 210 MT had been separated by reprocessing. The large stock of plutonium is a result of its production inside uranium-fueled reactors and of the reprocessing of weapons-grade plutonium during the weapons program. An option for getting rid of this plutonium is to use it as a fuel in a traditional light water reactor (LWR). Several fuel types with differing plutonium destruction efficiencies are under study.

Re-use of Waste

Another option is to find applications for the isotopes in nuclear waste so as to reuse them. Already, caesium-137, strontium-90 and a few other isotopes are extracted for certain industrial applications such as food irradiation and radioisotop thermoelectric generators. While reuse does not eliminate the need to manage radioisotopes, it reduces the quantity of waste produced. The Nuclear Assisted Hydrocarbon Production Method, Canadian patent application 2,659,302, is a method for the temporary or permanent storage of nuclear waste materials comprising the placing of waste materials into one or more repositories or boreholes constructed into an unconventional oil formation. The thermal flux of the waste materials fracture the formation alters the chemical and/or physical properties of hydrocarbon material within the subterranean formation to allow removal of the altered material. A mixture of hydrocarbons, hydrogen, and/or other formation fluids is produced from the formation. The radioactivity of high-level radioactive waste affords proliferation resistance to plutonium placed in the periphery of the repository or the deepest portion of a borehole. Breeder reactors can run on U-238 and transuranic elements, which comprise the majority of spent fuel radioactivity in the 1,000-1,00,000 year time span.

8.16 ADVANTAGES OF NUCLEAR ENERGY

Nuclear energy is the world's largest source of emission free energy. Nuclear power plants do not emit air pollutants, such as sulphur and particulates, or greenhouse gases. The use of nuclear energy in place of other energy sources helps to keep the air clean, preserve the Earth's climate, avoid ground-level ozone formation and prevent acid rain. Of all energy sources, nuclear energy has perhaps the lowest impact on the environment, including water, land, habitat, species, and air resources. Nuclear energy is the most eco-efficient of all energy sources because it produces the most electricity relative to its environmental impact.

1. Of all energy sources, nuclear energy has perhaps the lowest impact on the environment especially in relation to kilowatts produced because nuclear plants do not emit harmful gases, require a relatively small area, and effectively minimize or negate other impacts.

2. Nuclear energy is an emission-free energy source because it does not burn anything to produce electricity. Nuclear power plants produce no gases such as nitrogen oxide or sulphur dioxide that could threaten our atmosphere by causing ground-level ozone formation, smog, and acid rain. Nor does nuclear energy produce carbon dioxide or other greenhouse gases suspected to cause global warming. Throughout the nuclear fuel cycle, the small volume of waste byproducts actually created is carefully contained, packaged and safely stored. As a result, the nuclear energy industry is the only industry established since the industrial revolution that has managed and accounted for all of its wastes, preventing adverse impacts on the environment.

3. Nuclear power also provides water quality and aquatic life conservation. Water discharged from a nuclear power plant contains no harmful pollutants and meets regulatory standards for temperature designed to protect aquatic life. This water,

used for cooling, never comes in contact with radioactive materials. If the water from the plant is so warm that it may harm marine life, it is cooled before it is discharged to its source river, lake, or bay as it is either mixed with water in a cooling pond or pumped through a cooling tower.

4. Because the areas around nuclear power plants and their cooling ponds are so clean, that they are often developed as wetlands that provide nesting areas for waterfowl and other birds, new habitats for fish, and the preservation of other wildlife as well as trees, flowers, and grasses. Many energy companies have created special nature parks or wildlife sanctuaries on plant sites.

5. Nuclear waste is more easily accountable than fossil fuel waste coal for example leeches heavy metals, greenhouse gases, and toxic chemicals involved in mining.

8.17 DISADVANTAGES OF NUCLEAR ENERGY

The environmental impact of nuclear power results from the nuclear fuel cycle, operation, and the effects of nuclear accidents. Although power plants are regulated by federal and state laws to protect human health and the environment, there is a wide variation of environmental impacts associated with power generation technologies. The purpose of the following section is to give consumers a better idea of the specific air, water, land, and radioactive waste releases associated with nuclear power electricity generation.

1. **Radioactive waste:** The waste produced by nuclear reactors needs to be disposed off at a safe place since they are extremely hazardous and can leak radiations if not stored properly. Such kind of wastes can emit radiation up to 100 years. The storage of radioactive waste has been a major bottleneck for the expansion of nuclear programs. The nuclear wastes contain radio isotopes with long half-lives. This means that the radio isotopes stay in the atmosphere in some form or the other. These reactive radicals make the sand or the water contaminated. It is known as mixed waste. The mixed wastes cause hazardous chemical reactions and leads to dangerous complications. The radioactive wastes are usually buried under sand and are known as 'vitrification'. But these wastes can be used to make nuclear weapons.

2. **Nuclear accidents:** While so many new technologies have been put in place to make sure that such disasters should not happen again like the one in chernobyl or more recently Fukushima but the risk associated with them are relatively high. Even small radiation leaks can cause devastating effects. Some of the symptoms include nausea, vomiting, diarrhea and fatigue. People who work at nuclear power plants and live near those areas are at high risk of facing nuclear radiations, if it happens.

3. **Nuclear radiation:** There are power reactors called breeders. They produce plutonium. It is an element which is not found in the nature, however, it is a fissionable element. It is a byproduct of the chain reaction and is very harmful if introduced in the nature. It is primarily used to produce nuclear weapons. Most likely, it is named as dirty bomb.

4. **High cost:** Another practical disadvantage of using nuclear energy is that it needs a lot of investment to set up a nuclear power station. It is not always possible by the developing countries to afford such a costly source of alternative energy. Nuclear power plants normally take 5–10 years to construct as there are several legal formalities to be completed and mostly it is opposed by the people who live nearby.

5. **National risk:** Nuclear energy has given us the power to produce more weapons than to produce things that can make the world a better place to live in. We have to become more careful and responsible while using nuclear energy to avoid any sort of major accidents. They are hot targets for militants and terrorist organizations. Security is a major concern here. A little lax in security can prove to be lethal and brutal for humans and even for this planet.

6. **Impact on aquatic life:** Eutrophication is another result of radioactive wastes. There are many seminars and conferences being held every year to look for a specific solution. But there is no outcome as of now. Reports say that radioactive wastes take almost 10,000 years to get back to the original form.

7. **Major impact on human life:** We all remember the disaster caused during the Second World War after the nuclear bombs were dropped over Hiroshima and Nagasaki. Even after five decades of the mishap, children are born with defects. This is primarily because of the radiation effects. Do we have any remedy for this? The answer is still *no*.

8. **Fuel availability:** Unlike fossil fuels which are available to most of the countries, uranium is a very scarce resource and exists in only a few countries. Permission of several international authorities are required before someone can even thought of building a nuclear power plant.

9. **Non-renewable:** Nuclear energy uses uranium which is a scarce resource and is not found in many countries. Most of the countries rely on other countries for the constant supply of this fuel. It is mined and transported like any other metal. Supply will be available as long as it is there. Once all extracted, nuclear plants will not be of any use. Due to its hazardous effects and limited supply, it cannot be termed renewable.

Various nuclear energy programs are undergoing in developed as well as developing nations like India. Not to mention, nuclear energy advantages are far ahead of advantages of fossil fuels. For this it has considered the most favoured technology to produce energy.

10. **Air emissions:** Nuclear power plants do not emit carbon dioxide, sulphur dioxide, or nitrogen oxides as part of the power genera-tion process. However, fossil fuel emissions are associated with the uranium mining and uranium enrichment process as well as the transport of the uranium fuel to and from the nuclear plant.

11. **Water discharges:** Heavy metals and salts build up in the water used in all power plant systems, including nuclear ones. These water pollutants, as well as the higher temperature of the water discharged from the power plants, can negatively affect water quality and aquatic life. Nuclear power plants sometimes discharge small amounts of tritium and other radioactive elements as allowed by their individual wastewater permits. Waste generated from uranium mining operations and rainwater runoff can contaminate ground-water and surface water resources with heavy metals and traces of radioactive uranium.

12. **Spent fuel:** Every 18 to 24 months, nuclear power plants must shut down to remove and replace the "spent" uranium fuel. This spent fuel has released most of its energy as a result of the fission process and has become radioactive waste. Currently, the spent fuel is stored at the nuclear plants at which it is generated, either in steel-lined, concrete vaults filled with water or in above-ground steel or steel-reinforced concrete containers with steel inner canisters.

13. **Radioactive waste generation:** Enrich-ment of uranium ore into fuel and the operation of nuclear power plants generate wastes that contain low-levels of radioactivity. These wastes are shipped to a few specially designed and licensed disposal sites. When a nuclear power plant is closed, some equipment and structural materials become radioactive wastes. This type of radioactive waste is currently being stored at the closed plants until an appropriate disposal site is opened.

9

Solar Energy

9.1 SOLAR RADIATION AND ITS MEASUREMENT

9.1.1 Introduction

In general, the energy produced and radiated by the Sun, more specifically the term refers to the Sun's energy that reaches the Earth. Sunlight based energy, accepted as radiation, could be changed over specifically or in a round about way into different types of energy, for example, high temperature and power, which might be used by man. Since the Sun is relied upon to transmit at a basically consistent rate for a couple of billion years, it may be viewed as an inexpendable source of suitable energy. Sun oriented energy is the energy gained from the Sun that maintains life on Earth. For a long time, Sun oriented energy has been acknowledged as a gigantic source of energy and likewise a temperate source of energy on the grounds that it is unreservedly accessible. Then again, it is now after years of research that innovation has made it conceivable to harness solar energy. It is the essential source of the Earth's energy, giving about 99.97% of the high temperature, energy needed for chemophysical methods in the climate, sea area, and other water forms. It assumes an essential part as a renewable energy source as Sun powered energy estimations could be utilized to gauge potential force levels that might be created from photovoltaic cells and likewise important for deciding cooling burdens for structures. Different varieties of energy on Earth start in Sun oriented vitality. There are two courses in which Sun oriented energy might be utilized. The first is through the thermal mode, where hotness is utilized for cooking, warming or creating power, and the second one is through photovoltaic cells, where sunlight based vitality is changed over into power that could be utilized for a mixture of requisitions like lighting, moving or pumping. As solar energy is free of pollution, available in abundance and throughout the world, it is a highly interesting source of energy for everyone. Solar energy has the greatest potential of all the sources of renewable energy and if only a small amount of this form of energy could be used, it will be one of the most important supplies of energy especially when other sources in the country have depleted. It is of great importance to India since it lies in a temperate climate of the region of the world where sunlight is abundant for a major part of the year.

The applications of solar energy which are enjoying most success today are:

- Heating and cooling of residential building
- Solar water heating
- Solar drying of agricultural and animal products
- Solar distillation
- Salt production by evaporation of seawater or in inland brines
- Solar cookers
- Solar engines for water pumping
- Food refrigeration
- Bioconversion and wind energy, which are indirect sources of solar energy

- Solar furnaces
- Solar electric power generation by solar ponds: Steam generators heated by rotating reflectors with lenses and pipes for fluid circulation.

9.1.2 Solar Energy is Important in Nature

Solar energy is an important part of almost every life process. Plants and animals, alike, use solar energy to produce important nutrients in their cells. Plants use the energy to produce the green chlorophyll that they need to survive, while humans use the sun rays to produce vitamin-D in their bodies. However, when man learned to actually convert solar energy into usable energy, it became even more important.

9.1.3 Solar Energy is Important as a Clean Energy

Since Sun powered energy is totally common, it is viewed as a clean energy source. It doesn't disturb nature or make a risk to eco-frameworks the way oil and some other energy sources may. The little measure of effect it does have on nature is typically from the chemicals and solvents that are utilized throughout the production of the photovoltaic cells that are required to change over the Sun's energy into power. This is a little issue contrasted with the tremendous effect that one oil slick can have on the Earth.

Solar Energy is Versatile

Solar energy cells can be used to produce the power for a calculator or a watch. They can also be used to produce enough power to run an entire city. With that kind of versatility, it is a great energy source. Some of the ways solar energy is being used today are:

- Cars
- Cooking
- Coffee roasters
- Electricity for homes and businesses
- Thermal heating for homes and businesses
- Watches
- Water heaters
- Water treatment plants

There are many other things that are or can be powered by solar energy.

Why is Solar Energy Important?

The most important issue of all is probably why solar energy is important to you, personally the reason could be:

- Fossil fuels, like gas and oil, are not renewable energy sources. Once they are finished they can't be replenished. Someday these fuels will run out and then mankind will either need to come up with a new way to provide power or go back to life as it was prior to man's use of these things.
- Fossil fuels create massive pollution in the environment. This pollution affects waterways, the air you breathe, and even the meat and vegetables that you eat.
- These fuels are expensive to retrieve from the Earth and they are expensive to use. Other, more ecofriendly energy sources like wind and solar energies are relatively inexpensive and easy to produce.

In the future, solar energy may well be the primary form of energy. This could lead to a clean environment, cheaper utilities and a healthier world and it has the potential to allow technology and nature to coexist peacefully.

9.1.4 Solar Constant

The average amount of solar radiation received by the Earth's atmosphere, per unit area, when the Earth is at its mean distance from the Sun or when the Sun and Earth are spaced at 1 AU – the mean Earth/Sun distance of 1,49,597,890 km – is called the *solar constant*. It is equivalent to 1370 watts for every square meter. Sunlight based radiation differs with the Earth's separation from the Sun and with the manifestation or decay of sunspots. The world metrological organization (WMO) advertises a worth of 1367 Wm^{-2}. The solar constant is the aggregate coordinated irradiance over the whole range. The separation between the Earth and Sun shifts a bit as the year progresses. Due to this extra terrestrial flux likewise shifts. The Earth is closest to the Sun in summer and farthest in winter. This variation in distance prepares an almost sinusoidal variation in the force of sunlight based radiation (I) that achieves the Earth. This might be approximated by the mathematical statement

$$I/Isc = 1 + 0.033 \cos (360(n{-}2)/365)$$
$$= 1 + 0.033 \cos (360 \times n/365)$$

where 'n' is the day of the year.

The Sun is a spherical source of about 1.39 million km diameter, at an average distance (1 astronomical unit) of 149.6 million km from Earth. The direct portion of the solar radiation is collimated with an angle of approximately 0.53° (full angle), while the "diffuse" portion is incident from the hemispheric sky and from ground reflections and scatter. The "global" irradiation, the sum of the direct and diffuse components, is essentially uniform. Since there is a strong forward distribution in aerosol scattering, high aerosol loading of the atmosphere leads to considerable scattered radiation appears to come from a small annulus around the solar disk, the *solar aureole* or *the corona*. This radiation mixed with the direct beam is called circumsolar radiation. The Sun as a normal star is an average fundamental grouping midget of phantom class G-2. Its sweep is 6.960–108 m. The mean separation between the Sun and the Earth is 1.496–1011m and is known as the *cosmic unit* (AU). Sun based radiation is the electromagnetic radiation emitted by the Sun. Very nearly all known physical and natural cycles in the Earth system are determined by the Sun oriented radiation arriving at the Earth. Sun powered radiation is likewise the reason for environmental change that is sincerely outside to the Earth system. Of the 51 units of Sun powered radiation consumed by Earth's surface, 23 units are utilized to dissipate water, bringing about a misfortune of 23 units of hotness at Earth's surface. Seven units are utilized for the methods of conduction and convection, additionally creating a loss of heat at Earth's surface. The rate at which sunlight based vitality touches base at the highest point of the air is called Sun powered consistent Isc. This is the measure of energy gained in unit time on an unit territory perpendicular to the Sun's course at the mean distance of the Earth from the Sun.

9.1.5 Basics of Solar Radiation

Radiation from the Sun sustains life on Earth and determines climate. The energy flow within the Sun results in a surface temperature of around 5800 K, so the spectrum of radiation from the Sun is similar to that of a 5800 K blackbody with fine structure due to absorption in the cool peripheral solar gas. The solar radiation that penetrates the Earth's atmosphere and reaches the surface differs in both amount and character from the radiation at the top of the atmosphere Fig. 9.1. In the first place, part of the radiation is reflected back into the space, especially by clouds. Furthermore, the radiation entering the atmosphere is partly absorbed by molecules in the air. Oxygen and ozone, formed from oxygen, absorb nearly all the ultraviolet radiation, and water vapour and carbon dioxide absorb some of the energy in the infrared range. Solar radiation that has not been absorbed or scattered and reaches the ground directly from the Sun is called "direct radiation" or beam radiation. Solar radiation thus has many useful applications in architectural design, evapo-transpiration estimates, agriculture, and atmospheric, land, ocean, and hydrologic models . The acquisition and the development of database on the long term solar radiation will facilitate the evaluation of solar energy potential as an input to the country's energy budget Fig. 9.2.

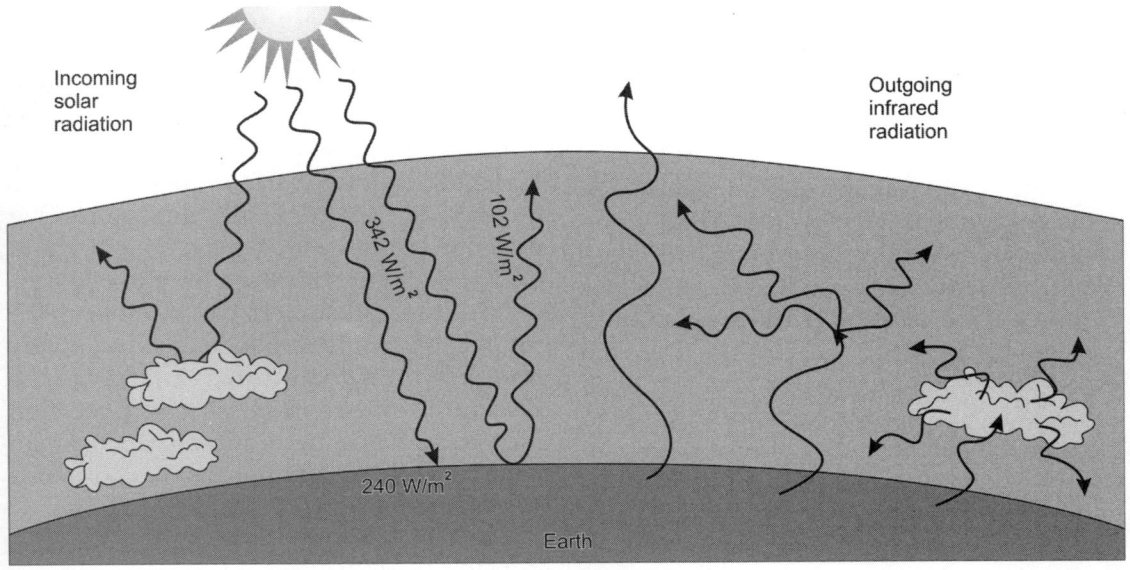

Fig. 9.1: Solar radiation and IR emission

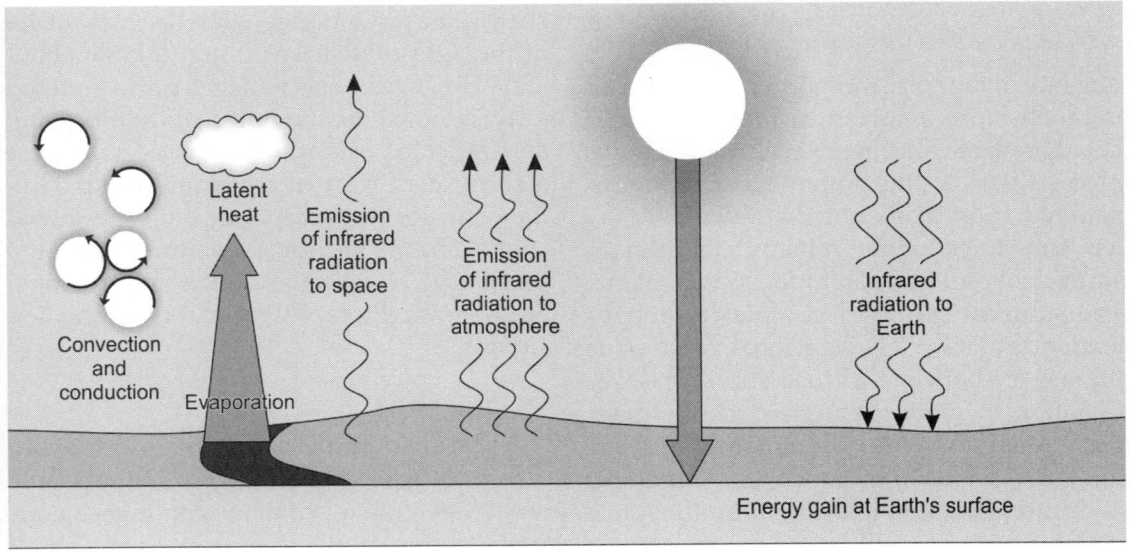

Fig. 9.2: Heat budget of the Earth surface

Direct and Diffuse Radiation

"Direct radiation" is likewise once in a while called "beam radiation" or "direct beam radiation". It is utilized to describe sunlight based radiation going on a straight line starting from the Sun to the surface of the Earth. "Diffuse radiation", then again, depicts the daylight that has been scattered by atoms and particles in the air, however, that has still made it down to the surface of the Earth. Direct radiation has a distinct course yet diffuse radiation is just going any way. Since when the radiation is administered, the beams are all going in the same course, an item can square all of them as soon as possible. This is the reason shadows are just handled when immediate radiation is blocked.

Ratio of Direct to Diffuse Radiation

When the sky is clear and the Sun is very high in the sky, *direct radiation* is around 85% of the total insolation striking the ground and *diffuse radiation* is about 15%. As the Sun goes lower in the sky, the percent of diffuse radiation keeps going up until it reaches 40% when the Sun is 10° above the horizon. Atmospheric conditions like clouds and pollution also increase the percentage of diffuse radiation. On an extremely overcast day, pretty much 100% of the solar radiation is diffuse radiation. Generally speaking, the larger the percentage of diffuse radiation, the less is the total insolation.

Direct/Diffuse Ratio Varies with Latitude and Climate

The rate of the solar radiation that is diffused is much more amazing in higher latitude, cloudier places than in lower altitude, sunnier places. Likewise, the precentage of the aggregate radiation that is diffuse radiation has a tendency to be higher in the winter than the spring in these higher latitude, cloudier places. The sunniest spots, by complexity, have a tendency to have less occasional variation in the degree between diffuse and immediate radiation. In London's sunniest month (June), the normal every day illumination is something like 5.5 kWh/m² and about half of that is diffuse. In December, the illumination is under 1 kWh/m² and by far the majority of that radiation is diffuse.

Tilted Solar Panels Gather Less Diffuse Radiation

When you tilt your Sun oriented boards, the Sun's beams are hitting them at a 90° plot, you are augmenting the measure of immediate radiation that they get. However, since diffuse radiation is for the most part pretty similarly appropriated all around the sky, the most diffuse radiation is accumulated when your Sun based boards are setting down evenly. The steeper your Sun oriented boards are tilted, the less of the sky they are confronting and the a greater amount of the sky's diffuse radiation they pass up for a great opportunity. Assuming that, for instance, the solar panels are tilted at a 45° edge, they are confronting far from around a quarter of the sky and would just gather something like three-fourths of the diffuse radiation in the sky. Still, on the ground that guide radiation is substantially more powerful than diffuse radiation, the measure of radiation missed by tilted Sun oriented boards is for the most part more than made up for by the additional radiation gained by tracking the Sun.

Reflected Radiation

Reflected radiation describes sunlight that has been reflected off of non-atmospheric things such as the ground. Asphalt reflects about 4% of the light that strikes it and a lawn about 25%. However, solar panels tend to be tilted away from where the reflected light is going and reflected radiation rarely accounts for a significant part of the sunlight striking their surface. An exception is in very snowy conditions which can sometimes raise the percentage of reflected radiation quite high. Fresh snow reflects 80% to 90% of the radiation striking it.

Global Insolation

"Global insolation" is the total insolation: direct + diffuse + reflected light. Often people use it to refer to the total insolation on a horizontal surface and if they want to talk about the total radiation striking a surface with some specific tilt, they will say something like "total insolation on an XYZ° tilt", etc. *"Normal radiation"* describes the radiation that strikes a surface that is at a 90° angle to the Sun's rays. Constantly keeping solar collectors (sensors) at a 90° angle with the Sun, than it maximize the direct radiation received on that day. Therefore, "normal global radiation" generally tells us the total solar radiation we could get.

9.1.6 Ground Measurements of Solar Radiation

Radiometry is the science of electromagnetic radiation measurement. The generic device is named *radiometer*. Each of the quantities are measured with a specific device; for instance, the *pyrheliometer* that measures the direct beam irradiance and the *pyranometer* that measures the horizontal beam and diffuse irradiances.

Solar Radiometers

Detection of the optical electromagnetic radiation is essentially performed by transformation of the shaft's energy in electric signs that consequently might be measured by routine methods. Because of their almost constant spectral affectability for the entire Sun oriented phantom reach, radiometers outfitted with warm sensors are generally used to measure broad and Sun based irradiance. Temperature variances wind, rain, and snow are elements that influence the estimations. The minimization of these perturbations is a difficult task in the engineering of solar radiometers.

Pyrheliometer

The pyrheliometer is a broadband instrument that measures the direct component Gn of solar radiation. A two-pivot Sun following instrument is frequently utilized for this reason. The detector is a multi-intersection thermopile set at the lowest part of a *collimating tube* (Fig. 9.3), gave a quartz window to ensure the instrument. The indicator is covered with optical dark paint (going about as a full absorber for Sun powered energy in the wavelength between range *0.280–3* lm). Its temperature is remunerated to minimize affectability of surrounding temperature change. The pyrheliometer aperture angle is *5°*. Consequently, radiation is received from the Sun and a limited circumsolar region, but all diffuse radiation from the rest of the sky is excluded. A readout device is used to give the instant value of the direct beam irradiance. Its scale

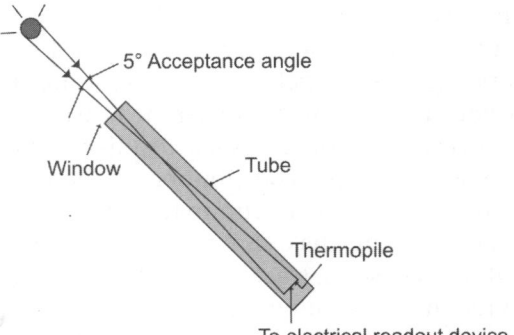

Fig. 9.3a: Schematic of a pyrheliometer

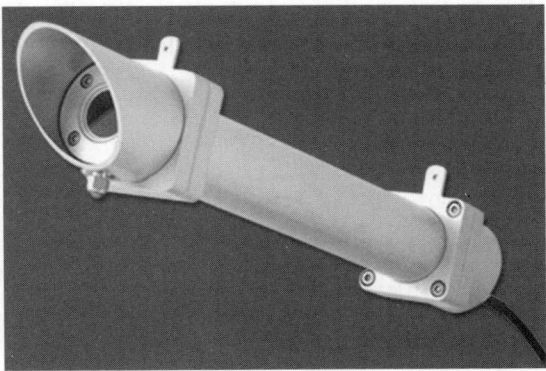

Fig. 9.3b: Photo of Hukseflux DR01 first class pyrheliometer (Hukseflux 2012). (Public license on Wikimedia commons).

is adapted to the sensitivity of the particular instrument in order to display the value in SI units, Wm^{-2}. For illustration, a picture of a Hukseflux DR01 first class pyrheliometer (Hukseflux 2012) is presented in Fig. 9.3b.

Pyranometer

Pyranometers are broadband instruments that measure global solar irradiance incoming from a $2p$ solid angle on a planar surface. A typical pyranometer is schematically represented in Fig. 9.4. It consists of a white disk for limiting the acceptance angle to **180°** and two concentric hemispherical transparent covers made of glass. The two domes shield the sensor from thermal convection, protect it against weather threat (rain, wind, and dust) and limit the spectral sensitivity of the instrument in the wavelength range 0.29–2.81 m. A cartridge of silica gel inside the dome absorbs water vapour. A pyranometer can also be used to measure the diffuse solar irradiance (Gd), provided that the contribution of the direct beam component is eliminated. For this, a small shading disk can be mounted on an automated solar tracker to ensure that the pyranometer is continuously shaded. Alternatively, a shadow ring may prevent the direct irradiance (Gb) from reaching the sensor whole day long (Fig. 9.4). As the daily maximum Sun elevation angle changes day by day, it is necessary to change periodically (days lag) the height of the shadow ring. Then

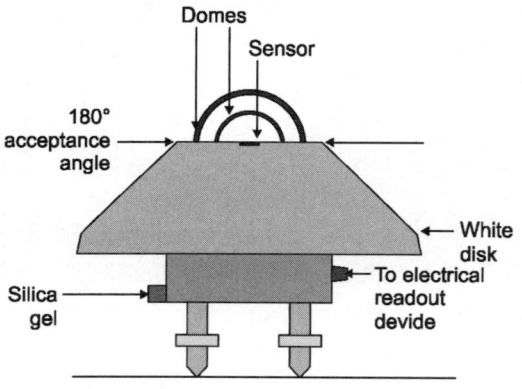

Fig. 9.4a: Schematic of a pyranometer

Fig. 9.4b: First class pyranometer LPPYRA 12 (Delta OHM 2012) equipped with shadow ring, mounted on the solar platform of the West University of Timisoara, Romania (SRMS 2012).

again, in light of the fact that the shadow ring likewise blocks a piece of the diffuse radiation, it is important to amend the measured qualities. The rate of diffuse radiation captured by the shadow ring fluctuates throughout the year with its position and atmospheric conditions. Self-calibrating absolute radiometers are used as primary standard, the other radiometers being calibrated against an absolute instrument. The uncertainty of the measured value depends on factors such as: resolution (the smallest change in the radiation quantity which can be detected by the instrument), nonlinearity of response (the change in sensitivity associated with incident irradiance level), deviation of the directional response (cosine response and azimuth response), time constant of the instrument (time to reach 95% of the final value), changes in sensitivity due to changes in weather variables (such as temperature, humidity, pressure, and wind), long-term drifts of sensitivity (defined as the ratio of electrical output signal to the irradiance applied). All the above uncertainties should be known for a well-characterized instrument. Certain instruments perform better for specific atmospheres, irradiances, and Sun oriented positions; accordingly, the instruments ought to be chosen as stated by their end utilization.

9.1.7 The Nature of Radiation

All objects over the temperature of absolute zero (–273.15 °C) radiate energy to their nature's turf. This energy, or radiation, is emitted as electromagnetic waves that go at the velocity of light. Numerous distinctive types of radiation have been distinguished. Each of these types is characterized by its wavelength. The wavelength of electromagnetic radiation can change from being interminably short to limitlessly long (Fig. 9.5).

Visible light is a type of electromagnetic radiation that might be observed by our eyes. Light has a wavelength of between 0.40 to 0.71 micrometers (μm). Fig. 9.5, represents the different otherworldly color groups that make

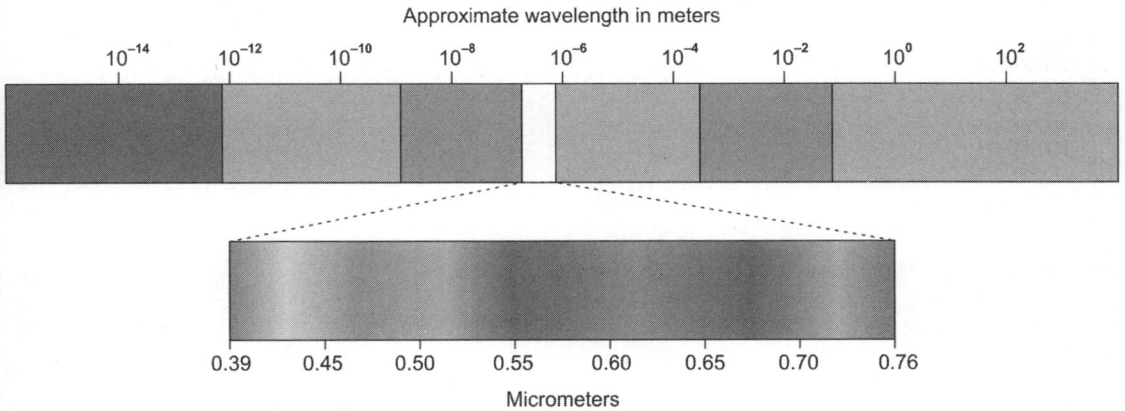

Approximate wavelength in meters

10^{-14} 10^{-12} 10^{-10} 10^{-8} 10^{-6} 10^{-4} 10^{-2} 10^{0} 10^{2}

0.39   0.45   0.50   0.55   0.60   0.65   0.70   0.76

Micrometers

Fig. 9.5: Electromagnetic spectrum (visible light ranges from 0.40 to 0.71 µm)

up light. The Sun transmits just a portion (44%) of its radiation in zone. Sun oriented radiation compasses a range from roughly 0.1 to 4.0 micrometers. The band from 0.1 to 0.4 micrometers is called ultraviolet radiation. Something like 7% of the Sun's discharge is in this wavelength band. About 48% of the Sun's radiation falls in the region between 0.71 to 4.0 micrometers. This band is known as visible band, close to 0.71 to 1.5 micrometers is called near infra red (NIR) and far infrared is 1.5 to 4.0 micrometers. The measure of electromagnetic radiation emitted by a form is specifically identified with its temperature. If the body is a perfect emitter (dark form), the measure of radiation given off is relative to the fourth power of its temperature as measured in Kelvin units. This natural phenomenon is described by the Stefan-Boltzmann, law. The following simple equation describes this law mathematically:

$$E^* = \sigma T^4$$

where, σ (sigma) = 5.67×10^{-8} Wm^{-2} K^{-4} and T is the temperature in Kelvin.

According to the Stephan-Boltzmann equation, a small increase in the temperature of a radiating body results in a large amount of additional radiation being emitted. In general, good emitters of radiation are also good absorbers of radiation at specific wavelength bands. This is especially true of gases and is responsible for the Earth's greenhouse effect. Likewise, weak emitters of radiation are also weak absorbers of radiation at specific wavelength bands. This fact is referred to as Kirchhoff's law. Some objects in nature behave like perfect black bodies, i.e. maintains thermal equilibrium, absorb and emit radiation. We call these objects black bodies. The radiation characteristics of the Sun and the Earth are very close to being black bodies. The wavelength of **maximum emission** of any body is inversely proportional to its absolute temperature. Thus, the higher the temperature, the shorter the wavelength of maximum emission. This phenomenon is often called Wien's law. The following equation describes this law:

$$\lambda_{max} = C/T$$

where, C is a constant equal to 2897 and T is the temperature in Kelvin.

Wien's law suggests that as the temperature of a body increases, the wavelength of maximum emission becomes smaller. According to the above equation the wavelength of maximum emission for the Sun (5800 °K) is about 0.5 micrometers, while the wavelength of maximum emission for the Earth (288 °K) is approximately 10.0 micrometers. A graph that describes the quantity of radiation that is emitted from a body at a particular wavelength is commonly called a spectrum. The following two graphs (Figs 9.6 and 9.7)

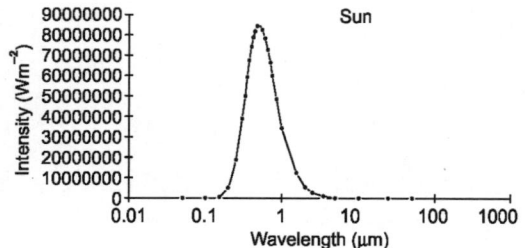

Fig. 9.6: Spectrum of the sun. The sun emits most of its radiation in a wavelength band between 0.1 and 4.0 micrometers (μm).

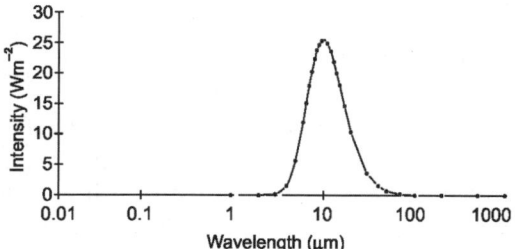

Fig. 9.7: Spectrum of the Earth. The Earth emits most of its radiation in a wavelength band between 0.5 and 30.0 micrometers (μm).

describe the spectrums for the Sun and Earth. The above graphs illustrate two important points concerning the relationship between the temperature of a body and its emissions of electromagnetic radiation given below:

1. The amount of radiation emitted from a body increases exponentially with a linear rise in temperature

2. The average wavelength of electromagnetic emissions becomes shorter with increasing temperature

Finally, the amount of radiation passing through a specific area is inversely proportional to the square of the distance of that area from the energy source. This phenomenon is called the inverse square law. Using this law we can model the effect that distance traveled has on the intensity of emitted radiation from a body like the Sun. Figure 9.8, suggests that the intensity of radiation emitted by a body quickly diminishes with distance in a nonlinear fashion.

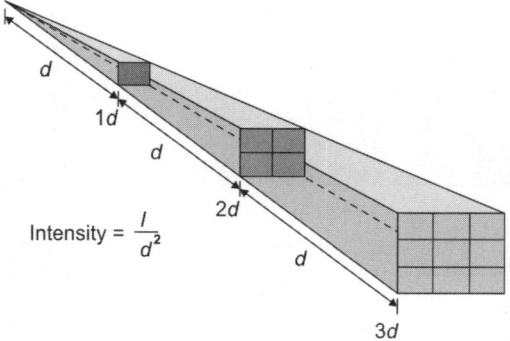

Fig. 9.8: Diagram illustrating the diffusion of radiation due to the inverse square law.

Mathematically, the **inverse square law** is described by the equation:

$$\text{Intensity} = I/d^2$$

where I is the intensity of the radiation at $1d$ (see above **diagram**) and d is the distance traveled.

9.1.8 Solar Spectrum

The distribution of solar radiation as a function of the wavelength is called the solar spectrum, which consists of a continuous emission with some superimposed line structures. The Sun's total radiation output is approximately equivalent to that of a blackbody at 5776 K. The solar radiation in the visible and infrared spectrum fits closely with the blackbody emission at this temperature. However, the ultraviolet (UV) region (0.4 μm) of solar radiation deviates greatly from the visible and infrared regions in terms of the equivalent blackbody temperature of the Sun. The deviations seen in the solar spectrum are a result of emission from the non-isothermal solar atmosphere. The solar constant is the amount of solar radiation received outside the Earth's atmosphere on a surface normal to the incident radiation per unit time and per unit area at the Earth's mean distance from the Sun. The solar constant is an important value for the studies of global energy balance and climate. Reliable measurements of solar constant can be made only from space and a more than 20 years record has been obtained based on overlapping satellite observations. The analysis of satellite data suggests a solar constant of 1366 Wm^{-2} with a measurement uncertainty of 73 Wm^{-2}. Of the radiant energy emitted from the Sun, approximately 50% lies in the infrared region about 40% in the visible region (0.4–0.7 µm), and about 10% in the UV region (0.4 µm). The solar constant is not in fact perfectly constant, but varies in relation to the solar activities. Past the exact moderate advancement of the Sun, a well-known Sun based movement is the sunspots, which are generally dark areas on the surface of the Sun. The occasional change in the amount of sunspots is alluded to as the sunspot cycle, and takes about 11 years, the alleged 11-year cycle. The cycle of sunspot maxima having the same attractive extremity is alluded to as the 22-year cycle. The Sun additionally pivots on its pivot once in about 27 days. Satellite perceptions recommend that the Sun oriented cycle variation of the Sun powered consistent is on the order of something like 0.1%, which could be so little there is no option specifically cause more than scarcely noticeable changes in the troposphere atmosphere. On the other hand, some circuitous proof demonstrates that the progressions in Sun oriented cycle consistently identified with sunspot action may have been essentially bigger throughout the last a few hundreds of years. Moreover, Sun oriented variability is much bigger (in relative terms) in the UV locale, and impels significant changes in the substance piece, temperature, and flow of the stratosphere, and in addition in the higher scopes of the upper climate.

Scattering and Absorption of Solar Radiation in the Earth-Atmosphere System

Sun oriented radiation entering the Earth's air is retained and scattered by air gasses, vapourizers, mists, and the Earth's surface. The absorbed radiation is added specifically to the hotness plan, though the scattered radiation is mostly come back to space and somewhat proceeds its way through the Earth-atmosphere framework, where it is liable to

further scrambling and retention. The division of the occurrence Sun based radiation that is reflected and backscattered to space is known as the albedo.We may talk about the albedo of the whole Earth or of distinct surfaces with reference either to monochromatic radiation or to the aggregate episode Sun powered radiation. Thus, the albedo of the Earth in general is something like 0.31.

Effects of Atmospheric Gases On Solar Radiation

The disseminating of Sun powered radiation via air particles might be portrayed by a hypothesis created by Rayleigh, who indicated that the measure of diffusing is contrarily relative to the fourth power of the wavelength, when the sizes of particles are much more modest than the wavelength of the occurrence radiation. We see blue sky in light of the fact that climatic atoms disseminate Sun powered radiation a great deal more in the blue than in the red part of the spectrum. Indeed, the sky is made noticeable through the scrambling methodology. Then again, sunrise and sunset seem reddish in light of the fact that the blue light in the immediate light is removed by dispersing throughout the long way through the air, leaving the remaining reddish shades of the range. Barometrical gasses likewise retain Sun oriented radiation in selected wavelength groups. The UV radiation with wavelengths shorter than 0.3 mm is deadly to the biosphere. The UV radiation in the interim 0.2–0.3 µm is chiefly consumed by O_3 in the stratosphere. The small amount of radiation with wavelengths shorter than 0.2 µm is absorbed at higher levels by O_2, N_2, O, and N. The photochemical processes due to absorption of solar UV radiation involving various forms of oxygen are critical in determining the amount of ozone in the stratosphere. The absorption spectrum of O_2 between 0.2 and 0.26 mm is weak, but of significance in the formation of ozone. In the troposphere, the absorption of solar radiation occurs in the visible and near-infrared regions, owing primarily to H_2O, CO_2, O_2, and O_3. The absorption in the visible, however, is very weak. The top curve is the solar spectrum at the top of the Earth's atmosphere and the lower curve represents the spectrum at sea level and the shaded area gives the combined effects of scattering and absorption of solar radiation by atmospheric gases. It is evident that the depletion of solar radiation is dominated by ozone absorption in the UV, Rayleigh's scattering is in both UV and visible, and water vapour absorption in the near infrared.

Effects of Aerosols on Solar Radiation

Aerosols are suspensions of liquid and solid particles in the atmosphere, excluding clouds and precipitation. The aerosol particle sizes range from 10^{-4} to 10 mm, falling under the following broad categories: sulfates, black carbon, organic carbon, dust, and sea salt. Aerosol concentrations and compositions vary significantly with time and location. Visibility measurements reflect the aerosol concentration at ground level. The visual range can vary from a few meters to 200 km, depending on the proximity to sources, the strength of the sources, and atmospheric conditions. Aerosols scatter and absorb solar radiation. Sulfate aerosols scatter primarily solar radiation and cause cooling of the Earth-atmosphere system. The increase in the reflected solar radiation at the top of the atmosphere due to such non-absorbing aerosols is nearly identical to the reduction in solar radiation at the surface. Carbonaceous aerosols (black carbon and organics) absorb and scatter solar radiation. The presence of black carbon aerosols results in the absorption of solar radiation, which reduces the solar radiation reaching the surface. At the same time, these aerosols absorb the upward solar radiation reflected from below and reduce the solar radiation reflected to space. Therefore, the effect of black carbon aerosols opposes the cooling effect of other aerosols at the top of the atmosphere, whereas at the surface all aerosols reduce solar radiation. The changes arising from the aerosol scattering and absorption of solar radiation are referred to as their *direct radiative forcing*. Aerosols can also modify solar radiation through their role in cloud condensation and as ice nuclei, an effect known as aerosol *indirect radiative forcing*. Aerosol particles in the atmosphere are produced both in nature and by people. A global aerosol optical depth of about 0.12, is suggested. These aerosols increase the reflected solar radiation at the top of the atmosphere by about 3 Wm^{-2} globally. Anthropogenic sources contribute significantly to the global aerosol optical depth. Global anthropogenic emissions of sulfates, organics, and black carbon even exceed natural sources. Such a large perturbation of the global aerosol loading due to human activities may significantly modify regional and global climates.

Effects of Clouds on Solar Radiation

Clouds regularly cover about 65% of the Earth, and occur in various types. Some, such as cirrus in the tropics and stratus near the coastal areas and in the arctic are climatologically persistent. Like aerosols, clouds show substantial spatial and temporal variations. Clouds are the most essential controller of sunlight based radiation. By reflecting sunlight based radiation again to space, they cool the Earth-atmosphere system—the alleged *cloud albedo impact*. Clouds additionally absorb solar radiation in the close infrared area. The cooling of the Earth-atmosphere system by the cloud albedo impact happens principally at the surface. The Sun powered albedo of clouds depends significantly on cloud sort and cloud structure, and the Sun oriented peak point. The most direct and straightforward analytic measure of the effect of mists on sunlight based radiation is the short-wave cloud constraining, which is characterized as the distinction of the net Sun powered irradiances at the highest point of the climate between all-sky and cloudless conditions. Here the net irradiance is the approaching Sun powered radiation minus the reflected radiation. Satellite estimations infer that the worldwide short-wave cloud compelling is about -45 Wm^{-2}. Short-wave cloud forcings are augmented (about -120 Wm^{-2}) in the middle of the *year side of the equator at* about *latitude 601°* where sunlight based information is vast and low clouds are abundant, with an optional most extreme in tropics. Note that the extent of short-wave cloud compelling is about ten times as expansive as those for a CO_2 multiplying. Hence small changes in the cloud-radiative forcing fields can play a significant role as a climate feedback mechanism.

Solar Radiation at the Earth's Surface

The radiation coming directly from the Sun received at the Earth's surface is called *direct solar radiation*. The amount of scattered radiation coming from all other directions is called *diffuse solar radiation*. The sum of both components as received on a horizontal surface is called *global solar radiation*. A significant fraction of the incoming solar radiation is reflected back by the surface. The surface albedo, defined as the ratio of the reflected over the incoming radiation, depends on the nature of the surface, solar Zenith angle, and wavelength. For a water surface the albedo is about 0.06, whereas for snow the albedo is about 0.6–0.8. The albedo of bare sea ice is about 0.4–0.6. Since large areas of the Earth are covered by water, snow and sea ice, changes in the snow and sea ice cover can have a significant impact on the global albedo. Bare land surfaces have typical surface albedo of 0.1–0.35, with the highest value for the desert sand. Albedos of most vegetation surfaces fall in the range 0.1–0.25. The albedo for green vegetation depends greatly on wavelength, reflecting strongly in the near infrared but absorbing in the ultraviolet and visible regions.

9.2 SOLAR ENERGY COLLECTORS

9.2.1 Introduction

Solar energy collectors are special kind of heat exchangers that transform solar radiation energy to internal energy of the transport medium. The major component of any solar system is the solar collector. This is a device which absorbs the incoming solar radiation, converts it into heat, and transfers this heat to a fluid (usually air, water, or oil) flowing through the collector. The solar energy thus collected is carried from the circulating fluid either directly to the hot water or space conditioning equipment, or to a thermal energy storage tank from which can be drawn for use at night and/or cloudy days. There are basically two types of solar collectors: non-concentrating or stationary and concentrating. A non-concentrating collector has the same area for intercepting and for absorbing solar radiation, whereas a Sun-tracking concentrating solar collector usually has concave reflecting surfaces to intercept and focus the Sun's beam radiation to a smaller receiving area, thereby increasing the radiation flux. A large number of solar collectors are available in the market. A Sun based collector is a gadget used to capture the high temperature energy of the Sun and convert it into a form all the more promptly usable by people. Not at all like a photovoltaic cell, a solar collector is moderately low-tech, and they could be made and introduced for an ease. The most fundamental kind of Sun powered authority includes a fluid medium that is warmed up by the Sun's beams and afterward transported to circulate heat somewhere else. A simple type of Sun powered authority might be seen simply by leaving a huge dark compartment loaded with water outside and underneath the hot Sun. After just a couple of hours, the water inside the holder will have ingested an extraordinary arrangement of the energy of the daylight, and will be exceptionally hot. This hot water could be utilized for a basic reason like cleaning up, to fill a heated water flask to high temperature, a sleeping bag, or some other type of hotness dispersion. More intricate collectors take this essential idea and extend it, making the high temperature gathering more effective, and the hotness appropriation more dynamic. A solar collector is essentially an even box and are made out of three fundamental parts, a transparent spread tubes which convey a coolant and a protected back plate. The solar collectors deals with the greenhouse impact principle; solar radiation occurrence upon the transparent surface of the solar collector is transmitted through this surface. The solar collector is typically cleared, the energy held inside solar collector is basically trapped and thus heats the coolant contained within the tubes. The tubes are usually made from copper, and the backplate is painted black to help absorb solar radiation. The solar collector is usually insulated to avoid heat losses.

9.2.2 Stationary Collectors

Solar energy collectors are basically distinguished by their motion, i.e. stationary, single axis tracking and two axes tracking, and the operating temperature. Initially, the stationary solar collectors are examined. These collectors are permanently fixed in position and do not track the Sun. Three type of collectors fall in this category:

1. Flat plate collectors (FPC);
2. Stationary compound parabolic collectors (CPC);
3. Evacuated tube collectors (ETC).

9.2.2.1 Flat-Plate Collectors

A typical flat-plate solar collector is shown in Fig. 9.9. When solar radiation passes through a transparent cover and impinges on the blackened absorber surface of high absorptivity, a large portion of this energy is absorbed by the plate and then transferred to the transport medium in the fluid tubes to be carried away for storage or use. The underside of the absorber plate and the side of casing are well insulated to reduce conduction losses. The liquid tubes can be welded to the absorbing plate, or they can be an integral part of the

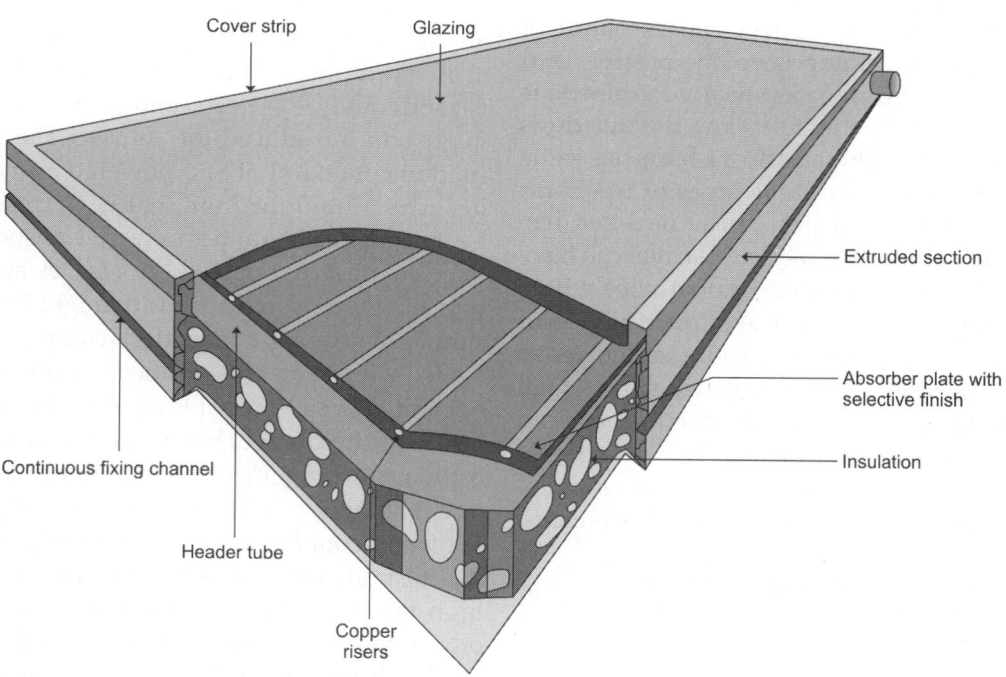

Fig. 9.9: Pictorial view of a flat-plate collector

plate. The liquid tubes are connected at both ends by large diameter header tubes. The transparent cover is used to reduce convection losses from the absorber plate through the restraint of the stagnant air layer between the absorber plate and the glass. It also reduces radiation losses from the collector as the glass is transparent to the shortwave radiation received by the Sun but it is nearly opaque to longwave thermal radiation emitted by the absorber plate (greenhouse effect). FPC are usually permanently fixed in position and require no tracking of the Sun. The collectors should be oriented directly towards the equator, facing south in the northern hemisphere and north in the southern. The optimum tilt angle of the collector is equal to the latitude of the location with angle variations of 10–158 more or less depending on the application.

1. **Glazing:** One or more sheets of glass or other diathermanous (radiation-transmitting) material.

2. **Tubes, fins, or passages:** To conduct or direct the heat transfer fluid from the inlet to the outlet.

3. **Absorber plates:** Flat, corrugated, or grooved plates, to which the tubes, fins, or passages are attached. The plate may be integral with the tubes.

4. **Headers or manifolds:** To admit and discharge the fluid.

5. **Insulation:** To minimise the heat loss from the back and sides of the collector.

6. **Container or casing:** To surround the aforementioned components and keep them free from dust, moisture, etc.

FPC have been built in a wide variety of designs and from many different materials. They have been used to heat fluids such as water, water plus antifreeze additive, or air. Their major purpose is to collect as much solar energy as possible at the lower possible total cost. The collector should also have a long effective life, despite the adverse effects of the Sun's ultraviolet radiation, corrosion and clogging because of acidity, alkalinity or hardness of the heat transfer fluid, freezing of water, or deposition of dust or moisture on the glazing, and breakage of the glazing because of thermal expansion, hail, vandalism or other causes. These causes can be minimised by the use of tempered glass.

Glazing materials: Glass has been widely used to glaze solar collectors because it can transmit as much as 90% of the incoming shortwave solar irradiation while transmitting virtually none of the longwave radiation emitted outward by the absorber plate. Glass with low iron content has a relatively high transmittance for solar radiation (approximately 0.85–0.90 at normal incidence), but its transmittance is essentially zero for the longwave thermal radiation (5.0–50 mm) emitted by Sun-heated surfaces. Plastic films and sheets also possess high shortwave transmittance, but because most usable varieties also have transmission bands in the middle of the thermal radiation spectrum, they may have longwave transmittances as high as 0.40. Plastics are also generally limited in the temperatures they can sustain without deteriorating or undergoing dimensional changes. Only a few types of plastics can withstand the Sun's ultraviolet radiation for long periods. However, they are not broken by hail or stones, and, in the form of thin films, they are completely flexible and have low mass. The commercially available grades of window and greenhouse glasses have normal incidence transmittances of about 0.87 and 0.85, respectively. For direct radiation, the transmittance varies considerably with the angle of incidence. Antireflective coatings and surface texture can also improve transmission significantly. The effect of dirt and dust on collector glazing may be quite small, and the cleansing effect of an occasional rainfall is usually adequate to maintain the transmittance within 2–4% of its maximum value. The glazing should admit as much solar irradiation as possible and reduce the upward loss of heat as much as possible. Although glass is virtually opaque to the longwave radiation emitted by collector plates, absorption of that radiation causes an increase in the glass temperature and a loss of heat to the surrounding atmosphere by radiation and convection. These are analysed in more details in Section-3. Various prototypes of transparently insulated FPC and CPC have been built and tested in the last decade. Low cost and high temperature resistant transparent insulating (TI) materials have been developed so that the commercialisation of these collectors becomes feasible. It was experimentally proved that the efficiency of the collector was comparable with that of ETC. However, no commercial collectors of this type are available in the market.

9.2.2.2 Collector Absorbing Plates

The collector plate absorbs as much of the irradiation as possible through the glazing, while loosing as little heat as possible upward to the atmosphere and downward through the back of the casing. The collector plates transfer the retained heat to the transport fluid. The absorptance of the collector surface for short-wave solar radiation depends on the nature and colour of the coating and on the incident angle. Usually black colour is used, however various colour coatings have been proposed mainly for aesthetic reasons. By suitable electrolytic or chemical treatments, surfaces can be produced with high values of solar radiation absorptance and low values of long-wave emittance. Essentially, typical selective surfaces consist of a thin upper layer, which is highly absorbent to shortwave solar radiation but relatively transparent to longwave thermal radiation, deposited on a surface that has a high reflectance and a low emittance for longwave radiation. Selective surfaces are particularly important when the collector surface temperature is much higher than the ambient air temperature. Lately, a low-cost mechanically manufactured selective solar absorber surface method has been proposed. An energy efficient solar collector should absorb incident solar radiation, convert it to thermal energy and deliver the thermal energy to a heat transfer medium with minimum losses at each step. It is possible to use several different design principles and physical mechanisms in order to create a selective solar absorbing surface. Solar absorbers are based on two layers with different optical properties, which are referred as *tandem absorbers*. A semiconducting or dielectric coating with high solar absorptance and high infrared transmittance on top of a non-selective highly reflecting material such as metal, constitutes one type of tandem absorber. Another alternative is to coat a non-selective highly absorbing material with a heat mirror having a high solar transmittance and high infrared reflectance. Today, commercial solar absorbers are made by electroplating, anodization, evaporation, sputtering and by applying solar selective paints. Much of the progress during recent years has been based on the implementation of vacuum techniques for the production of fin type absorbers used in low temperature applications. The chemical and electro-chemical processes used for their commercialization were readily taken over from the metal finishing industry. The requirements of solar absorbers used in high temperature applications, however, namely extremely low thermal emittance and high temperature stability, were difficult to fulfil with conventional wet processes. Therefore, largescale sputter deposition was developed in the late 70s. The vacuum techniques are now-a-days mature, characterized by low cost and have the advantage of being less environmentally polluting than the wet processes. For fluid-heating collectors, passages must be integral with or firmly bonded to the absorber plate. A major problem being a good thermal bond between tubes and absorber plates without incurring excessive costs for labour or materials. Material most frequently used for collector plates are copper, aluminium, and stainless steel. UV-resistant plastic extrusions are used for low temperature applications. If the entire collector area is in contact with the heat transfer fluid, the thermal conductance of the material is not important. Copper tubes are used most often because of their superior resistance to corrosion. Thermal cement, clips, clamps, or twisted wires have been tried in the search for low-cost bonding methods. Mechanical pressure, thermal cement, or brazing may be used to make the assembly. Soft solder must be avoided because of the high plate temperature encountered at stagnation conditions.

Air or other gases can be heated with FPC, particularly if some type of extendor is used to counteract the low heat transfer coefficients between metal and air. Metal or fabric matrices or thin corrugated metal sheets may be used, with selective surfaces applied to the latter when a high level of performance is required. The principal requirement is a large contact area between the absorbing surface and the air.

Various applications of solar air collectors are reported. Reduction of heat loss from the absorber can be accomplished either by a selective surface to reduce radiative heat transfer or by suppressing convection. These are usually low-cost units which can offer cost effective solar thermal energy in applications such as water preheating for domestic or industrial use, heating of swimming pools, space heating and air heating for industrial or agricultural applications. FPC are by far the most used type of collector. FPC are usually employed for low temperature applications up to 100 °C, although some new types of collectors employing vacuum insulation and/ or TI can achieve slightly higher values . Due to the introduction of highly selective coatings actual standard FPC can reach stagnation temperatures of more than 200 °C. With these collectors good efficiencies can be obtained up to temperatures of about 100 °C.

9.2.2.3 Compound Parabolic Collectors (CPC)

CPC are non-imaging concentrators. These have the capability of reflecting to the absorber all of the incident radiation within wide limits. Their potential as collectors of solar energy was pointed out by Winston. The necessity of moving the concentrator to accommodate the changing solar orientation can be reduced by using a trough with two sections of a parabola facing each other, as shown in Fig. 9.10. Compound parabolic concentrators can accept incoming radiation over a relatively wide range of angles. By using multiple internal reflections, any radiation that is entering the aperture, within the collector acceptance angle, finds its way to the absorber surface located at the bottom of the collector. The absorber can take a variety of configurations. It can be cylindrical as shown in Fig. 9.10 or flat. In the CPC shown in Fig. 9.10, the lower portion of the reflector (AB and AC) is circular, while the upper portions (BD and CE) are parabolic. As the upper part of a CPC contribute little to the radiation reaching the absorber, they are usually truncated thus forming a shorter version of the CPC, which is also cheaper. CPCs' are usually covered with glass to avoid dust and other materials from entering the collector and thus reducing the reflectivity of its walls. These collectors are more useful as linear or trough-type concentrators. The acceptance angle is defined as the angle through which a source of light can be moved and still converge at the absorber. The orientation of a CPC collector is related to its acceptance angle. Also depending on the collector acceptance angle, the collector can be stationary or tracking. A CPC concentrator can be orientated with its long axis along either the north-south or the east-west direction and its aperture is tilted directly towards the equator at an angle equal to the local latitude. When orientated along the north-south direction the collector must track the Sun by turning its axis so as to face the Sun continuously. As the acceptance angle of the concentrator along its long axis is wide, seasonal tilt adjustment is not necessary. It can also be stationary but radiation will only be received for hours when the Sun is within the collector acceptance angle. When the concentrator is orientated

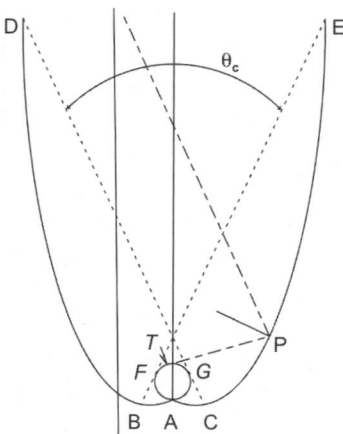

Fig. 9.10: Schematic diagram of compound parabolic collectors

with its long axis along the east-west direction, with a little seasonal adjustment in tilt angle the collector is able to catch the Sun's rays effectively through its wide acceptance angle along its long axis. The minimum acceptance angle in this case should be equal to the maximum incidence angle projected in a north-south vertical plane during the times when output is needed from the collector. For stationary CPC collectors mounted in this mode the minimum acceptance angle is equal to 478. In practice bigger angles are used to enable the collector to collect diffuse radiation at the expense of a lower concentration ratio. Smaller (less than 3) concentration ratio CPCs' are of greatest practical interest. In particular, a simple analytic technique was developed for the calculation of the average number of reflections for radiation passing through a CPC, which is useful for computing optical loses. Many numerical examples are presented which are helpful in designing a CPC. Two basic types of CPC collectors have been designed; *the symmetric and asymmetric*. These usually employ two main types of absorbers; fin type with pipe and tubular absorbers. Advanced flat plate collector fixing of risers on the absorber plate embedded ultrasonically welded absorber coating black mat paint chromium selective coating.

9.2.2.4 *Evacuated Tube Collectors*

Conventional simple flat-plate solar collectors were developed for use in Sunny and warm climates. Their benefits however are greatly reduced when conditions become unfavourable during cold, cloudy and windy days. Furthermore, weathering influences such as condensation and moisture will cause early deterioration of internal materials resulting in reduced performance and system failure. Evacuated heat pipe solar collectors (tubes) operate differently than the other collectors available on the market. These solar collectors consist of a heat pipe inside a vacuum-sealed tube, as shown in Fig. 9.11. ETC have demonstrated that the combination of a selective surface and an effective convection suppressor can result in good performance at high temperatures. The vacuum envelope reduces convection and conduction losses, so the collectors can operate at higher temperatures than FPC. Like FPC, they collect both direct and diffuse radiation. However, their efficiency is higher at low incidence angles. This effect tends to give ETC an advantage over FPC in day-long performance. ETC use liquid-vapour phase change materials to transfer heat at high efficiency. These collectors feature a

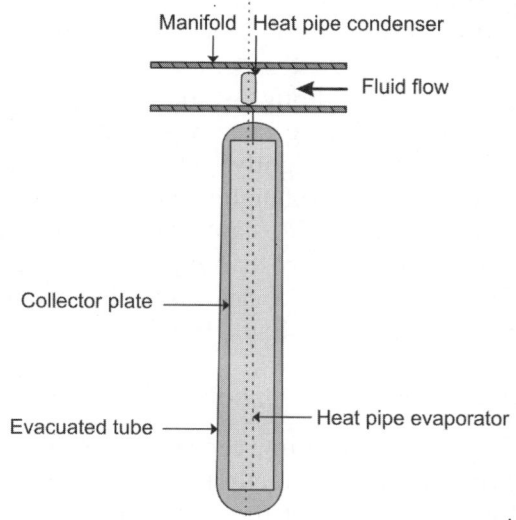

Fig. 9.11: Schematic diagram of an evacuated tube collector

heat pipe (a highly efficient thermal conductor) placed inside a vacuum-sealed tube. The pipe, which is a sealed copper pipe, is then attached to a black copper fin that fills the tube (absorber plate). Protruding from the top of each tube is a metal tip attached to the sealed pipe (condenser). The heat pipe contains a small amount of fluid (e.g. methanol) that undergoes an evapourating-condensing cycle. In this cycle, solar heat evapourates the liquid, and the vapour travels to the heat sink region where it condenses and releases its latent heat. The condensed fluid return back to the solar collector and the process is repeated. When these tubes are mounted, the metal tips up, into a heat exchanger (manifold) as shown in Fig. 9.11. Water or glycol, flows through the manifold and picks up the heat from the tubes. The heated liquid circulates through another heat exchanger and gives off its heat to a process or to water that is stored in a solar storage tank. Because no evaporation or condensation above the phase-change temperature is possible, the heat pipe offers inherent protection from freezing and overheating. This selflimiting temperature control is a unique feature of the evacuated heat pipe collector. ETC basically consist of a heat pipe inside a vacuum sealed tube. A large number of variations of the absorber shape of ETC are on the market. Evacuated tubes with CPC-reflectors are also commercialised by several manufacturers. One manufacturer recently presented an all-glass ETC, which may be an important step to cost reduction and increase of lifetime. Another variation of this type of collector is what is called Dewar tubes. In this two concentric glass tubes are used and the space in between the tubes is evacuated (vacuum jacket). The advantage of this design is that it is made entirely of glass and it is not necessary to penetrate the glass envelope in order to extract heat from the tube thus leakage losses are not present and it is also less expensive than the single envelope system.

Another type of collector developed recently is the integrated compound parabolic collector (ICPC). This is an ETC in which at the bottom part of the glass tube a reflective material is fixed . The collector combines the vacuum insulation and non-imaging stationary concentration into a single unit. In another design a tracking ICPC is developed which is suitable for high temperature applications.

9.2.2.5 Sun Tracking Concentrating Collectors

Energy delivery temperatures can be increased by decreasing the area from which the heat losses occur. Temperatures far above those attainable by FPC can be reached if a large amount of solar radiation is concentrated on a relatively small collection area. This is done by interposing an optical device between the source of radiation and the energy absorbing surface. Concentrating collectors exhibit certain advantages as compared with the conventional flat-plate type. The main ones are:

1. The working fluid can achieve higher temperatures in a concentrator system when compared to a flat-plate system of the same solar energy collecting surface. This means that a higher thermodynamic efficiency can be achieved.

2. It is possible with a concentrator system, to achieve a thermodynamic match between temperature level and task. The task may be to operate thermionic, thermodynamic, or other higher temperature devices.

3. The thermal efficiency is greater because of the small heat loss area relative to the receiver area.

4. Reflecting surfaces require less material and are structurally simpler than FPC. For a concentrating collector the cost per unit area of the solar collecting surface is therefore less than that of a FPC.

5. Owing to the relatively small area of receiver per unit of collected solar energy, selective surface treatment and vacuum insulation to reduce heat losses and improve the collector efficiency are economically viable.

Their disadvantages are:

1. Concentrator systems collect little diffuse radiation depending on the concentration ratio.
2. Some form of tracking system is required so as to enable the collector to follow the Sun.
3. Solar reflecting surfaces may loose their reflectance with time and may require periodic cleaning and refurbishing.

Many designs have been considered for concentrating collectors. Concentrators can be reflectors or refractors, can be cylindrical or parabolic and can be continuous or segmented. Receivers can be convex, flat, cylindrical or concave and can be covered with glazing or uncovered. Concentration ratios, i.e. the ratio of aperture to absorber areas, can vary over several orders of magnitude, from as low as a unity to high values of the order of 10,000. Increased ratios mean increased temperatures at which energy can be delivered but consequently these collectors have increased requirements for precision in optical quality and positioning of the optical system. Because of the apparent movement of the Sun across the sky, conventional concentrating collectors must follow the Sun's daily orientation. There are two methods by which the Sun's motion can be readily tracked. The first is the altazimuth method which requires the tracking device to turn in both altitude and azimuth, i.e. when performed properly, this method enables the concentrator to follow the Sun exactly. Paraboloidal solar collectors generally use system. The second one is the one-axis tracking in which the collector tracks the Sun in only one direction either from east to west or from north to south. Parabolic trough collectors (PTC) generally use this system. These systems require continuous and accurate adjustment to compensate for the changes in the Sun's orientation. The first type of a solar concentrator, shown in Fig. 9.12, is effectively a FPC fitted with simple flat reflectors which can markedly increase the amount of direct radiation reaching the

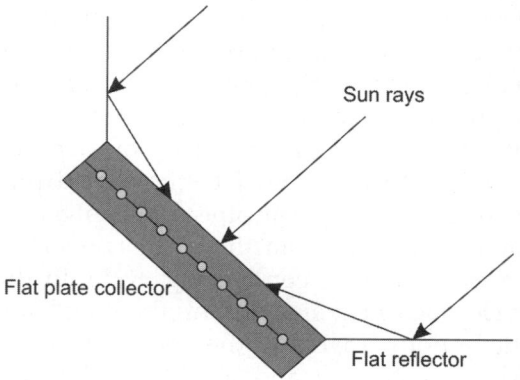

Fig. 9.12: Flat plate collector with flat reflectors

collector. This is a concentrator because the aperture is bigger than the absorber but the system is stationary. The model facilitates the prediction of the total energy absorbed by the collector at any hour of the day for any latitude for random tilt angles and azimuth angles of the collector and reflectors. type of collector, already covered under the stationary collectors, the CPC is also classified as concentrator. This, depending on the acceptance angle, can be stationary or tracking. When tracking is used this is very rough or intermitted as concentration ratio is usually small and radiation can be collected and concentrated by one or more reflections on the parabolic surfaces. As was seen above one disadvantage of concentrating collectors is that, except at low concentration ratios, they can use only the direct component of solar radiation, because the diffuse component cannot be concentrated by most types. However, an additional advantage of concentrating collectors is that, in summer, when the sunrises well to the north of the east-west line, the sun-follower, with its axis oriented north-south, can begin to accept radiation directly from the Sun long before a fixed, south-facing flatplate can receive anything other than diffuse radiation from the portion of the sky that it faces. Thus, in relatively cloudless areas, the concentrating collector may capture more radiation per unit aperture area than a FPC. In concentrating collectors solar energy is optically concentrated before being transferred into heat. Concentration can be obtained by reflection or refraction of solar radiation by the use of mirrors or lens. The reflected or refracted light is concentrated in a focal zone, thus increasing the energy flux in the receiving target. Concentrating collectors can also be classified into non-imaging and imaging depending on whether the image of the Sun is the concentrator belonging in the first category is the CPC, whereas all the other types of concentrators belong to the imaging type. The collectors falling in this category are parabolic trough collectors, linear fresnel reflector, parabolic dish and central receiver.

9.2.3 ADVANTAGES AND DISADVANTAGES OF SOLAR COLLECTORS

Advantages

- They have advantage of using both beam and diffuse solar radiation.

- They do not require orientation towards Sun.

- They require little maintenance.

Disadvantages

- Additional requirements of maintenance particular to retain the quality of reflecting surface against dirt, weather, oxidation etc.

- Non-uniform flux on the absorber whereas flux in flat-plate collectors is uniform.

- High initial cost.

9.3 SOLAR ENERGY STORAGE

9.3.1 Introduction

Thermal mass systems can store solar energy in the form of heat at domestically useful temperatures for daily of interseasonal durations. Thermal storage systems generally use readily available materials with high specific heat capacities such as water, Earth and stone. Well-designed systems can lower peak demand, shift time-of-use to off-peak hours and reduce overall heating and cooling requirements. Energy storage became a dominant factor in economic development with the widespread introduction of electricity. Unlike other common energy storage in prior use such as wood or coal, electricity must be used as it is being generated, or converted immediately into another form of energy such as potential, kinetic or chemical. Energy storage may be in the form of sensible heat of solids or liquid medium, as heat of fusion in chemical systems or as chemical energy of products in the reversible chemical reaction. Mechanical energy can be converted to potential energy and stored in elevated fluids. Product of solar processes other than energy may be stored in tanks until needed. The choice of media for energy storage depends on the nature of processes. For water heating, energy storage as sensible heat of stored water is logical. If air heating collectors are used, storage is sensible or latent heat effects in particular storage units are indicated, such as sensible heat in a pebble bed heat exchanger. If photovoltaic or photo chemical processes are used, storage mostly logical in the form of chemical energy or energy is stored in the batteries. Thus the energy may be stored in a variety of forms, e.g. As heat, electrical, chemical, mechanical and magnetic. Measurements of solar radiation are important because of the increasing number of solar heating and cooling applications, and the need for accurate solar irradiation data to predict performance. Experimental determination of the energy transferred to a surface by solar radiation required instruments which will measure the heating effect of direct solar radiation and diffuse solar radiation. Measurements are also made of beam radiation, which respond to solar radiation received from a very small portion of the circum solar sky. The need for energy storage of some kind is almost immediate evident for a solar-electric system. An optimally designed solar-electric system will collect and convert when the insolation is available during the day. Unfortunately the time when solar energy is most available will rarely coincide exactly with the demand for electrical energy.

9.3.2 Solar Pond

Natural or artificial waterway for gathering and retaining Sun oriented radiation energy and putting away it as high temperature. In this way sunlight based lake consolidates solar energy accumulation and sensible high temperature stockpiling. The most simplest kind of solar pond is extremely shallow, something like 5 to 10 cm profound, with a radiation engrossing lowest part. A bed of protecting material under the lake minimizes misfortune of high temperature to the ground. A bended spread, made of transparent fiber glass, over the pond licenses passage of Sun powered radiation yet decreases misfortunes by radiation and convection. In a suitable atmosphere, all the pond water can get hot enough for utilization in space warming and farming and different procedures. The solar pond works on a very simple principle. It is well-known that water or air is heated they become lighter and rise upward, e.g. a hot air balloon. Similarly, in an ordinary pond, the Sun's rays heat the water and the heated water from within the pond rises and reaches the top but loses the heat into the atmosphere. The net result is that the pond water remains at the atmospheric temperature. The solar pond restricts this tendency by dissolving salt in the bottom layer of the pond making it too heavy to rise.

A solar pond has three zones. The top zone is the surface zone, or UCZ (upper convective zone), which is at atmospheric temperature and has little salt content. The bottom zone is very hot, 70°–85°C, and is very salty. It is this zone that collects and stores solar energy in the form of heat, and is, therefore, known as the storage zone or LCZ (lower convective zone), Fig. 9.13. Separating these crops on two zones is the important gradient zone or NCZ (non-convective zone).

Applications of solar pond:
- Heating and cooling of buildings
- Production of power
- Industrial process of heat
- Desalination
- Heating animal housing and drying crops on farms
- Heat for biomass conversion

Solar ponds have several advantages. They have a low cost per unit area of collection and an inherent storage capacity. Also, they can be easily constructed over large areas, enabling the diffuse solar resource to be concentrated on a grand scale. Solar ponds address three environmental issues arising from the use of conventional fuels. First, heat energy is provided without burning fuel, thus reducing pollution. Second, conventional energy

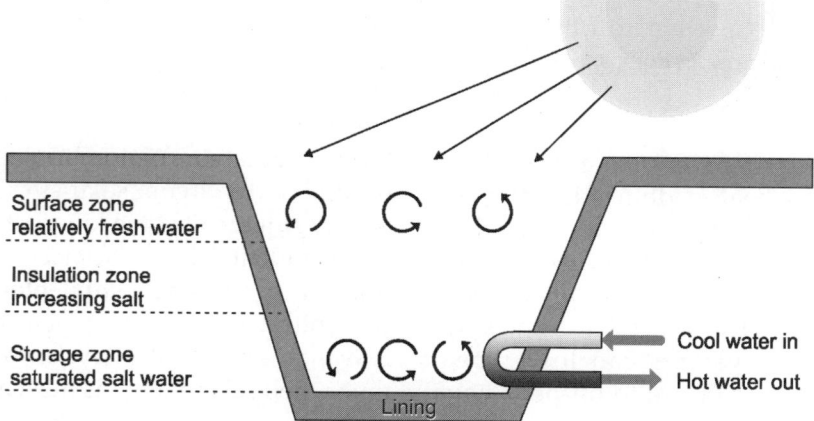

Fig. 9. 13: Zonation of solar pond

resources are conserved. Third, solar ponds coupled with desalting units can be used to purify contaminated or minerally-impaired water, and the pond itself can become the receptacle for the waste products.

9.4 APPLICATIONS OF SOLAR ENERGY

9.4.1 Introduction

The actual and proposed applications of solar energy are discussed below:

- Solar water heating
- Space heating
- Space cooling
- **Solar energy:** Thermal electric conversion
- **Solar energy:** Photovoltaic electric conversion
- Solar distillation
- Solar pumping
- Agriculture and industrial process heat
- Solar furnace
- Solar cooking
- Solar greenhouses

9.4.2. Active solar water heating

The main components on an active solar water heating system are:

- Solar collector to capture the Sun's energy and to transfer is to the coolant medium
- A circulation system that moves the fluid between the solar collector and the storage tank
- Storage tank
- Back up heating system
- Control system to regulate the system operation

The two main types of solar water heating systems are the closed loop system and the open loop system. The open loop system used water as the coolant, the water circulates between the solar collector and the storage tank. There are two main types of open loop system these are the draindown system and the recirculating system, the main principle behind both systems is the activation of circulation from the collector to the storage tank when the temperature within the solar collector reaches a certain value.

i. In the draindown system a valve is used to allow the solar collector to fill with water when the collector reaches a certain temperature.

ii. In the recirculating system water is pumped through the collector when the temperature in the storage tank reaches a certain critical value.

In applications where there is likely to be a temperature drop below zero degrees then it is necessary to use a closed loop system. The main difference between the open loop systems is the water is replaced with a coolant which will not freeze in the tempeture range which the solar collector may be subject to. The coolant will usually be a refrigerant, oil or distilled water. Closed loop systems are generally more costly than their open loop counter parts and great care must be taken to avoid contamination of the water with refrigerant. The energy captured by the coolant is then transferred to the hot water via a heat exchanger. In a drain back system the coolant may be distilled water. The system works on the principle that there is only water in the collector when the pump is operating. This has the benefit that the coolant used in the system will not have the chance to cool down during the night when temperature may drop to a level which may cause the coolant to increase in density and thus perhaps cause is not be as free flowing as it should. The only necessary feature on the drain back system is that the solar collectors are elevated from the heat exchanger or drain back tank so that the coolant can flow out of the collector. This system again works on the principle that the water will circulate between the collector and the drain back tank when the designated temperature is reached between the solar collector and the hot water (Fig. 9.14).

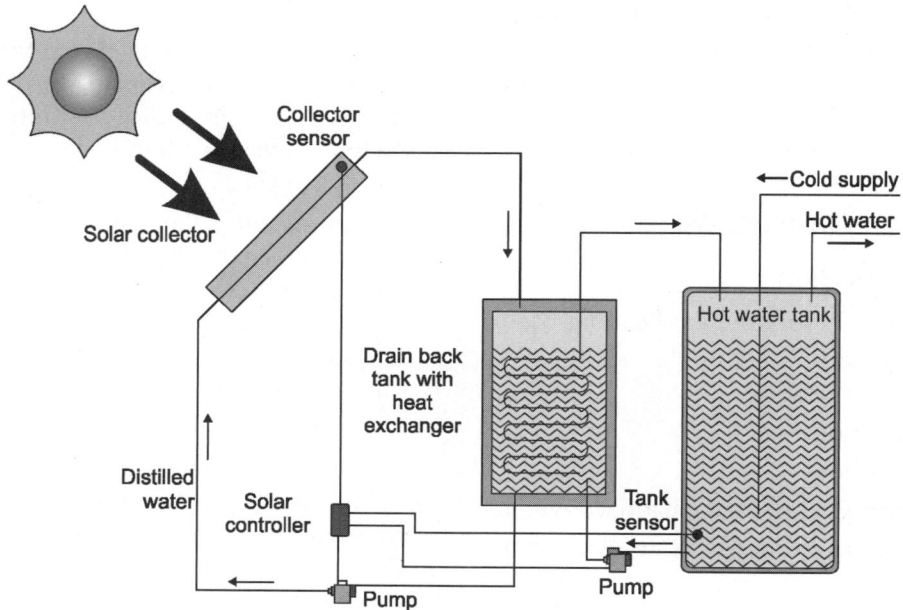

Collector sensor

Solar collector

Cold supply

Hot water

Hot water tank

Drain back tank with heat exchanger

Distilled water

Solar controller

Tank sensor

Pump

Pump

Fig. 9.14: Schematic diagram of a solar water heater

9.4.3 Solar Space Heating

The system components in a space heating application are the same for water heating with the addition of radiators for space heating or underfloor heating coils or even forced air systems. A radiator system will generally work in a very similar manner to the hot water application, the main difference is the inclusion of a boiler, heated water from the collector is passed through the heat exchanger or drain back tank and is then passed to a boiler which is used to supplement the water bearing requirements before passing into the radiators to be used for space heating (Fig. 9.15). Solar process heating systems are designed to provide large quantities of hot water or space heating for nonresidential buildings. A typical system includes solar collectors that work along with a pump, a heat exchanger, and/or one or more large storage tanks. The two main types of solar collectors used are—*an evacuated-tube collector and a parabolic-trough collector*—can operate at high temperatures with high efficiency. An evacuated-tube collector is a shallow box full of many glass,

double-walled tubes and reflectors to heat the fluid inside the tubes. A vacuum between the two walls insulates the inner tube, holding in the heat. Parabolic troughs are long, rectangular, curved (U-shaped) mirrors tilted to focus sunlight on a tube, which runs down the center of the trough. This heats the fluid within the tube. Many different concepts have been proposed for using solar energy in space heating of buildings. There are two primary categories into which virtually all solar heating systems may be divided. The first is passive systems, in which solar radiation is collected by some element of the structure itself, or admitted directly into building through large, south facing windows. The second is the active systems which generally consists of: (a) separate solar collectors, which may heat either water or air, (b) storage devices which can accumulate the collected energy for use at nights and during inclement days, and (c) a backup system to provide heat for protected periods of bad weather. The heat from a solar collector can also be used to cool a building. It may seem impossible to use heat to cool a building, but it makes more sense if you just

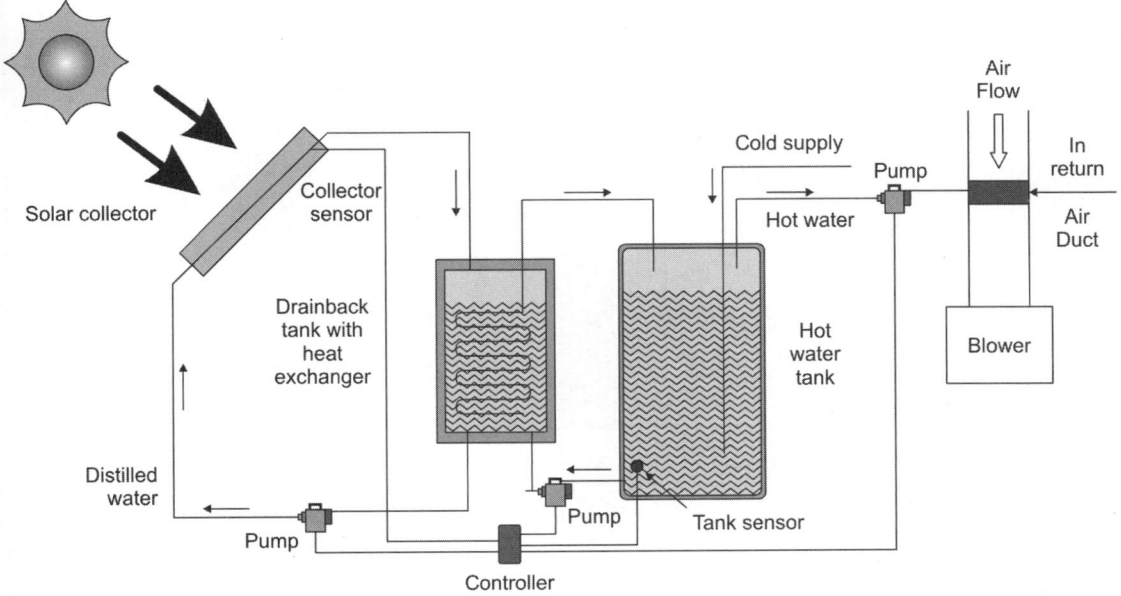

Fig. 9.15: Schematic diagram of a solar space heater

think of the solar heat as an energy source. Your familiar home air conditioner uses an energy source, electricity, to create cool air. Solar absorption coolers use a similar approach, combined with some very complex chemistry tricks, to create cool air from solar energy. Solar energy can also be used with evapourative coolers (also called "swamp coolers") to extend their usefulness to more humid climates, using another chemistry trick called *desiccant cooling*.

Air Distribution Systems

Again the air appropriation system works in a way very much alike to the boiling point water system, the primary distinction is the incorporation of a blower and an air pipe. The framework utilizes an extra controlled pump which will permit wind stream over the coil when the temperature in the stockpiling tank is sufficiently high that passing air over the loops in the return conduit of the device will permit the system to make a positive contribution to the heating space warming interest. In large commercial or industrial applications, system configuration is somewhat not the

same as residential applications. It is worth noting that the temperature rise across a collector is fairly constant to use an example if the temperature of supply to the collector is around 60°F and the temperature of return is around 73°C or the return is 173°F and the supply is 160°C, this basically means that high and low temperature applications should not be put in series inside a loop. The low temperature application would basically drag down the higher temperature application. Vacuum collectors are excellent performers in high temperature applications, the collector loop should be dedicated to the higher temperature application until the load is satisfied. In applications such as for hospitals, hotels or commercial office blocks it might be necessary for the installation of two or more tanks connected in series (Fig. 9.16). (1) storage tank, (2) preheat tank, (3) cold feed, (4) mixing valve, (5) supply and return to collector, (6) hot water out system operation: hot water from the collector passes through the coil in tank one (1), then, depending on it's temperature, it is diverted by a three way valve (4) to either: the coil in tank (2) if it is above the set temperature,

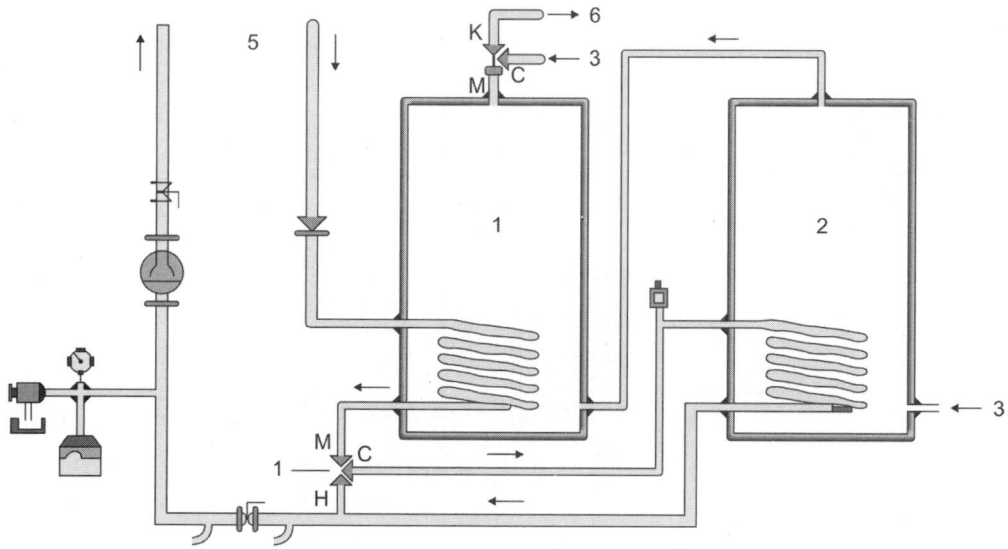

Fig. 9.16: Flow diagram of space heating system

(meaning tank (1) is hot) or the collector, if it is below the set temperature of the mixing valve. Commercial and industrial design considerations: the system can be expanded to include more than one preheat tanks, the heat exchange coils are linked by three way valves and the water which is to be heated runs in series through the tanks in the opposite direction. The three way valve can either be thermally controlled or electrically operated. No more than 100 tubes should be plumbed in series. Care must be taken when designing the pipe work in each section to ensure that each section receives equal flow.

9.4.4 Residential Application

Use of solar energy for homes has number of advantages. The solar energy is used in residential homes for heating the water with the help of solar heater. The photovoltaic cell installed on the roof of the house collects the solar energy and is used to warm the water. Solar energy can also be used to generate electricity. Batteries store energy captured in day time and supply power throughout the day. The use of solar appliances is one of the best ways to cut the expenditure on energy.

9.4.5 Industrial Application

Sun's thermal energy is used in office, warehouse and industry to supply power. Solar energy is used to power radio and TV stations. It is also used to supply power to lighthouse and warning light for aircraft.

9.4.6 Remote Application

Solar energy can be used for power generation in remotely situated places like schools, homes, clinics and buildings. Water pumps run on solar energy in remote areas. Large scale desalination plant also use power generated from solar energy instead of electricity.

Solar heating systems can effectively reduce a household's energy bill by over 1/2. Hot water requirements and home/space heating in North America represents over 1/2 of our annual energy costs. By using solar energy we can capture the Sun's radiation and use it for both hot water heating and as well as home heating by simply piping hot water through traditional or modern radiators, furnaces, or use it in hydronic system for infloor radiant heat (Fig. 9.17).

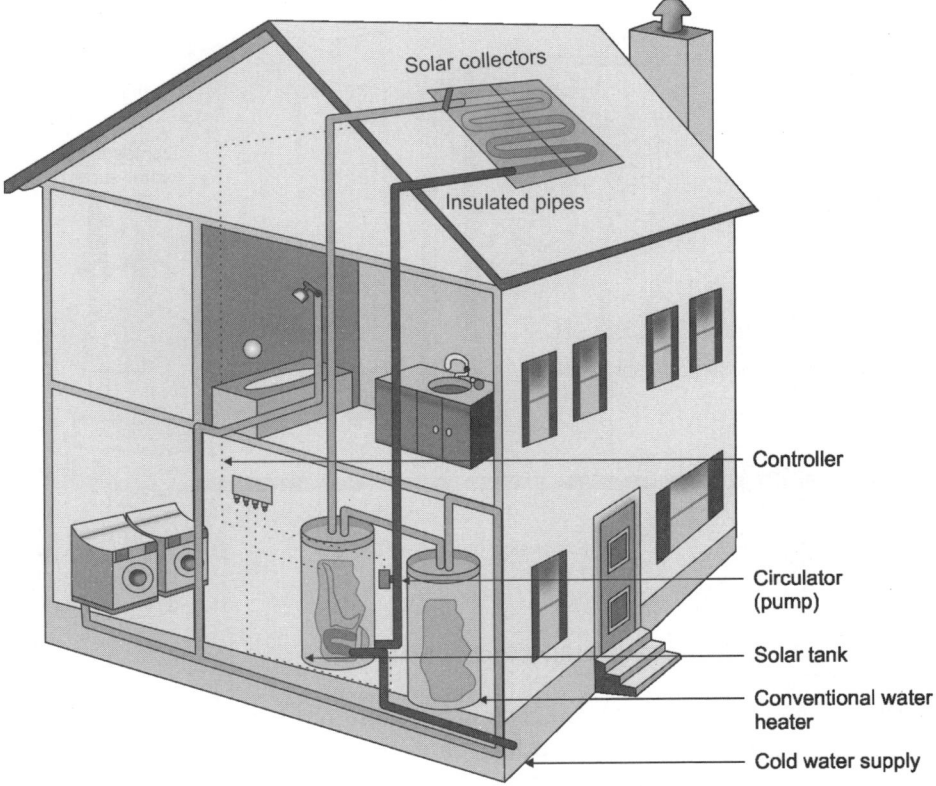

Solar collectors

Insulated pipes

Controller

Circulator (pump)

Solar tank

Conventional water heater

Cold water supply

Fig. 9.17: Utilization of solar energy

9.4.7 Solar Distillation

Fresh water and coast is a need for the sustenance of life and likewise the way to man's thriving. It is for the most part watched that in some arid, semi arid and seaside regions which are thinly populated and scattered, one or two relatives are constantly occupied in bringing new water from a long distance. In these ranges solar energy is copious and might be utilized for changing over saline water into refined water. The unadulterated water could be acquired by refining in the least difficult sunlight based still, by and large known as the basin type solar still. The Sun powered refining methodology produces consumable water much the way that nature produces rain. The Sun's hotness causes water to evapourate, differentiating the water vapour from salt or polluting influences. This water vapour gathers on the still for accumulation for utilization.

Sun based refining is an old method for decontaminating water and making saltwater consumable; Aristotle depicted the methodology as right on time as the fourth century BC. The principal advanced, huge scale sunlight based still, implicit Chile in 1872, comprised of 64 basins that supplies up to 20,000 liters of water every day to a mining group in the area.

Solar hot water system (Fig. 9.18), turns cold water into hot water with the help of Sun rays. Around 60°–80 °C temperature can be attained depending on solar radiation, weather conditions and solar collector system efficiency. Hot water for homes, hostels, hotels, hospitals, restaurants, dairies, industries etc. can be installed on roof-tops, building terrace and open ground and where there is no shading, south orientation of collectors and over-head tank above SWH system.

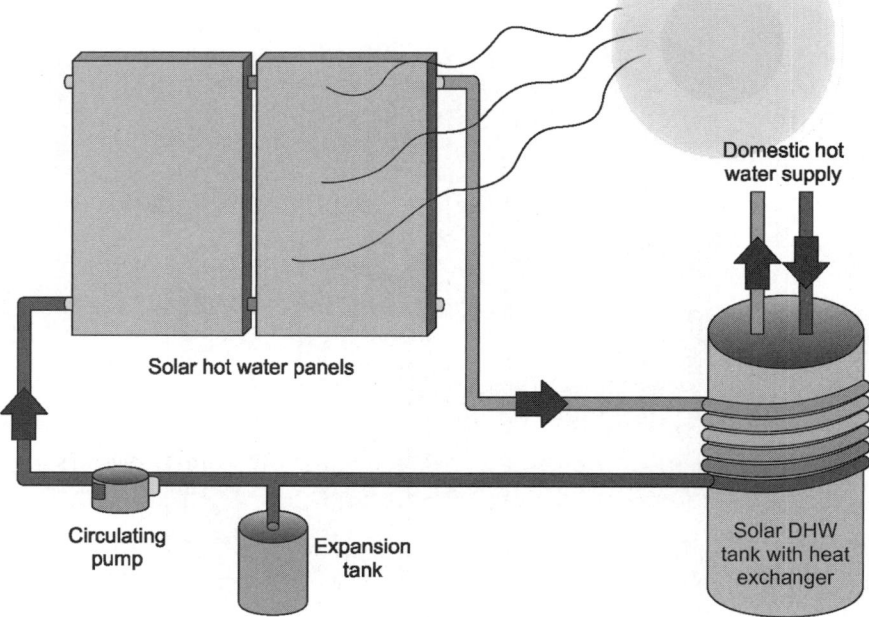

Fig. 9.18: Solar hot water system

SWH system generates hot water on clear sunny days (maximum), partially clouded (moderate) but not in rainy or heavy overcast day, only soft and potable water can be used . Stainless steel is used for small tanks, whereas, mild steel tanks with anticorrosion coating inside are used for large tanks. Solar water heaters (SWHs) of 100–300 litres capacity are suited for domestic application. Larger systems can be used in restaurants, guest houses, hotels, hospitals and industries.

Sometimes the only water available to a community is not just a health hazard, but is completely undrinkable. In the case of brackish or salty water, the compounds dissolved into the water cannot be removed with a simple filter. This solution highlights a method of separating pure water from the salty source water using a solar still. The process of separating water from salts and other impurities by heating the source water until the pure water becomes a vapour is called *distillation*, and the device that performs this task is called a *still*. When the salty or otherwise contaminated water is heated, pure water changes

from a liquid form into a gas. Only the water makes this change, leaving salt, minerals and other contaminants behind. The water vapour is then collected and condensed back into its liquid form, where it can be stored for drinking or manufacturing purposes. This seems like a simple process, but it takes huge amount of energy. Evapourating water consumes 10,000 times more energy per kilogram than pumping water (upto 20 meters). For this reason, distillation should only be used when pure water that is free of any mineral content is required, such as in lead-acid car batteries, or if the only drinking water source is salty. In fact, to make water taste good, and to provide humans with the nutrients we need to be healthy, a small amount of salt and minerals must be added back to the distilled water. The cost of transporting water by boat or vehicle is moreover equal to the cost of producing clean water with a still. To build an energy efficient still, the Sun can be harnessed as a heat source and used to evapourate water. Stills come in many different shapes and designs, which are discussed in the technical brief, but they all have some sort of basin for the source water, a slanted clear glass or plastic cover and a collection point for the condensed pure water. When the Sun heats the basin containing the source water it forms vapour which rises and hits the relatively cool glass or plastic cover. When the vapour hits the cover it turns back into a liquid and trickles down the slanted surface into the collection tank. The cover works just like a glass of ice-water on a hot day. The condensation that forms on the outside of the glass was initially vapour in the air. For a still to be efficient, the source water must get very hot, the temperature difference between the source water and the glass or plastic cover must be large, and the system must be sealed tightly so the water vapour does not leak out of the still before it condenses. The minimum daily requirement for distilled water per person per day is 5 liters. (People require 20 liters of water per day, but salty water can be used for some purposes.) A solar still will produce 5 liters of distilled water each day for every two square meters of condensing surface. This is about the size of a common doorway. For stills to be useful, they must be quite large. In India, a solar still of 15 square meters, which could be used by a maximum of 7 people, costs approximately $575. Costs per liter of water produced will vary greatly based on the following factors: still material and construction cost, cost of the land the still occupies, the life of the still, maintenance costs, and source water cost.

9.4.8 SOLAR THERMAL POWER GENERATION

Solar power is the transformation of sunlight into power, either specifically utilizing photovoltaics (PV), or in a roundabout way utilizing concentrated Sun oriented force (CSP). Concentrated solar power systems use lenses or mirrors and following frameworks to center a substantial range of sunlight into a small beam. Photovoltaics change over light into electric current utilizing the photoelectric

impact. Concentrating solar power (CSP) frameworks use lenses or mirrors and following systems to center a huge territory of daylight into a little bar. The concentrated hotness is then utilized as a heat source for conventional power plant. Different procedures are utilized to track the Sun and center light. In these systems a working liquid is warmed by the concentrated sunlight, and is then utilized for energy generation or energy storage. Warm capacity productively permits up to 24 hour power era. Heat might be changed over directly into electrical energy by Sun powered cell or thermionic or thermoelectric strategies however, these procedures may not be suitable for utilization to high temperature working liquid. The high temperature energy is then changed over into mechanical energy in a turbine and then into electrical energy by method for an conventional generator coupled to the turbine. This mechanical force processing system is known as the solar thermal power production system (Fig. 9.19).

Solar thermal power generation employs power cycles which are broadly classified as low, medium and high temperature cycles. Low temperature cycles generally use flat-plate collectors so that maximum temperatures are limited to about 100°C. Medium temperature cycles work at maximum temperatures ranging from 150° to 300°C, while high temperature cycles work at temperatures above 300 degrees celcius. Over the last few years, a few experimental tower power plants have been built or under construction in USA, France, Italy and Japan. The largest of these is the solar thermal test facility at Albuquerque in New Mexico. In solar thermal power production system the energy is first collected by using a solar pond, a flat plate collector or a focusing collector. This energy is used to increase the internal energy or temperature of the fluid. This fluid may be directly using in any of the common or known cycles such as Rankine, Brayton or Stirling or passes through a heat exchanger to heat a secondary fluid

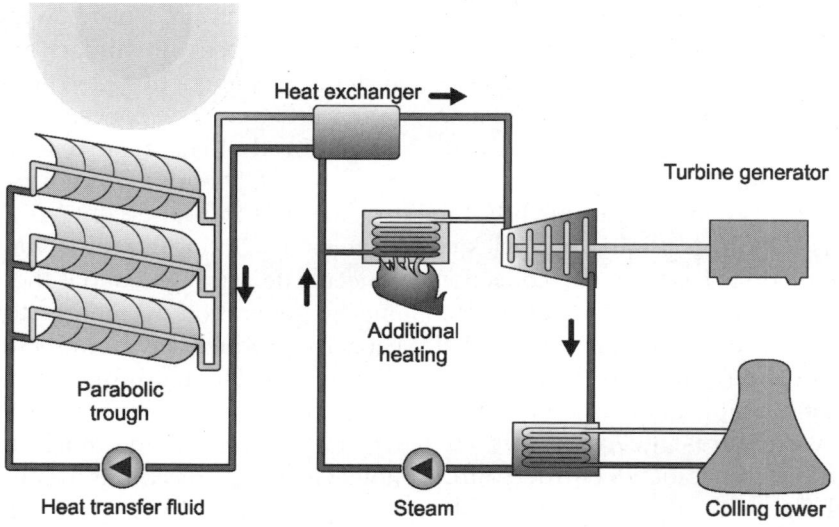

Heat exchanger ➡

Turbine generator

Parabolic
trough

Additional
heating

Heat transfer fluid   Steam   Colling tower

Fig. 9.19: Solar thermal power generation

which is being used in a cycle to produce mechanical power from which electrical power can be produced. Functioning of solar thermal system is shown in Fig. 9.20.

9.4.9 Solar Photovoltaics

Solar Electric Power Generation

Solar panels turn energy from the Sun's rays directly into useful energy that can be used in homes and businesses. There are two main types: solar thermal and photovoltaic, or PV. Solar thermal panels use the solar energy to heat water that can be used in washing and heating. PV panels use the photovoltaic effect to turn the solar energy directly into electricity, which can supplement or replace a building's usual power supply.

Figure 9.21, demonstrates the working procedure of photovoltaic panel for the production of solar energy. A PV panel is made up of a semiconducting material, usually silicon-based, sandwiched between two electrical contacts. To generate as much electricity as possible, PV panels need to spend as much time as possible in direct sunlight (1a). A sloping, south-facing roof is the ideal place to mount a solar panel. A sheet of glass (1b)

protects the semiconductor sandwich from hail, grit blown by the wind, and wildlife. The semiconductor is also coated in an antireflective substance (1c), which makes sure that it absorbs the sunlight it needs instead of scattering it uselessly away.

When sunlight strikes the panel and is absorbed, it knocks loose electrons from some of the atoms that make up the semiconductor (1d). The semiconductor is positively charged on one side and negatively charged on the other side, which encourages all these loose electrons to travel in the same direction, creating an electric current. The contacts (1e and 1f) capture this current (1g) in an electrical circuit. The electricity PV panels (2) generate direct current (DC). Before it can be used in homes and businesses, it has to be changed into alternating current (AC) using an inverter (3). The inverted current then travels from the inverter to the building's fuse box (4), and from there to the appliances that need it. PV systems installed in homes and businesses can include a dedicated metering box (5) that measures how much electricity the panels are generating. As an incentive to generate renewable energy, energy suppliers pay the system's owner a fixed rate for every unit of

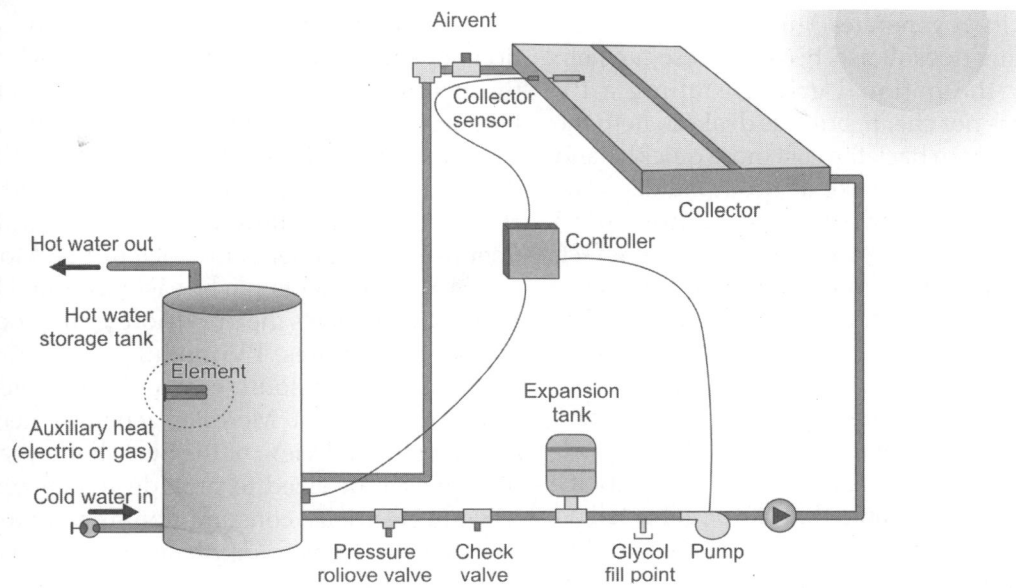

Fig. 9.20: Solar thermal system—how it works

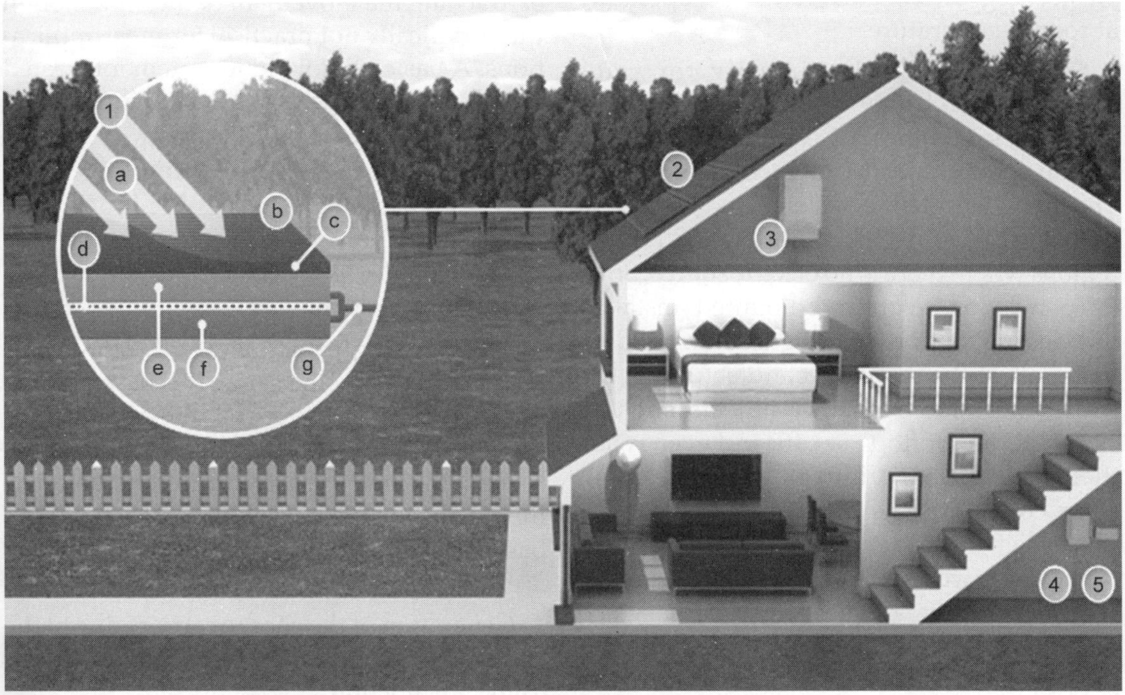

1 Solar panel detail   2 Solar panel   3 Inverter   4 Fuse box   5 Metering box

a Sunlight   b Cover glass   c Antireflective coating   d Semiconductor   e Front contact   f Back contact   g Current

Fig. 9.21: Schematic view of photovoltaic panels

electricity it generates plus a bonus for units the owner doesn't use, because these can help supply the national grid. Installing a PV system is not cheap, but this deal can help the owner to earn back the cost more quickly- and potentially even make a profit one day. The direct conversion of solar energy into electrical energy by means of the photovoltaic effect, that is, the conversion of light into electricity. Energy conversion devices which are used to convert sunlight to electricity by the use of the photovoltaic effect are called solar cells. Photovoltaic cells are made of semiconductors that generate electricity when they absorb light. As photons are received, free electricity charges are generated that can be collected on contacts applied to the surfaces of the semiconductors. Solar cells are not heat engines, and therefore do not need to operate at high temperatures, they are adopted to the weak energy flux of solar radiation, operating at room temperature.

Solar PV is used primarily for grid-connected electricity to operate residential appliances, commercial equipment, lighting and air conditioning for all types of buildings. Through stand-alone systems and the use of batteries, it is also well suited for remote regions where there is no electric source. Solar PV panels can be ground mounted, installed on building rooftops or designed into building materials at the point of manufacturing. The future will see everyday objects such as clothing, the rooftops of cars and even roads themselves turned into power-generating solar collectors.The efficiency of solar PV increases in colder temperatures and is particularly well-suited for Canada's climate. A number of technologies are available which offer different solar conversion efficiencies and pricing. Solar PV modules can be grouped together as an array of series and parallelly connected modules to provide any level of power requirements, from mere watts (W) to kilowatt (kW) and megawatt (MW) size. The size of the solar array, battery bank, and AC inverter required for a typical solar PV application depends on a number of factors, such as the amount of electricity, the amount of sunlight at the site, the number of days without backup that require, and the peak electricity demand at any given time. Sufficient battery storage can easily allow a solar PV system to operate fully independently of a utility. Most solar PV equipments can be easily checked to ensure that it meets the provisions of the code for safety purposes. PV modulës should be oriented between south-east and south-west (due south is best). Modules generally need an unobstructed view of the Sun all the year. Systems can be sized to provide 100 percent of your electricity consumption at a cottage or campsite, or as a supplement to conventional utility electricity. A tracking system can orient the solar array to maximize its electricity production throughout the day and the year by tracking the movement of the Sun, though this is typically not practical for most applications. A basic photovoltaic system integrated with the utility grid is shown in Fig. 9.22.

A photovoltaic system consists of:

- Solar cell array
- Load leveler
- Storage system
- Tracking system

The working principle of all today solar cells is essentially the same. It is based on the *photovoltaic effect*. In general, the photovoltaic effect means the generation of a potential difference at the junction of two different materials in response to visible or other radiation. The basic processes behind the photovoltaic effect are:

1. Generation of the charge carriers due to the absorption of photons in the materials that form a junction.

2. Subsequent separation of the photo-generated charge carriers in the junction.

3. Collection of the photo-generated charge carriers at the terminals of the junction.

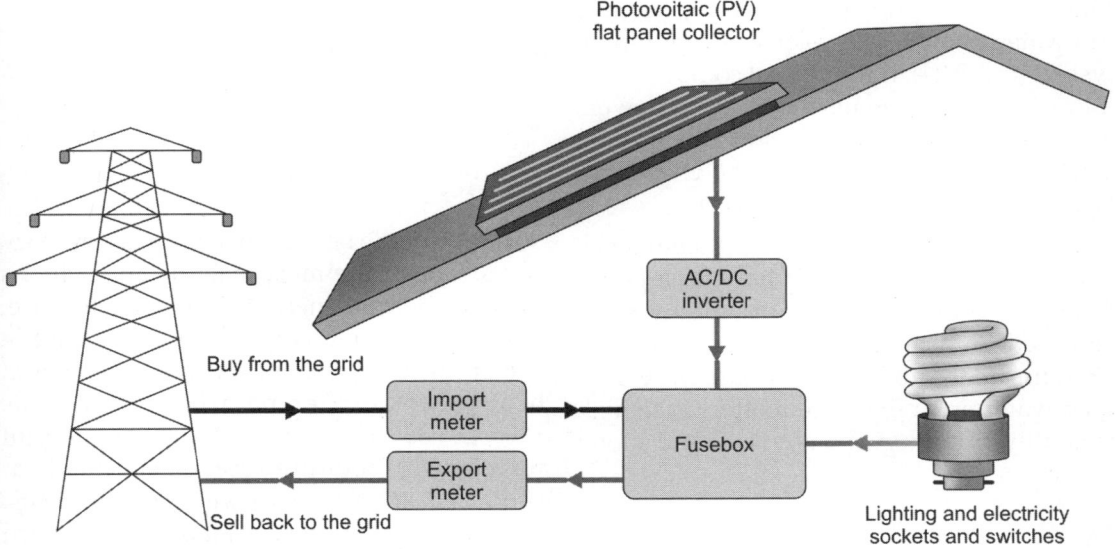

Fig. 9.22: Basic photovoltaic system integrated with power grid

The photo-voltaic effect can be described easily for p-n junction in a semiconductor. In an intrinsic semiconductor such as silicon, each one of the four valence electrons of the material atom is tied in a chemical bond, and there are no free electrons at absolute zero. Such a piece of semiconductor with one side of p-type semiconductor and the other of the n-type semiconductor is called a p-n junction. In this junction after the photons are absorbed, the free electrons of the n-side will tend to flow to the p-side, and the holes of respective deficiencies. This diffusion will create an electric field 'E' from the n-region to the p-region. The combination of n-type and p-type semiconductors thus constitutes a photovoltaic cell or solar cell. All such cells generate direct current which can be converted into alternating current if desired. Figure 9.23, demonstrates how this p-n junction provides an electric field that sweeps the electrons in one direction and the positive holes in the other.

Concentrated PV (CPV) systems concentrate sunlight on solar cells, greatly increasing the efficiency of the cells. The PV cells in a CPV system are built into concentrating collectors that use a lens or mirrors to focus the sunlight

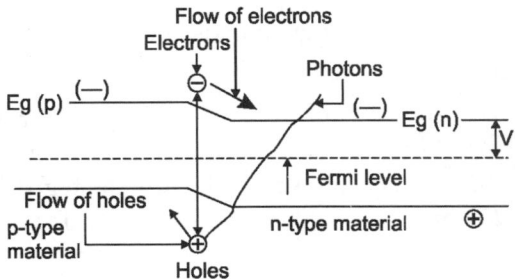

Fig. 9.23: p-n junction electric fields

onto the cells. CPV systems must track the Sun to keep the light focused on the PV cells. The primary advantages of CPV systems are high efficiency, low system cost, and low capital investment to facilitate rapid scale-up; the systems use less expensive semiconducting PV material to achieve a specified electrical output. Reliability, however, is an important technical challenge for this emerging technological approach; the systems generally require highly sophisticated tracking devices.

9.4.10 Solar Pumping

A solar-powered pump is a pump running on electricity generated by photovoltaic panels or the thermal energy available from collected sunlight as opposed to grid electricity or diesel run water pumps. The solar water pumping system (Fig. 9.24), is the perfect compact solution for pumping water from the borewell, openwell, lake, river or stream to the ground level. The pumped water can be used right away or stored in remote locations.

The solar PV modules in this system generate DC electricity which is fed into a pump through a controller. The solar modules are mounted on a manually-operated tracking structure on the ground using hardware. This solar water pumping system offers very high reliability, minimum maintenance and a long service life.

Features

- Can be used with a surface or submersible pump
- Durable and rugged construction
- Environmental friendly
- DC submersible pump—total lift up to 300 ft (92 m)
- DC surface pump—Total lift up to 70 ft (21 m)

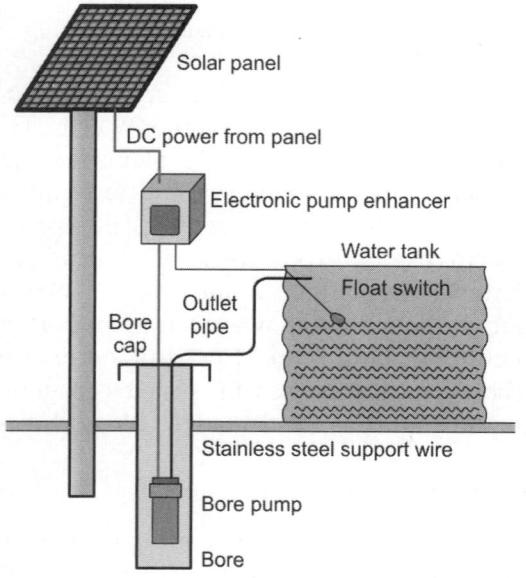

Fig. 9.24: Solar pump

Applications

- Village/drinking water supply
- Residential/industrial well pumping
- Irrigation pumps

9.4.11 SOLAR FURNACE

Solar power can be used in many extraordinary ways. One of the most majestic applications of solar thermal energy is the solar furnace. These are enormous installations that make use of solar thermal energy for extreme high heat processes (temperatures as high as 3500°C/6330°F). The technology is wonderful for high temperature researchers; the heat produced is very clean with no pollutants. There are a variety of uses for this energy, such as hydrogen fuel production, foundry applications and high temperature materials testing. Thus science can advance without enormous energy bills. Research can take place in areas previously deemed too costly or polluting to be worthwhile. The largest solar furnace currently in existence is at Odeillo in the Pyrenees-Orientales in France; it has been operational since 1970. This area boasts extremely high air quality and approximately 300 days of sunlight a year, making it a perfect spot for a solar furnace. This is also the same area in which the first solar furnace was built; this solar furnace was put in place at Mont-Louis in 1949 by Professor Felix Trombe. Odeillo and Mont-Louis are within 15 km of each other. The furnace makes use of a large parabolic reflector concentrating the Sun into an area, the size of the common cooking pot. The reflector is discrete; 63 individual flat mirrors track the Sun in unison and redirect the solar thermal energy towards the crucible. Another application of similar technology is the solar beam parabolic concentrator from solartron energy systems. It uses a single dish, tracking the Sun on two axes, to focus the solar thermal energy onto an absorber approximately 10″ × 10″. The device is significantly smaller than industrial solar furnaces, but operates on the same concept. By focusing, and thereby multiplying solar power, one may yield great efficiencies in solar thermal energy. Due to its size, the Solar beam operates at a lower temperature than the industrial furnace at Odeillo. The solar thermal energy harvested by the solar beam concentrator can be used for hot water, space heating, air conditioning or even process heat within many different types of industry. People with variety of goals can take advantage of the solar power available each and every day. This solar thermal energy is cheaper than using grid sources, all the while shielded from changes in prices or disruptions in service. Even during a cloudy day, the solar beam tracks the Sun with its mathematical algorithms and concentrates the energy diffused through the clouds. A typical solar furnace system is shown in Fig. 9.25.

9.4.12 Solar Cooking

Using stoves and ovens, we can cook foods like meat, vegetables, beans, rice, bread and fruit in just about any way. We can bake, stew, steam, fry and braise. Using a solar cooker, we can do the same things, but by using sunlight instead of gas or electricity. Sunlight isn't hot in and of itself. It's just radiation, or light waves—basically energy generated by fluctuating electric and magnetic fields. It feels warm on your skin, but that's because of what happens when those light waves hit the molecules in your skin. This interaction is similar to the concept that makes one form of solar cooker, the box cooker, generate high temperature from sunlight. At its simplest, the sunlight-to-heat conversion occurs when

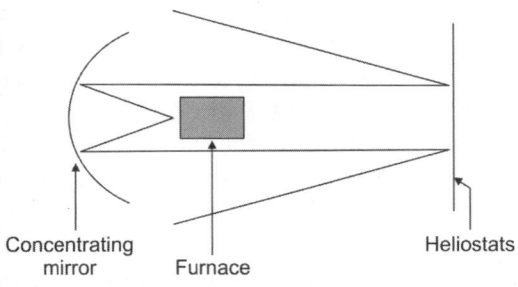

Concentrating mirror   Furnace   Heliostats

Fig. 9.25: Schematic of a solar furnace

photons (particles of light) moving around within light waves interact with molecules moving around in a substance. The electromagnetic rays emitted by the Sun have a lot of energy in them. When they strike matter, whether solid or liquid, all of this energy causes the molecules in that matter to vibrate. They get excited and start jumping around. This activity generates heat. Solar cookers use a couple of different methods to harness this heat. The *box cooker* is a simple type of solar cooker. At maybe 3 to 5 feet (1 to 1.5 meters) across, it's essentially a sun-powered oven— an enclosed box that heats up and seals in that heat. At its most basic, the box cooker consists of an open-topped box that's black on the inside, and a piece of glass or transparent plastic that sits on top. It often has several reflectors (flat, metallic or mirrored surfaces) positioned outside the box to collect and direct additional sunlight onto the glass. To cook, you leave this box in the Sun with a pot of food inside, the pot sitting on top of the black bottom of the box. When sunlight enters the box through the glass top, the light waves strike the bottom, making it scorching hot. Dark colors are better at absorbing heat, that's why the inside is black. The molecules that make up the box get excited and generate more heat. The box traps the heat, and the oven gets hotter and hotter. The effect is the same as what goes on in a standard oven: the food cooks. A *parabolic cooker* can get even hotter, up to 400 °F (204 °C), which is hot enough to fry food or bake bread. This slightly more complicated design uses curved, reflective surfaces to focus lots of sunlight into a small area. It works a lot like a stove, and it's big, sometimes up to several feet across. Both parabolic and box cookers are quite large, making them difficult to carry around. And box cookers are heavy because of the glass. A *panel cooker*, which uses parabolic reflectors positioned above a box-type oven, tends to be smaller and lighter Fig. 9.26. The cooking pot goes in a plastic bag while it cooks, which acts as a heat trap (like the transparent top on a

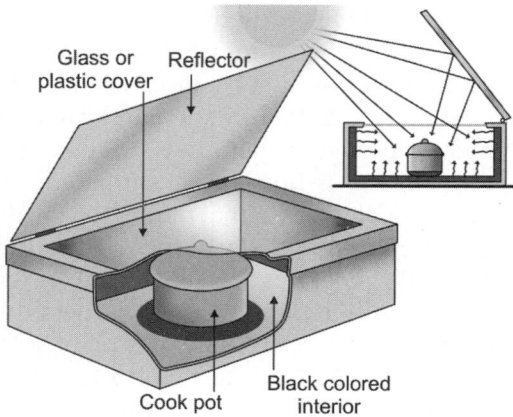

Glass or plastic cover   Reflector

Cook pot   Black colored interior

Fig. 9.26: Details of a box type solar cooker

box cooker). People sometimes use these types of cookers in camping.

9.4.13 Solar Greenhouses

The solar greenhouse differs from conventionally designed and operated greenhouses in that it does not rely on outside sources of energy for winter heating and summer cooling. A conventional greenhouse is usually all glass and pays no attention to direction of the Sun. The solar greenhouse tries to get as much solar energy as possible by using glazing on surfaces with southern exposure to permit the entry of heat and light. The northern walls are insulated to reduce heat loss at night. Vents promote natural circulation to help keep the interior cool. A storage mass, usually in the form of steel drum filled with water, helps to reduce the difference between day and night temperatures. The south side of the solar greenhouse is calculated to permit the winter Sun, but exclude the summer Sun.

The solar greenhouse relies on passive solar energy. The advantage of passive solar heat is that it can be built right into a freestanding or attached solar greenhouse. The passive design will use very little mechanical equipment, extra piping or special maintenance, as active systems often do.

In more general terms, any solar greenhouse must contain the following parts to be considered a complete passive solar heating system:

A. A collector, such as the double layer of greenhouse window glazing (glass or plastic).

B. An absorber, usually the darkened surfaces of the walls, floors, and water-filled containers inside the greenhouse,

C. A storage mass, normally the concrete, brick, and/or water that retains the heat after it has been absorbed.

D. A distribution system, which is the means of getting heat into and around the house fans, natural circulation flows.

E. A control system (or heat regulation device), such as a movable insulation used to prevent heat loss from the greenhouse at night. Roof overhangs that block the summer Sun and thermostats that activate fans are also controls. Some controls may be operated by the occupant.

The following principles briefly explain the basics of understanding how a solar greenhouse operates:

A. The sunshines through the clear areas in shortwaves.

B. These waves strike objects in the greenhouse and are reradiated as longwaves, the longwaves do not readily return through the glazing. This is known as the greenhouse effect. The greenhouse effect is similar to hot air trapped in a car on a sunny day with the windows closed. The inside air becomes warmer than the outside air .

C. Massive (heavy) objects in the greenhouse such as masonry walls, rocks, water drums, concrete, etc., absorb heat during the day and return heat to the structure at night. Pound for pound, the most efficient heat store you can get is enclosed water. It is necessary that the greenhouse have considerable mass in order to perform properly (about 2 gallons of water or 80 pounds of concrete per square foot of glazing). If this is done, the greenhouse will maintain temperature as high as 30 °F above outdoor lows in winter.

D. The warm air (80°–90 °F) from the greenhouse goes directly into the adjoining structure. This works best if there are high and low openings. The vents establish a natural air circulation system that benefits the home and the greenhouse. At night the openings can either be left open or closed, at the occupant's option, If open, the greenhouse will draw on some home heat and will keep higher temperature.

E. The partially shaded and insulated greenhouse roof will keep it warmer in winter and cooler in summer. Notice that the south face of the unit is tilted 60°, according to the latitude on which the greenhouse is located. The tilt maximizes winter Sun and reflects a large percentage of the summer Sun off the front of the greenhouse. Thus, overheating is less of a problem. How well the greenhouse keeps warm is largely determined by how well it is constructed and sealed. All cracks and joints in the greenhouse must be insulated and caulked to prevent infiltration heat losses. The greatest heat lose area for the greenhouse is through the clear wall portions. A moveable insulator to cover these areas would greatly increase the winter performance of the greenhouse.

Greenhouse provides crop cultivation under controlled environment. A greenhouse is a structure covered with a transparent material that utilizes solar radiant energy to grow plants and may have heating, cooling and ventilating equipments for temperature control. It controls:

- Soil temperature
- Air temperature
- Air humidity
- Soil moisture
- Light
- Air composition
- Exposure to rain
- Hail storm
- Protection from plant enemies
- Root medium composition

9.5 SOLAR POWER PLANTS IN INDIA

India is densely populated and has high solar *isolation*, an ideal combination for using solar power in India. India is already a leader in *wind power generation*. In the solar energy sector, some large projects have been proposed, and a 35,000 km² area of the *Thar desert* has been set aside for solar power projects, sufficient to generate 700–2100 GW. Many states have contributed significantly in solar arena but needless to say the solar state of India "Gujarat" still leads the solar sector and is way-way ahead due to its much favorable Gujarat State Solar Policy of phase I and II among all the states by housing almost 710 MW of solar power plants out of about 1060 MW of solar power installed till August 2012 in India. Of the total in renewable, India has witnessed an unprecedented growth in solar installations over past 1 year from showcasing approximately 56 MW in 2011 to around 1060 MW of solar power in 2012 (Table 9.1).

Many states have contributed significantly in solar arena but needless to say that the solar state of India "Gujarat" still leads the solar sector and is way ahead due to its much favorable Gujarat State Solar Policy of Phase I and II among all the states by housing almost 710 MW of solar power plants out of about 1060 MW of solar power plants installed till August 2012 in India. India will build the world's largest solar plant to generate 4,000 MW from sunlight near the Sambhar Lake in Rajasthan that will sell electricity at an estimated rate of ₹ 5.50 per unit. The proposed solar project's capacity is about three times India's total solar power capacity and comparable with coal-fired ultra mega power projects of Tata power and Reliance power. Being the first project of this scale anywhere in the world this project is expected to set a trend for large scale solar power development in the world. It would be set up and run by a joint venture of five public sector utilities Bhel, Powergrid corporation of India, Solar Energy Corporation of India, Hindustan Salts limited and Rajasthan

Table 9.1 Statewise distribution of solar energy till August 2012

State	Capacity in MW
Gujarat	709.54
Rajasthan	198.7
Andhra Pradesh	21.8
Chattisgarh	4.0
Punjab	9.3
Tamil Nadu	15.1
Haryana	7.8
Uttar Pradesh	12.4
Jharkhand	16.0
Uttarakhand	5.1
Karnataka	14.0
West Bengal	2.1
Madhya Pradesh	7.4
Maharashtra	20.0
Delhi	2.5
Odisha	13.0
Lakshadweep	0.8
Andaman and Nicobar	0.1
Total	**1059.64 MW**

Electronics and Instruments Limited, the statement said.

The first phase of the project, which would be 1000 MW is expected to be commissioned in 2016. "Based on the experience gained during implementation of the first phase of the project, the remaining capacity would be implemented through a variety of models," it said. The project would have to acquire 23000 acre of land out of which 18000 acre would be provided by Hindustan Salts limited. The tariff is expected to be competitive.

9.6 SITE SELECTION CRITERIA FOR SOLAR POWER PLANTS

The following are the important key factors to be considered before setting up a solar captive power plant:

1. **Load supported:** They are usually the largest single influence on the size and cost of a PV system. A PV system designer can minimize a PV system's cost by efficiently using the energy available. The first step is to estimate the average daily power demand of each load to be used. It is important to note that one should be thorough, but realistic, while estimating the load. A 25% safety factor can cost a great deal of money.

2. **Type of loads:** While estimating the load, it is necessary to calculate for both AC and DC loads.

3. **Hours of operation:** An hour of operation is an important factor. This value helps us determine the exact consumption of electricity (kWh) of each appliance. Calculating this value will help the designer in the first level assessment of the size of the solar system that will be needed to power the site under consideration. More importantly the time of operation during the day will enable a designer to do a more accurate sizing of the PV system. For example, a refrigerator runs for 24 hours in a day. Other appliances like washing machine will run 2 hours a day in the afternoon. So a PV system can be designed to supplement grid electricity by providing electricity during peak hours. It is also possible, if the user wishes, to design a complete solar system to provide

electricity throughout the day with battery backup.

4. Days of autonomy: Autonomy refers to the number of days a battery system will provide a given load without being recharged by the PV array or another source. General weather conditions determine the number of "no Sun" days which is a significant variable in determining the autonomy. Local weather patterns and microclimates must also be considered. Cross-check weather sources because errors in solar resource estimates can cause disappointing system performance.

The most important factors in determining an appropriate autonomy for a system are the size and type of loads that the system services. The general range of autonomy is as follows:

- 2–3 days for non-essential uses or systems with a generator back up
- 5–7 days for critical loads with no other power source

5. Space available: For setting up 1 kW SPV system without batteries, the required shade free area is 100 sqft.

Knowing the installation site before designing the system is recommended for good planning of component placement, wire runs, shading, and terrain peculiarities. The primary requirement in selecting the space is that, it should be shade free. Shading critically affects a PC array's performance. Even a small amount of shade on a PV panel can reduce its performance significantly. For this reason, minimizing shading is much more important in PV system design. Carefully determining solar access or shade-free location is fundamental to cost effective PV performance. When a site is selected, be sure that the following parameters are met and tasks completed:

- Be sure that the array is not shaded from 9 AM to 3 PM.
- Identify the obstacles, if any, that shade the array between 9 AM and 3 PM
- Make recommendations to eliminate any shading, move the array to avoid shading, or increase the array size to offset losses due to shading.

- The vital consideration in designing solar greenhouses is the availability of direct sunlight. In order to learn about the availability of direct sunlight, it is necessary to understand how the Sun moves across the sky. The sun rises in the east, follows an arc through the southern sky, and sets in the west. The Sun travels a different path in the winter than in summer, thus solar radiation falls on the Earth's surfaces (and on your greenhouse) at different angles during each season. During the cold part of the year the Earth's tilt causes the Sun to appear low in the southern sky, rather than overhead as in summer. Since the Sun rises and sets in the southern sky, east and west walls can see very little sunlight in the phase of winter. The north wall loses out completely. But a double-glazed south-facing window gains more solar heat in eight hours than it loses over the entire 24-hour day.

9.7 ADVANTAGES AND DISADVANTAGES

Advantages

- **Solar power helps to slow/stop global warming:** Global warming threatens the survival of human society, as well as the survival of countless species. Luckily, decades (or even centuries) of research have led to efficient solar panel systems that create electricity without producing global warming pollution. Solar power is now very clearly one of the most important solutions to the global warming crisis.

- **Solar power saves society, billions or trillions of dollars:** Even long before society's very existence is threatened by global warming, within the coming decades, global warming is projected to cost society trillions of dollars, if left unabated. So, even ignoring the very long-term threat of societal suicide, fighting global warming with solar power will likely save society billions or even trillions of dollars.

- **Solar power provides energy reliability:** The rising and setting of the Sun is extremely consistent. All across the world, we know exactly when it will rise and set every day of the year. While clouds may be a bit less predictable, we do also have fairly good seasonal and daily projections for the amount of sunlight that will be received in different locations. All in all, this makes solar power an extremely reliable source of energy

- **Solar power provides energy security.** On top of the above reliability benefits, no one can go and buy the Sun or turn sunlight into a monopoly. Combined with the simplicity of solar panels, this also provides the notable solar power advantage of energy security, something the US military has pointed out for years, and a major reason why it is also putting a lot of its money into the development and installation of solar power systems.

- **Solar power provides energy independence:** Similar to the energy security boost, solar power provides the great benefit of *energy independence*. Again, the "fuel" for solar panels cannot be bought or monopolized. It is free for all to use. Once you have solar panels on your roof, you have an essentially independent source of electricity that is all yours. This is important for individuals, but also for cities, counties, states, and even companies.

- **Solar power creates jobs:** As a source of energy, solar power is a job-creating powerhouse. Money invested in solar power creates two to three times more jobs than money invested in coal or natural gas. Solar jobs come in many forms, from manufacturing, installing, monitoring and maintaining solar panels, to research and design, development, cultural integration, and policy jobs.

Disadvantages

Solar power disadvantages are actually not so plentiful. In fact, there's only one notable disadvantage to solar power that I can think of. That disadvantage is that the Sun doesn't shine 24 hours a day. When the Sun goes down or is heavily shaded, solar PV panels stop producing electricity. If we need electricity at that time, we have to get it from some other source. In other words, we couldn't be 100% powered by solar panels. At the very least, we need batteries to store electricity produced by solar panels for use sometime later.

However, there are a couple of key things to note regarding this solar power disadvantage. Firstly, the Sun actually does shine when we need electricity most. As humans (not vampires), our days more or less follow the movement of the Sun. Society more or less wakes up when the Sun rises. At the time of the sun's greatest height and visibility, humans tend to be most active. At this time, we are of course using much more electricity than in the middle of the night, so electricity is in greater demand. (This also makes electricity more expensive in the middle of the day, making electricity produced from solar panels more valuable).

10
Environmental Aspects and Protocols

10.1 SUSTAINABLE DEVELOPMENT

Sustainable development has been defined in many ways, but the most frequently quoted definition is from *our common future*, also known as the Brundtland report:

Sustainable development means "the development that meets the needs of the present without compromising the ability of future generations to meet their own needs."

All definitions of sustainable development require that we see the world as a system that connects space; and a system that connects time. Sustainable development ties together concern for the carrying capacity of natural systems with the social challenges faced by humanity. As early as the 1970s, "sustainability" was employed to describe an economy "in equilibrium with basic ecological support systems." Ecologists have pointed to *The Limits to Growth*, and presented the alternative of a "steady state economy" in order to address environmental concerns.

In 1987, the United Nations released the Brundtland report, which included what is now one of the most widely recognised definitions. According to the same report, the definition contains within it two key concepts:

- The concept of 'needs', in particular the essential needs of the world's poor, to which overriding priority should be given; and

- The idea of limitations imposed by the state of technology and social organization on the environment's ability to meet present and future needs.

The concept of sustainable development is rooted in this sort of systems thinking. It helps us understand ourselves and our world. It is an organizing principle for human life on a finite planet. Sustainable development ties together concern for the carrying capacity of natural systems with the social and economic challenges faced by humanity. As early as the 1970s, 'sustainability' was employed to describe an economy "in equilibrium with basic ecological support systems." Scientists in many fields have highlighted the *limits to growth*, and economists have presented alternatives, for example a 'steady state economy', to address concerns over the impacts of expanding human development on the planet. The concept of sustainable development has in the past most often been broken out into three constituent domains: environmental domain, economic domain and social domain (Fig. 10.1).

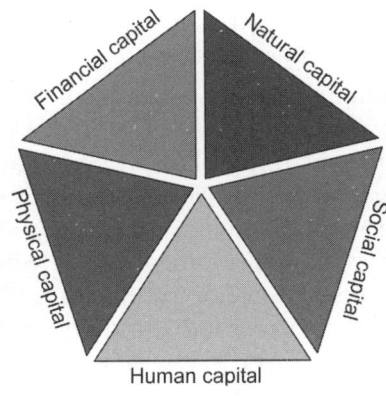

Fig. 10.1: Sustainable development

10.1.1 History of the Concept

The concept of sustainable development was originally synonymous with that of sustainability and is often still used in that way. Both terms derive from the older forestry term "sustained yield", which in turn is a translation of the German term "nachhaltiger Ertrag" dating from 1713. According to different sources, the concept of sustainability in the sense of a balance between resource consumption and reproduction was however applied to forestry already in the 12th to 16th century. 'Sustainability' is a semantic modification, extension and transfer of the term 'sustained yield'. This had been the doctrine and, indeed, the 'holy grail' of foresters all over the world for more or less two centuries. The essence of 'sustained yield forestry' was described for example by William A. Duerr, a leading American expert on forestry: "To fulfill our obligations to our descendents and to stabilize our communities, each generation should sustain its resources at a high level and hand them along undiminished. The sustained yield of timber is an aspect of man's most fundamental need: to sustain life itself." A fine anticipation of the Brundtland formula.

Technically it started in 1969 with the national environmental policy act (NEPA). It was centered to "foster and promote the general welfare, to create and maintain conditions under which man and nature can exist in productive harmony and fulfill the social, economic and other requirements of present and future generations."

Stockholm meeting was a big event of the 1970's, where the UN meet on the Human Environment in Stockholm, Sweden. This meeting is where developed countries voiced concern about the environmental implications of worldwide development, while countries that were still developing raised their own continuing need for industrial development. Therefore the idea of "sustainable development" was born out of an effort to find an understanding between the development requirements of the countries in the Southern hemisphere and the conservation demands of the developed states in the North. The meeting increased awareness of the world environmental issues and set in motion events which lead to the general acknowledgment of the concept of "sustainable development" as a method of realizing the development requirements of all folks without having to sacrifice the Earth's capacity to sustain life.

Out of the Stockholm meeting, the UN environmental program was formed to license the concept of environmentally-sound development. Based in Nairobi, Kenya, UNEP provided the UN with an agency to look at the planet's growing environmental and development issues with a view to recommending to nation-wide states and world bodies on suitable actions. The work of the UNEP helped launch, among other stuff, the World Environmental Academic Programme (IEEP) in 1975 and the World Conservation Technique in 1980.

The first use of the term "sustainable" in the modern sense was by the Club of Rome in March 1972 in its epoch-making report on the "Limits to Growth", written by a group of scientists led by Dennis and Donella Meadows of the Massachusetts Institute of Technology. Describing the desirable "state of global equilibrium", the authors used the word "sustainable": "We are searching for a model output that represents a world system that is:

1. Sustainable without sudden and uncontrolled collapse; and

2. Capable of satisfying the basic material requirements of its entire people.

On December 1983, Gro Harlem Brundtland, the PM of Norway, was asked by the Secretary General of the UN to chair a special independent commission, the World Commission on Environment and Development called the WCED. Its mission: to examine the vital environmental and development issues around the planet and fashion practical suggestions to address them. A second target was to bolster global cooperation on environmental and development issues. And, eventually, the commission wanted to raise the level of knowledge of and dedication to viable development on the side of people, associations, companies and govertments.

When the commission was organised, some wanted it to be restricted to environmental problems only. Nevertheless they suspected that environmental quality and supportable development were two inseparable ideas which should be linked in compound, a world technique. With this established, the commission therefore outlined viable development as "development that fulfills the requirements of the present without risking the capability of generations to come to meet their own needs."

A further end result of the WCED report, was the UN meeting on environment and development. A two year series of preparatory conferences finished in the Earth Summit in Rio de Janeiro, June 1992. This marked the second meeting of global leaders to talk about environmental and development issues and was significantly bigger than its precedent the Stockholm meeting held twenty years before. The Earth Summit was bigger not just in the level of collaboration by the states of the Earth Summit, but also in the extent of issues it tried to address. Over a hundred heads of state and central authority attended the Earth Summit and 170 countries sent delegations. As an element of the Earth Summit, countrywide leaders had a chance to sign world conventions on global warming and biodiversity, a "declaration of environment and development" and an agenda for the 21st century, which looked to create a strong effort to teach folks about the state of both environment and development, and to help them to make calls which can lead to supportability.

On the anniversary of the Earth Summit in June 1993, President Clinton signed an executive order creating the president's council on tolerable development. In his address to the country he revealed. "Every country faces a challenge to spot and implement policies that may meet the requirements of the present without sacrificing the future. America will face that test with the assistance of this Council and the concepts and experience that its members bring to this crucial task."The twenty-five member Council built new partnerships among delegates from industry, administration (including US the Cupboard members) and environmental, work and civil rights associations to develop bold fresh approaches to integrate business and environmental policies. Their charge: to seriously change the president's vision of tolerable development into a concrete plan.

Their first work concluded in Febuary 1996, with the publishing of their report titled, "sustainable America: a new understanding for wealth, opportunity, and a good environment for the future." In January 1997, the commission issued its 2nd major report titled, "building on understanding: a progress report on supportable America." Secretary general of UNCED, Maurice powerful, summarised *Agenda Twenty-one* as, a "program of action for a viable future for the human family and a primary step towards making sure the world will change into a more just, secure and wealthy habitat for all humanity."

Agenda Twenty-one called on all nations of the Earth to do a complete process of planning and action to reach supportability. As well as worldwide agenda, this document also detailed a role for towns and counties. Chapter twenty-eight of *Agenda Twenty-one* (known as local *Agenda Twenty-one*) states: "local authorities construct, operate and maintain commercial, social and environmental sub-structure, oversee planning processes, build local environmental policies and rules, and … as the level of state nearest the folk, they play a crucial role in teaching, mobilizing and replying to the general public to push viable development."

10.2 CLIMATE CHANGE

10.2.1 Introduction

A long term change in the Earth's climate, especially a change due to an increase in the average atmospheric temperature is called climate change. Climate means the average pattern in which weather varies in time. The climate of region depends on the presence or absence of water, the reflection of solar radiation or albedo, the ability to transfer water to the atmosphere (evaporation), the capacity to store heat, topography and texture of the region. Although they constitute only a fraction of the total land area of the Earth, metropolitan areas emit the bulk of all air pollutants. These air pollutants influence temperature, visibility and precipitation as well as other climatic elements.

Climatic change is a significant and lasting change in the statistical distribution of weather patterns over periods ranging from decades to millions of years. It may be a change in average weather conditions, or in the distribution of weather around the average conditions, (i.e. more or fewer extreme weather events). Climate change is caused by factors that include oceanic processes (such as oceanic circulation), biotic processes, variations in solar radiation received by Earth, plate tectonics and volcanic eruptions, and human-induced alterations of the natural world; these human-induced effects are currently causing global warming, and "climate change" is often used to describe human-specific impacts.

Things get cooler, or warmer, or wetter, or drier. Climate change refers to any long term alteration or disruption of typical seasonal weather patterns. Climate change is the overall change in our environment. It is a long-term change in the statistical distribution of weather patterns over periods of time that range from decades to millions of years. It may be a change in the average weather conditions or a change in the distribution of weather events with respect to an average, for example, greater or fewer extreme weather events. Climate change may be limited to a specific region, or may occur across the whole Earth. It refers to any distinct change in measures of climate lasting for a long period of time. In other words, "climate change" means major changes in temperature, rainfall, snow, or wind patterns lasting for decades or longer. Climate change may result from:

- Natural factors, such as changes in the sun's energy or slow changes in the earth's orbit around the sun (Fig. 10.2).
- Natural processes within the climate system (e.g. changes in ocean circulation).
- Human activities that change the atmosphere's makeup (e.g. burning fossil fuels) and the land surface (e.g. cutting down forests, planting trees, building developments in cities and suburbs, etc.).

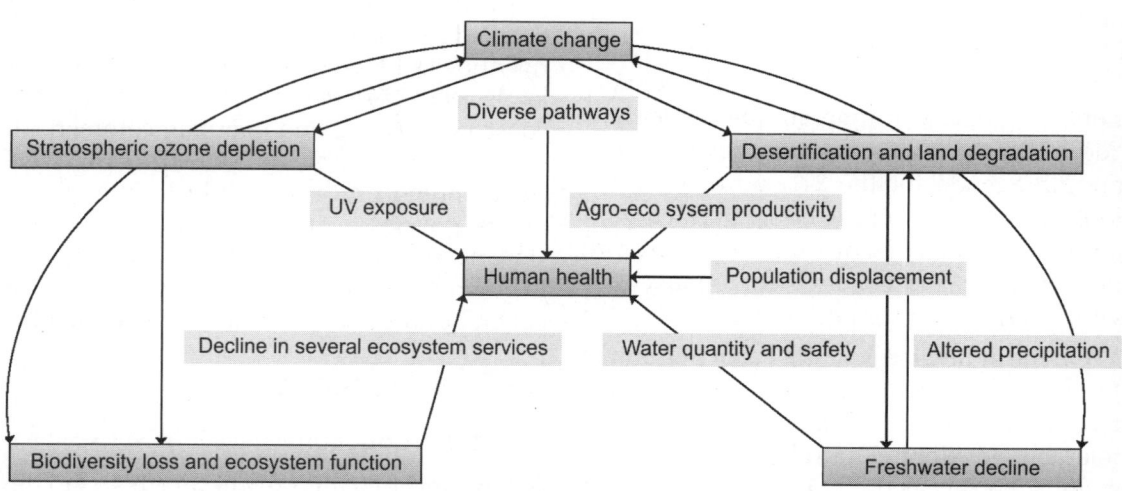

Fig.10.2: Climate change

The United Nations Framework Convention on Climate Change (UNFCCC) defines climatic change as a change of climate which is attributed directly or indirectly to human activity that alters the composition of the global atmosphere and which is in addition to natural climate variability observed over comparable time period. Projections of future climate change are derived from a series of experiments made by computer based global climate models. These are calculated based on factors like future population growth and energy use. Climatologists of the intergovernmental panel on climate change (IPCC) have reviewed the result of several experiments in order to estimate changes in climate in the course of this century. These studies have shown that in the near future the global mean surface temperature will rise by 1.4°–5.8 °C. this 'warming' will be greatest over land areas, and at high latitudes. The projected rate of warming is greater than has occurred in the last 10000 years. The frequency of weather extremes is likely to increase, leading to floods or drought. There will be fewer cold spells but more heat waves. The frequency and intensity of the El Nino is likely to increase.

10.2.2 Sources

Internal Forcing Mechanisms

Natural changes in the components of earth's climate system and their interactions are the cause of internal climate variability, or "internal forcings." Scientists generally define the five components of earth's climate system to include atmosphere, hydrosphere, cryosphere, lithosphere (restricted to the surface soils, rocks, and sediments), and biosphere.

1. **Ocean variability:** The ocean is a fundamental part of the climate system, some changes in it occurring at longer timescales than in the atmosphere. Short-term fluctuations (years to a few decades) such as the El Nino-Southern oscillation, the Pacific decadal oscillation, the North Atlantic oscillation, and the Arctic oscillation, represent climate variability rather than climate change. On longer time scales, alterations to ocean processes such as thermohaline circulation play a key role in redistributing heat by carrying out a very slow and extremely deep movement of water and the long-term redistribution of heat in the world's oceans.

2. **Life:** Life affects climate through its role in the carbon and water cycles and such mechanisms as albedo, evapo-transpiration, cloud formation, and weathering.

External Forcing Mechanisms

1. **Orbital variations:** Slight variations in earth's orbit lead to changes in the seasonal distribution of sunlight reaching the Earth's surface and how it is distributed across the globe. There is very little change to the area-averaged annually averaged sunshine; but there can be strong changes in the geographical and seasonal distribution. The three types of orbital variations are variations in earth's eccentricity, changes in the tilt angle of earth's axis of rotation, and precession of earth's axis. Combined together, these produce Milankovitch cycles which have a large impact on climate and are notable for their correlation to glacial and interglacial periods, their correlation with the advance and retreat of the Sahara, and for their appearance in the stratigraphic record.

2. **Solar output:** The sun is the predominant source for energy input to the earth. Both long- and short-term variations in solar intensity are known to affect global climate. Three to four billion years ago the Sun emitted only 70% as much power as it does today. If the atmospheric composition had been the same as today, liquid water should not have existed on earth. However, there is evidence for the presence of water on the early earth, in the Hadean and Archeaneons, leading to what is known as the faint young sun paradox. Interestingly, a 2010 study suggests, "that the effects of solar variability on temperature throughout the atmosphere may be contrary to current expectations."

3. **Volcanism:** Volcanic eruptions release gases and particulates into the atmosphere. Eruptions large enough to affect climate occur on average several times per century, and cause cooling for a period of a few years. When a volcano erupts it throws out large volumes of sulphur dioxide (SO_2), water vapour, dust, and ash into the atmosphere. Although the volcanic activity may last only a few days, yet the large volumes of gases and ash can influence climatic patterns for years. Millions of tonnes of sulphur dioxide gas can reach the upper levels of the atmosphere (called the stratosphere) from a major eruption. The gases and dust particles partially block the incoming rays of the sun, leading to cooling. Sulphur dioxide combines with water to form tiny droplets of sulphuric acid. These droplets are so small that many of them can stay aloft for several years. They are efficient reflectors of sunlight, and screen the ground from some of the energy that it would ordinarily receive from the sun.

4. **Greenhouse gases and their sources:** Carbon dioxide is undoubtedly, the most important greenhouse gas in the atmosphere. Changes in land use pattern, deforestation, land clearing, agriculture, and other activities have all led to a rise in the emission of carbon-dioxide. Methane is another important greenhouse gas in the atmosphere. About ¼ of all methane emissions are said to come from domesticated animals such as dairy cows, goats, pigs, buffaloes, camels, horses, and sheep. These animals produce methane during the cud-chewing process. Methane is also released from rice or paddy fields that are flooded during the sowing and maturing periods. When soil is covered with water it becomes anaerobic or lacking in oxygen. Under such conditions, methane-producing bacteria and other organisms decompose organic matter in the soil to form methane.

Deforestation by means of cutting down and burning these tropical rainforests usually pave the way for agriculture and industry which often produce even more CO_2. Forests reduce greenhouse gas emissions to combat global warming. 20% of global greenhouse gas emissions result from deforestation and degradation of forest, more than all the world's cars, trucks, ships and planes combined. Fossil fuels release carbon dioxide into the atmosphere contributing to global warming and climate change. Forest alleviates this change by converting carbon dioxide to carbon during photosynthesis. The world's forests contain about 125% of the carbon found in the atmosphere. This carbon is stored in the form of wood and vegetation through "carbon sequestration".

5. **Agriculture:** Agriculture has been shown to produce significant effects on climate change, primarily through the production and release of greenhouse gases such as carbon dioxide, methane, and nitrous oxide. Another contributing cause of climate change is when agriculture alters the earth's land cover, which can change its ability to absorb or reflect heat and light. Land use change such as deforestation and desertification, together with use of fossil fuels, are the major anthropogenic sources of carbon dioxide.

6. **Human activities:** Human activities contribute to climate change by causing changes in earth's atmosphere in the amounts of greenhouse gases, aerosols (small particles), and cloudiness. The largest known contribution comes from the burning of fossil fuels, which releases carbon dioxide gas to the atmosphere. Greenhouse gases and aerosols affect climate by altering incoming solar radiation and out-going infrared (thermal) radiation that are part of earth's energy balance. Changing the atmospheric abundance or properties of these gases and particles can lead to a warming or cooling of the climate system.

10.2.3 Effects

The potential future effects of global climate change include more frequent wildfires, longer periods of drought in some regions and an increase in the number, duration and intensity of tropical storms. Global climate change has already had observable effects on the environment. Glaciers have shrunk, ice on rivers and lakes is breaking up earlier, plant and animal ranges have shifted and trees are flowering sooner.

1. Heat-trapping gases emitted by power plants, automobiles, deforestation and other sources are warming up the planet. High temperatures are to blame for an increase in heat-related deaths and illness, rising seas, increased storm intensity, and many of the other dangerous consequences of climate change.

2. Rising temperatures are changing weather and vegetation patterns across the globe, forcing animal species to migrate to new, cooler areas in order to survive. The rapid nature of climate change is likely to exceed the ability of many species to migrate or adjust.

3. Climate change is intensifying the circulation of water on, above and below the surface of the earth–causing drought and floods to be more frequent, severe and widespread. Higher temperatures increase the amount of moisture that evaporates from land and water, leading to drought in many areas. Lands affected by drought are more vulnerable to flooding once rain falls. As temperatures rise globally, droughts will become more frequent and more severe, with potentially devastating consequences for agriculture, water supply and human health.

4. Hot temperatures and dry conditions also increase the likelihood of forest fires.

5. As the earth heats up, sea level rises because warmer water takes up more room than colder water, a process known as thermal expansion. Melting glaciers compound the problem by dumping even more fresh water into the oceans. Rising seas threaten to inundate low-lying areas and islands, threaten dense coastal populations, erode shorelines, damage property and destroy ecosystems such as mangroves and wetlands that protect coasts against storms.

6 Climate change will cause hurricanes and tropical storms to become more intense— lasting longer, unleashing stronger winds, and causing more damage to coastal ecosystems and communities.

7. The decline in Arctic sea ice, both in extent and thickness, over the last several decades is further evidence for rapid climate change. Sea ice is frozen seawater that floats on the ocean surface. It covers millions of square miles in the polar regions, varying with the seasons. In the Arctic, some sea ice remains year after year, whereas almost all Southern Ocean or Antarctic sea ice melts away and reforms annually. Satellite observations show that Arctic sea ice is now declining at a rate of 11.5 percent per decade, relative to the 1979 to 2000 average.

8. **Vegetation:** A change in the type, distribution and coverage of vegetation may occur given a change in the climate. Some changes in climate may result in increased precipitation and warmth, resulting quaternary glaciations in improved plant growth and the subsequent sequestration of airborne CO_2. A gradual increase in warmth in a region will lead to earlier flowering and fruiting times, driving a change in the timing of life cycles of dependent organisms. Larger, faster or more radical changes, however, may result in vegetation stress, rapid plant loss and desertification in certain circumstances. An example of this occurred during the Carboniferous rainforest collapse (CRC), an extinction event 300 million years ago.

9. **Pollen analysis:** Palynology is the study of contemporary and fossil palynomorphs, including pollen. Palynology is used to infer the geographical distribution of plant species, which vary under diffe-

rent climate conditions. Different groups of plants have pollen with distinctive shapes and surface textures, and since the outer surface of pollen is composed of a very resilient material, they resist decay. Changes in the type of pollen found in different layers of sediment in lakes, bogs, or river deltas indicate changes in plant communities. These changes are often a sign of a changing climate. As an example, palynological studies have been used to track changing vegetation patterns throughout the and especially since the last glacial maximum.

10. **Precipitation:** Past precipitation can be estimated in the modern era with the global network of precipitation gauges. Surface coverage over oceans and remote areas is relatively sparse, but, reducing reliance on interpolation, satellite data has been available since 1970. Quantification of climatological variation of precipitation in prior centuries and epochs is less complete but approximated using proxies such as marine sediments, ice cores, cave stalagmites, and tree rings. Climatological temperatures substantially affect precipitation. For instance, during the last glacial maximum of 18,000 years ago, thermal-driven evaporation from the oceans onto continental landmasses was low, causing large areas of extreme desert, including polar deserts (cold but with low rates of precipitation). In contrast, the world's climate was wetter than today near the start of the warm Atlantic period of 8000 years ago.

10.3 GREEN HOUSE EFFECT

10.3.1 Introduction

It is the trapping of the sun's warmth in a planet's lower atmosphere, due to the greater transparency of the atmosphere to visible radiation from the sun than to infrared radiation emitted from the planet's surface. The greenhouse effect is a process by which thermal radiation from a planetary surface is absorbed by atmospheric greenhouse gases, and is re-radiated in all directions. Since part of this re-radiation is back towards the surface and the lower atmosphere, it results in an elevation of the average surface temperature above what it would be in the absence of the gases (Fig.10.3).

There are two common meanings of the term "greenhouse effect". There is a "natural" greenhouse effect that keeps the earth's climate warm and habitable. There is also the "man-made" greenhouse effect, which is the enhancement of earth's natural greenhouse effect by the addition of greenhouse gases from the burning of fossil fuels (mainly petroleum, coal, and natural gas). The industrial activities that our modern civilization depends upon have raised atmospheric carbon dioxide levels from 280 parts per million to 379 parts per million in the last 150 years. The panel also concluded; there's more than 90 percent probability that human-produced greenhouse gases such as carbon dioxide, methane and nitrous oxide have caused much of the observed increase in Earth's temperature over the past 50 years.

Gases that contribute to the greenhouse effect include:

1. **Water vapour:** The most abundant greenhouse gas, but importantly, it acts as a feedback to the climate. Water vapour increases as the earth's atmosphere warms, but so does the possibility of clouds and precipitation, making these some of the most important feedback mechanisms to the greenhouse effect.

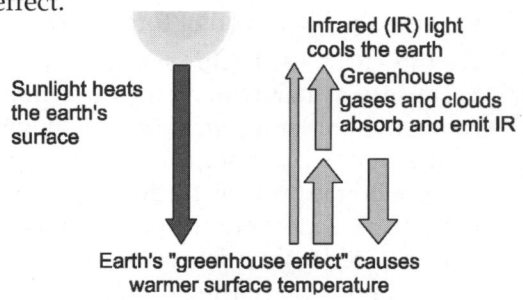

Sunlight heats the earth's surface

Infrared (IR) light cools the earth

Greenhouse gases and clouds absorb and emit IR

Earth's "greenhouse effect" causes warmer surface temperature

Fig.10.3: Greenhouse effect

2. **Carbon dioxide (CO_2):** A minor but very important component of the atmosphere, carbon dioxide is released through natural processes such as respiration and volcanic eruptions and through human activities such as deforestation, land use changes, and burning fossil fuels. Humans have increased atmospheric CO_2 concentration by one-third since the industrial revolution began. This is the most important long-lived "forcing" of climate change.

3. **Methane:** A hydrocarbon gas produced both through natural sources and human activities, including the decomposition of wastes in landfills, agriculture, and especially rice cultivation, as well as ruminant digestion and manure management associated with domestic livestock. On a molecule-for-molecule basis, methane is a far more active greenhouse gas than carbon dioxide, but also one which is much less abundant in the atmosphere.

4. **Nitrous oxide:** A powerful greenhouse gas produced by soil cultivation practices, especially the use of commercial and organic fertilizers, fossil fuel combustion, nitric acid production, and biomass burning.

5. **Chlorofluorocarbons (CFCs):** Synthetic compounds entirely of industrial origin used in a number of applications, but now largely regulated in production and release to the atmosphere by international agreement for their ability to contribute to destruction of the ozone layer. They are also greenhouse gases.

10.3.2 Sources

1. **Electricity production** (33% of 2011 greenhouse gas emissions): Electricity production generates the largest share of greenhouse gas emissions. Over 70% of our electricity comes from burning fossil fuels, mostly coal and natural gas.

2. **Transportation** (28% of 2011 greenhouse gas emissions): Greenhouse gas emissions from transportation primarily come from burning fossil fuel for our cars, trucks, ships, trains, and planes. Over 90% of the fuel used for transportation is petroleum based, which includes gasoline and diesel.

3. **Industry** (20% of 2011 greenhouse gas emissions): Greenhouse gas emissions from industry primarily come from burning fossil fuels for energy as well as greenhouse gas emissions from certain chemical reactions necessary to produce goods from raw materials.

4. **Commercial and residential** (11% of 2011 greenhouse gas emissions): Greenhouse gas emissions from businesses and homes arise primarily from fossil fuels burned for heat, the use of certain products that contain greenhouse gases, and the handling of waste.

5. **Agriculture** (8% of 2011 greenhouse gas emissions): Greenhouse gas emissions from agriculture come from livestock such as cows, agricultural soils, and rice production.

6. **Land use and forestry** (offset of 14% of 2011 greenhouse gas emissions): Land areas can act as a sink (absorbing CO_2 from the atmosphere) or a source of greenhouse gas emissions.

10.3.3 Effects

1. When the weather gets warmer, evaporation from both land and sea increases. This can cause drought in areas of the world where the increased evaporation is not compensated by more precipitation. In some regions of the world this will result in crop failure and famine especially in areas where temperatures are already high. The extra water vapour in the atmosphere will fall again as extra rain, which can cause flooding in other places of the world.

2. Worldwide, glaciers are shrinking rapidly at present. Ice appears to be melting faster than previously estimated. In areas that are dependent on meltwater from mountain areas, this can cause drought and lack of domestic water supply.

3. The warmer climate will probably cause more heatwaves, more violent rainfall and also an increase in the number and/or severity of storms.

4. Sea level rises because of melting ice and snow and because of the thermal expansion of the sea (water expands when warmed). Areas that are just above sea level now, may become submerged.

10.3.4 Control

Besides, the burning of fossil fuels like petroleum has greatly raised the levels of carbon dioxide. For this, alternative sources of energy such as geothermal energy, biomass energy and wind energy should be utilized so that the environment is not affected by this combustion activity.

The causes of greenhouse effect include deforestation. Our population is increasing and to accommodate to needs for amenities and economic stability, more and more forests are being cut down or burnt. This in turn heightens the percentage of carbon dioxide as there are not enough plants to absorb carbon dioxide for photosynthesis. To overcome this problem, the authority has to implement regulations to address this issue by constraining the law and penalizing those who do not comply with the law. Aforestation will lead to restoration of forest areas which inturn will keep atmospheric conditions in check. More trees means more absorption of CO_2. Moreover, electrical appliances are a contributor to this issue too. Electrical appliances like refrigerator and air conditioner releases chlorofluorocarbons (CFCs) which have added to the greenhouse effects. Therefore, hydrofluorocarbons (HFCs) should substitute CFCs as these gases break down in the atmosphere and return to the Earth. CFC's are thus banned.

10.4 GLOBAL WARMING

10.4.1 Introduction

Atmospheric carbon dioxide, and a host of other gases, such as methane, nitrous oxide, tropospheric ozone and chlorofluorocarbons (CFCs), transmit short wavelength radiations from the sun; whilst at the same time absorb long wavelength infrared radiations from the earth. As the concentration of these gases increases, less of the earth's long wavelength radiations escape into space. The result is that the earth's atmosphere is warming up. This global warming called 'greenhouse effect' is said to be one of the most important environmental problem (Fig. 10.4).

10.4.2 Sources of Global Warming

The following factors are largely responsible for the increase in atmospheric carbon dioxide concentrations.

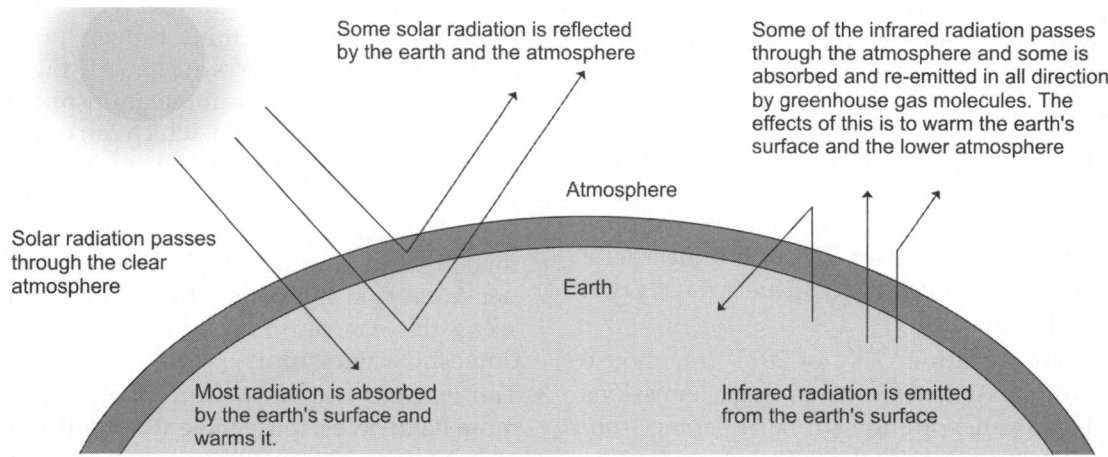

Fig. 10.4: The greenhouse effect (*Source:* US department of state, 1992).

1. **Burning of fossil fuels:** The most important and significant source of atmospheric carbon dioxide is the burning of fossil fuels. With the escalation of population and increase in industrial growth, the demand for fossil fuel has greatly increased.

2. **Deforestation:** Deforestation adds carbon dioxide to the atmosphere in two ways: Firstly, most of the trees are either burned or decomposed by bacteria, emitting carbon dioxide directly to the air. Secondly, the deforested land is unable to sequester carbon dioxide through photosynthesis. As a result of these two phenomena, deforestation contributes 10 to 30% as much carbon dioxide to the atmosphere as fossil fuel emissions do (Fig. 10.5).

3. **Volcanoes:** Volcanoes emit huge amount of carbon dioxide approximately 25 million tons, therefore the entire region around the volcano is enriched in carbon dioxide.

4. **Greenhouse role of trace gases:** The important trace gases which contribute to the greenhouse effect are methane, nitrous oxide, ozone and chlorofluorocarbons. Addition of one molecule of CFC can have the same greenhouse effect as the addition of 104 molecules of carbon dioxide to the atmosphere. Since their atmospheric levels are rising rapidly and since each molecule of these gases absorbs more infrared radiations than a carbon dioxide molecule, their combined greenhouse effect is almost equal to that of carbon dioxide.

5. **Methane:** The principal source of methane are:

1. Action of anaerobic bacteria on rice paddies and wetlands.
2. Leakage from coal mines and natural gas pipeline.
3. Decomposition of organic matter in landfills.
4. Incomplete combustion of forest or range fires.

Methane contributes to the greenhouse effect to an extent of 19%.

6. **Nitrous oxide:** The major sources of nitrous oxide are:

1. Microbial action on nitrogenous fertilizers in soil.
2. Burning of biomass, fossil fuels and forests.

It's contribution to greenhouse effect is about 4%.

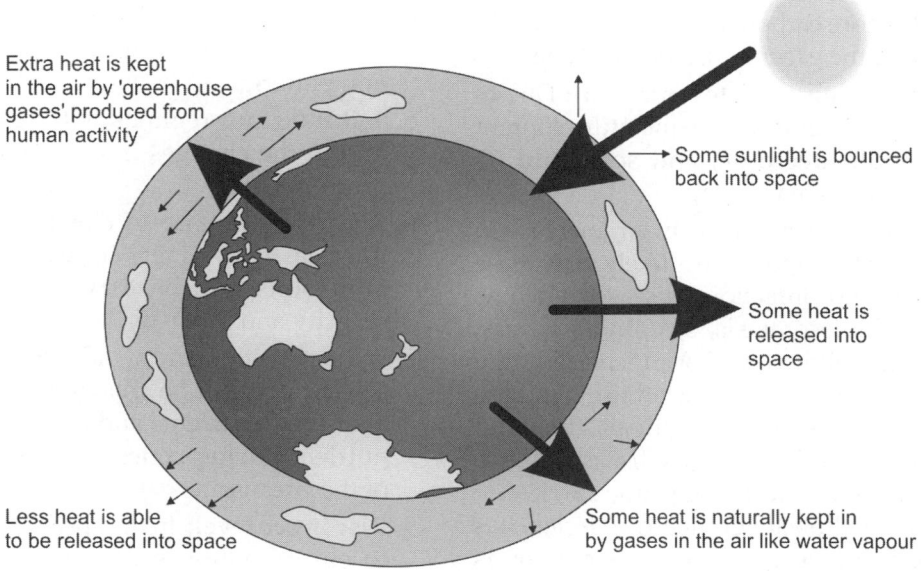

Fig. 10.5: Global warming

10.4.3 Consequences of Global Warming

1. **Changing patterns of rainfall:** Patterns of rainfall would change worldwide causing large shifts in agriculturally productive areas.

2. **Carbon dioxide fertilization:** An elevated carbon dioxide levels might seem an agricultural blessing, accelerating the pace of photosynthesis. The increase in rate of photosynthesis as a result of enhanced carbon dioxide levels is called carbon dioxide fertilization.

3. **Lower nitrogen content:** The dead plant material, such as fallen leaves and twigs, are rich in nitrogen. These act as natural fertilizers, providing nitrogen-based nutrients to the soil and thereby increasing soil productivity. However, plants growing in higher concentrations of carbon dioxide have less nitrogen and more carbon content. Less nitrogen in plants means less protein content. Insect pests that feed on carbon dioxide fertilized plants, therefore, would eat more leaf to obtain enough nitrogen.

4. **Increased rate of decomposition:** As a result of increased global temperature due to greenhouse effect, the dead plant matter and soil organic matter would decompose at the rate higher than normal. The decomposition shall yield more carbon dioxide, which would supplement the greenhouse phenomena.

5. **Evaporation of water from soil:** Due to increase in temperature, the moisture content of soil would decrease, and so would its fertility towards many crops.

6. **Effect on human health:** An increase in average global temperature is likely to increase the incidence of infectious diseases, such as malaria, schistosomiasis, sleeping sickness, dengue and yellow fever. An increase in the global temperature is suspected to extend the range of vectors the mosquitoes, flies and snails that transmit infectious disease. Due to global warming, one of the prime carriers of dengue and yellow fever the *Aedes aegypti* has extended its range in such diverse regions as Coasta Rica, Colombia, Kenya and India.

7. **Effects on wildlife:** With every rise of 1°C, plant and tree species will have to move about 90 kilometres polewards to survive. Changing rainfall patterns will compound the ecological disaster, while rise in sea levels will swamp coastal habitats. As trees and plants die out and habitats disappear, so will the animals that depend on them. And as the world continues to get warmer there will be nowhere for habitats to re-establish themselves.

8. **Climate effects:** Work done with different climatic models shows there is scientific uncertainty about the effects of global change. However, work on these simulation models has agreed on many common things, including:

1. There will be a warming of the earth's surface and lower atmospheric and a cooling of stratosphere.

2. The warming trend over the earth's surface is varied. Warming in the tropics is smaller than the global mean by about 2°–3°C depending on seasonal changes, which in other latitudes, the average warming might account for 5°–10°C increase in temperatures.

3. Precipitation patterns will be changed. Some areas will become wetter and some areas dryer.

4. Seasonal patterns will change due to the changing of temperature and precipitation patterns.

5. Soil moisture regimes will be changed due to the changes in evaporation and precipitation.

6. With the decrease in cloud cover over Eurasia in summer, contrast, tropical monsoons will be driven with more severity and intensity.

7. Wind direction and wind stress over the sea surface will be changed, which will alter ocean currents and cause change in nutrient mixing zones and productivity of the oceans.

9. **Rise in sea level:** In the absence of efforts to cut greenhouse gas emissions, sea levels will rise by between 10 and 30 cm

by the year 2030 and 30 to 100 cm by the end of next century. The direct effects are:

1. Recession of shorelines and wetlands.
2. Increased tidal range and estuarine salt-front intrusion.
3. An increase in salt-water contamination of coastal fresh-water aquifers. All the above effects have profound implication on human society, especially in many coastal areas that are densely populated.

10.5 ACID RAIN

10.5.1 Introduction

The term 'acid rain' was first used by Robert A. Smith in 1872 from his studies of air in Manchester, England. The widespread occurrence of acid rain was recognized only in 1980. Acid rain is a rain or any other form of precipitation that is unusually acidic, i.e. elevated levels of hydrogen ions (low pH). What we call acid rain is the oxides of sulphur and nitrogen originating from industrial exhausts and fossil fuel combustion, the major sources of acid forming gases combine with the water in the air. Acid forming gases are oxidized over several days by which time they travel several thousand kilometres. In the atmosphere these gases are ultimately converted into sulphuric and nitric acids. This acidic mixture then falls as rain, sleet, mist or snow or as solid flakes. Hydrogen chloride emission forms hydrochloric acid. These acids cause acidic rain.

Rain water is turned acidic when its pH falls below 5.6. In fact clean or natural rain water has a pH of 5.6 at 20°C because of formation of carbonic acid due to dissolution of CO_2 in water. Parts of India such as North-East, Coastal regions of Kerela, Odisha, Bihar, and West Bengal have reported decline in fertility of soil due to reduced pH of soil (increased in acidity).

Acid rain is the phenomenon of wet and dry acidic deposition.

Wet deposition: Wet deposition of acids occurs when any form of precipitation (rain, snow, and so on) removes acids from the atmosphere and delivers it to the Earth's surface. This can result from the deposition of acids produced in the raindrops (see aqueous phase chemistry above) or by the precipitation removing the acids either in clouds or below clouds. Wet removal of both gases and aerosols are both of importance for wet deposition (Fig. 10.6).

Dry deposition: Acid deposition also occurs via dry deposition in the absence of precipitation. This can be responsible for as much as 20% to 60% of total acid deposition. This occurs when particles and gases stick to the ground, plants or other surfaces. During the last few decades acid rain occurred within the downwind of areas of major industrial areas in Europe and America. Emissions of sulphur dioxide (SO_2) into sulphate or NO_2 into nitrate particles and by combining with water vapour into mild sulphuric or nitric acids and return to earth as dew, drizzle, fog,

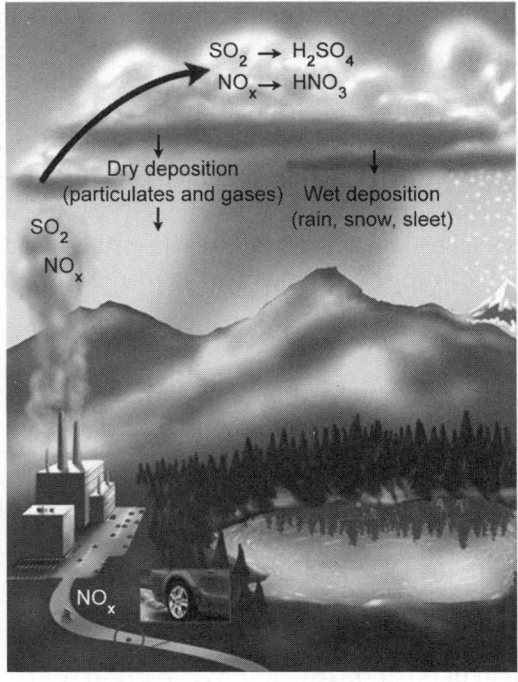

Fig. 10.6: Acid rain formation

sleet, snow, and rain is the mechanism of acid rain. Acid rain is a rain or any other form of precipitation that is unusually toxic, meaning that it possesses elevated levels of hydrogen ions (low pH). It can have harmful effects on plants, aquatic animals, and infrastructure. Acid rain is caused by emissions of sulphur dioxide and nitrogen acids, which react with the water, small molecules in the atmosphere to produce acids. Governments have made efforts since the 1970s, to reduce the release of sulphur dioxide into the atmosphere with positive results. Nitrogen oxides can also be produced naturally by lightning strikes and sulphur dioxide is produced by volcanic eruptions. The chemicals in acid rain can cause paint to peel, corrosion of steel structures such as bridges, and erosion of stone statues. "Acid u" is a popular term referring to the deposition of wet (rain, snow, sleet, fog, cloud water, and dew) and dry (acidifying particles and gases) acidic components. Distilled water, once carbon dioxide is removed, has a neutral pH of 7. Liquids with a pH less than 7 are acidic, and those with a pH greater than 7 are alkaline. "Clean" or unpolluted rain has an acidic pH, but usually no lower than 5.7, because carbon dioxide and water in the air react together to form carbonic acid, a weak acid. However, unpolluted rain can also contain other chemicals which affect its pH. A common example is nitric acid produced by electric discharge in the atmosphere such as lightning. Carbonic acid is formed by the reaction:

$$H_2O \text{ (l)} + CO_2 \text{ (g)} \rightleftharpoons H_2CO_3 \text{ (aq)}$$

Carbonic acid then can ionize in water forming low concentrations of hydronium and carbonate ions:

$$H_2O \text{ (l)} + H_2CO_3 \text{ (aq)} \rightleftharpoons HCO_3^- \text{ (aq)} + H_3O^+ \text{ (aq)}$$

Acid deposition as an environmental issue would include additional acids to H_2CO_3.

10.5.2 Causes of Acid Rain

Acid deposition can occur via natural sources like volcanoes but it is mainly caused by the release of sulphur dioxide and nitrogen oxide during fossil fuel combustion. When these gases are discharged into the atmosphere they react with the water, oxygen, and other gases already present there to form sulphuric acid, ammonium nitrate, and nitric acid. These acids then disperse over large areas because of wind patterns and fall back to the ground as acid rain or other forms of precipitation. Acid deposition can occur via natural sources like volcanoes but it is mainly caused by the release of sulfur dioxide and nitrogen oxide during fossil fuel combustion. When these gases are discharged into the atmosphere they react with the water, oxygen, and other gases already present there to form sulfuric acid, ammonium nitrate, and nitric acid. These acids then disperse over large areas because of wind patterns and fall back to the ground as acid rain or other forms of precipitation.

Man-made sources include emission of sulfur dioxide and nitrogen oxides due to combustion of fossil fuels. Roughly two-thirds of all sulfur dioxide and one-fourth of all nitrogen oxides come from generation of electricity through burning of fossil fuels such as coal. These gases react in the atmosphere with water, oxygen, and other chemicals to form various acidic compounds such as sulfuric acid, ammonium nitrate, and nitric acid. The existing wind blows these acidic compounds over large areas across borders and they fall back to the ground in the form of acid rain or other forms of precipitation. Upon reaching the Earth, it flows across the surface, absorbs into the soil and enters into lakes and rivers and finally gets mixed up with sea water.

The gases responsible for acid deposition are normally a by-product of electric power generation and the burning of coal. As such, it began entering the atmosphere in large amounts during the Industrial Revolution and was first discovered by a Scottish chemist, Robert Angus Smith, in 1852. In that year, he discovered the relationship between acid rain and atmospheric pollution in Manchester,

England. Although it was discovered in the 1800s, acid deposition did not gain significant public attention until 1960 and the term acid rain was coined in 1972.

10.5.3 Adverse Effects

Acid rain has been shown to have adverse impacts on forests, freshwaters and soils, killing insects and aquatic life-forms as well as causing damage to buildings and having impacts on human health.

1. **Surface waters and aquatic animals:** Both the lower pH and higher aluminium concentrations in surface water that occur as a result of acid rain can cause damage to fish and other aquatic animals. At pH lower than 5 most fish eggs will not hatch and lower pH can kill adult fish. As lakes and rivers become more acidic, biodiversity is reduced. Acid rain has eliminated insect life and some fish species, including the brook trout in some lakes, streams, and creeks in geographically sensitive areas, such as the Adirondack mountains of the United States. However, the extent to which acid rain contributes directly or indirectly via runoff from the catchments to lake and river acidity, (i.e. depending on characteristics of the surrounding watershed) is variable. The United States environmental protection agency's (EPA) website states: "Of the lakes and streams surveyed, acid rain caused acidity in 75% of the acidic lakes and about 50% of the acidic streams." As the lake becomes more acidic the fish find it more difficult to reproduce successfully. It is not only the acid in the water that kills them, but also poisonous minerals like aluminium that are washed out of the surrounding ground into the water. The birds that eat the fish also begin to suffer as the harmful minerals build up inside.

2. **Soils:** Soil biology and chemistry can be seriously damaged by acid rain. Some microbes are unable to tolerate changes to low pH and are killed. The enzymes of these microbes are denatured (changed in shape so they no longer

function) by the acid. The hydronium ions of acid rain also mobilize toxins such as aluminium, and leach away essential nutrients and minerals such as magnesium.

$$2H^+ (aq) + Mg^{2+} (clay) \rightleftharpoons 2H^+ (clay) + Mg^{2+} (aq)$$

Soil chemistry can be dramatically changed when base cations, such as calcium and magnesium, are leached by acid rain thereby affecting sensitive species, such as sugar maple (acer saccharum).

Acid rain can damage soil by destroying many vital substances and washing away the nutrients. Soils naturally contain small amounts of poisonous minerals such as mercury and aluminium. Normally these minerals do not cause serious problems, but when acid rain falls on the ground and the acidity of the soil increases, chemical reactions occur allowing the poisonous minerals to be taken up by the plant roots. The trees and plants are then damaged and any animals eating them will absorb the poisons, which will stay in their bodies.

3. **Forests and other vegetation:** The acid takes important minerals away from the leaves and the soil. Without these minerals, trees and plants cannot grow properly. Damaged trees lose their leaves, have stunted growth and damaged bark. This makes it easier for fungi and insects to attack the tree, and as a result the tree may die. Acid rain not only damages soil but can also affect the trees directly. Pollutants can block or damage the little pores on the leaves through which the plant takes in the air it needs to survive. High altitude forests are especially vulnerable as they are often surrounded by clouds and fog which are more acidic than rain.

Other plants can also be damaged by acid rain, but the effect on food crops is minimized by the application of lime and fertilizers to replace lost nutrients. In cultivated areas, limestone may also be added to increase the ability of the soil to keep the pH stable, but this tactic is largely unusable in the case of wilderness lands. When calcium is leached from the needles of red spruce, these trees become less cold tolerant and exhibit winter injury and even death.

4. **Human health effects:** Acid rain does not directly affect human health. The acid in the rainwater is too dilute to have direct adverse effects. However, the particulates responsible for acid rain (sulphur dioxide and nitrogen oxides) do have an adverse effect. Increased amounts of fine particulate matter in the air do contribute to heart and lung problems including asthma and bronchitis.

5. **Other adverse effects on monuments:** Acid rain can also damage buildings and historic monuments and statues, especially those made of rocks, such as limestone and marble that contain large amounts of calcium carbonate. When sulphur pollutants fall on to buildings made from limestone and sandstone they react with minerals in the stone to form a powdery substance that can be washed away by rain. Acids in the rain react with the calcium compounds in the stones to create gypsum, which then flakes off.

$$CaCO_3 (s) + H_2SO_4 (aq) \rightleftharpoons CaSO_4 (aq) + CO_2 (g) + H_2O (l)$$

The effects of this are commonly seen on old gravestones, where acid rain can cause the inscriptions to become completely illegible. Acid rain also increases the corrosion rate of metals, in particular iron, steel, copper and bronze. Famous buildings like the Statue of Liberty in New York, the Taj Mahal in India and St. Paul's Cathedral in London, all been damaged by this sort of air pollution.

Acid rain can also damage stained glass windows in churches, railway lines and steel bridges. The acid rain slowly eats away them all. Building materials crumble away, metals are corroded.

10.5.4 Control of Acid Rain

Because of these problems and the adverse effects air pollution has on human health, a number of steps are being taken to reduce sulfur and nitrogen emissions. Most notably, many governments are now requiring energy producers to clean smoke stacks by using scrubbers which trap pollutants before they are released into the atmosphere and catalytic converters in cars to reduce their emissions. Additionally, alternative energy sources are gaining more prominence today and funding is being given to the restoration of ecosystems damaged by acid rain worldwide.

The strategy to curb acid deposition is quite uncomplicated–reduce the sulfur and nitrogen oxide emissions by mitigating the sources. Installation of scrubbers in the smokestacks of coal-fired power plants and use of clean coal-burning technologies can effectively reduce sulfur emissions. Similarly, alterations in car engine designs and installation of emission control devices such as catalytic converters can be useful, however, the gains are likely to be offset by the growing number of vehicles.

1. Emission of SO_2 and NO_2 from industries and power plants should be reduced by using pollution control equipments such as scrubbers in the smokestacks of factories. These spray a mixture of water and lime-stone into the polluting gases, recapturing the sulphur.

2. Liming of lakes and soils should be done to correct the adverse effects of acid rain.

3. A coating of protective layer of inert poly-mer should be given in the interior of water pipes for drinking water.

4. In catalytic converters, the gases are passed over metal coated beads that convert harmful chemicals into less harmful ones.

10.6 OZONE LAYER DEPLETION

10.6.1 Introduction

Ozone depletion is the term commonly used to describe the thinning of the ozone layer in the stratosphere. Ozone depletion occurs when the natural balance between the production and destruction of ozone in the stratosphere is tipped in favour of destruction.

Ozone depletion is the seasonal loss of a large swath of our stratospheric ozone above Antartica, as well as the general degradation of this protective layer around the globe. With less ozone in the atmosphere, more ultraviolet radiation strikes Earth, causing more skin cancer, eye damage, and possible harm to crops. The ozone layer is a belt of naturally occurring ozone gas that sits 9.3 to 18.6 miles (15 to 30 kilometers) above Earth and serves as a shield from the harmful ultraviolet B radiation emitted by the sun. Ozone is a highly reactive molecule that contains three oxygen atoms. It is constantly being formed and broken down in the upper atmosphere, 6.2 to 31 miles (10 to 50 kilometers) above Earth, in the region called the stratosphere. The earth's stratospheric ozone layer plays a critical roll in absorbing ultraviolet radiation emitted by our sun. In the last thirty years, it has been discovered that stratospheric ozone is depleting as a result of anthropogenic pollutants. There are a number of chemical reactions that can deplete stratospheric ozone, however, some of the most significant depletion comes from the catalytic destruction of ozone by freed halogen radicals like chlorine and bromine.

Ozone depletion is largely a result of man-made substances. Humans have introduced gases and chemicals into the atmosphere that have rapidly depleted the ozone layer in the last century. This depletion makes humans more vulnerable to the UVB rays which are known to cause skin cancer as well as other genetic deformities.

Ozone Layer Depletion

Ozone depletion describes two distinct, but related observations: a slow, steady decline of about 4% per decade in the total volume of ozone in earth's stratosphere (ozone layer) since the late 1970s, and a much larger, but seasonal decrease in stratospheric ozone over earth's polar regions during the same period. The latter phenomenon is commonly referred to as the ozone hole. In addition to this well-known stratospheric ozone depletion, there are also topospheric ozone depletion events, which occur near the surface in polar region during spring.

Ozone is formed in the atmosphere when ultraviolet radiation from the sun strikes the stratosphere, splitting oxygen molecule (O_2) into atomic oxygen (O). The atomic oxygen quickly combines with further oxygen molecules to form ozone.

$$O_2 + h\nu \rightarrow O + O$$
$$O + O_2 \rightarrow O_3$$

Ozone is also destroyed by the following reaction with atomic oxygen.

$$O + O_3 \rightarrow O_2 + O_2$$

All the three reactions are known as *champan reaction*. Ozone formation reaction becomes slower, with increase in altitude while reaction of decomposition of ozone becomes faster. The concentration of ozone is a balance between these two complimentary reactions.

The most common stratospheric ozone measuring unit is *dobson unit (DU)*. The average amount of ozone in the stratosphere across the globe is about 300 DU. Highest level of ozone are usually found in the mid to high latitude. When stratospheric ozone falls below 200 DU, this is considered low enough to represent the beginning of an ozone hole.

10.6.2 Sources

Several catalytic chemical reactions have been identified as ozone destruction mechanisms. The chemicals that start these reactions are called catalysts because they are not used up by the reaction. Rather, they are regenerated by the reaction and therefore are capable of reacting with ozone over and over again. Each of them can destroy thousands or even hundred's of thousands of ozone molecules before being destroyed itself by some other process. The chemicals involved in these catalytic reactions include chlorine oxide, hydrogen oxide, and nitrogen oxide. Relatively recently, human activities have introduced large quantities of these catalysts into the atmosphere.

The Main Ozone-Depleting Substances (ODS)

1. **Chlorofluorocarbons (CFCs):** The most widely used ODS, accounting for over 80% of total stratospheric ozone depletion.Used as coolants in refrigerators, freezers and air conditioners in buildings and cars manufactured before 1995. Found in industrial solvents, dry-cleaning agents and hospital sterilants. Also used in foam products - such as soft-foam padding, (e.g. cushions and mattresses) and rigid foam, (e.g. home insulation).

2. **Halons:** Used in some fire extinguishers, in cases where materials and equipment would be destroyed by water or other fire extinguisher chemicals. In brominated fluorocarbons, halons cause greater damage to the ozone layer than do CFCs from automobile air conditioners.

3. **Methyl chloroform:** Used mainly in industry—for vapour degreasing, some aerosols, cold cleaning, adhesives and chemical processing.

4. **Carbon tetrachloride:** Used in solvents and some fire extinguishers.

5. **Hydrofluorocarbons (HFCs):** HFCs have become major, "transitional" substitutes for CFCs. They are much less harmful to stratospheric ozone than CFCs are. But HFCs still cause some ozone destruction and are potent greenhouse gases (Fig. 10.7).

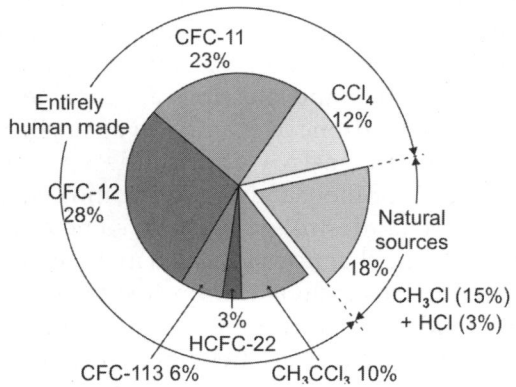

Fig. 10.7: Types of Ozone layer depletion gases

10.6.3 Consequences of Ozone Layer Depletion

Stratospheric ozone filters out most of the sun's potentially harmful shortwave ultraviolet (UV) radiation. If this ozone becomes depleted, then more UV rays will reach the earth (Fig. 10.8). Exposure to higher amounts of UV radiation could have serious impacts on human beings, animals and plants, such as the following:

1. **On vegetation:** An increase of UV radiation would be expected to affect crops. A number of economically important species of plants, such as rice, depend on cyanobacteria residing on their roots for the retention of nitrogen. Cyanobacteria are sensitive to UV light and they would be affected by its increase.

2. **On plankton:** There is a difference in the orientation and motility of planktons when excess of UV rays reach earth. Researchers speculate that the extinction was caused by a significant weakening of the ozone layer at that time when the radiation from the supernova produced nitrogen oxides that catalyzed the destruction of ozone (planktons are particularly susceptible to effects of UV light, and are vitally important to marine food webs). Decrease in plankton could disrupt the fresh and saltwater food chains, and lead to a species shift in Canadian waters. Loss of biodiversity in our oceans, rivers and lakes could reduce fish yields for commercial and sport fisheries.

3. **On man:** UVB (the higher energy UV radiation absorbed by ozone) is generally accepted to be a contributory factor to skin cancer. In addition, increased surface UV leads to increased tropospheric ozone, which is a health risk to humans. More skin cancers, sunburns and premature aging of the skin. More cataracts, blindness and other eye diseases: UV radiation can damage several parts of the eye, including the lens, cornea, retina and conjunctiva. Cataracts (clouding of the lens) are the

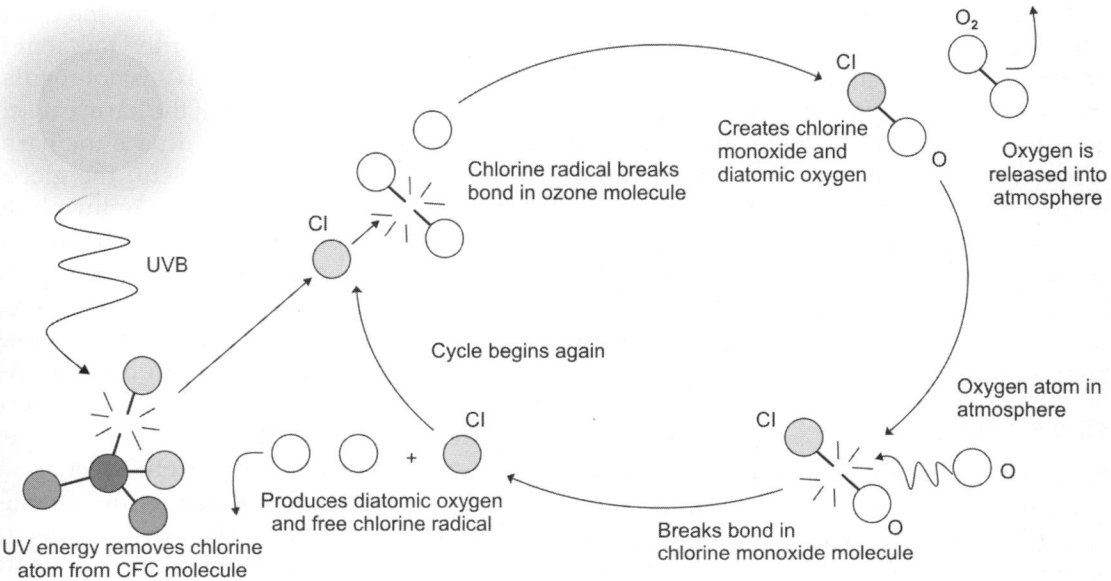

Fig. 10.8: Ozone layer depletion

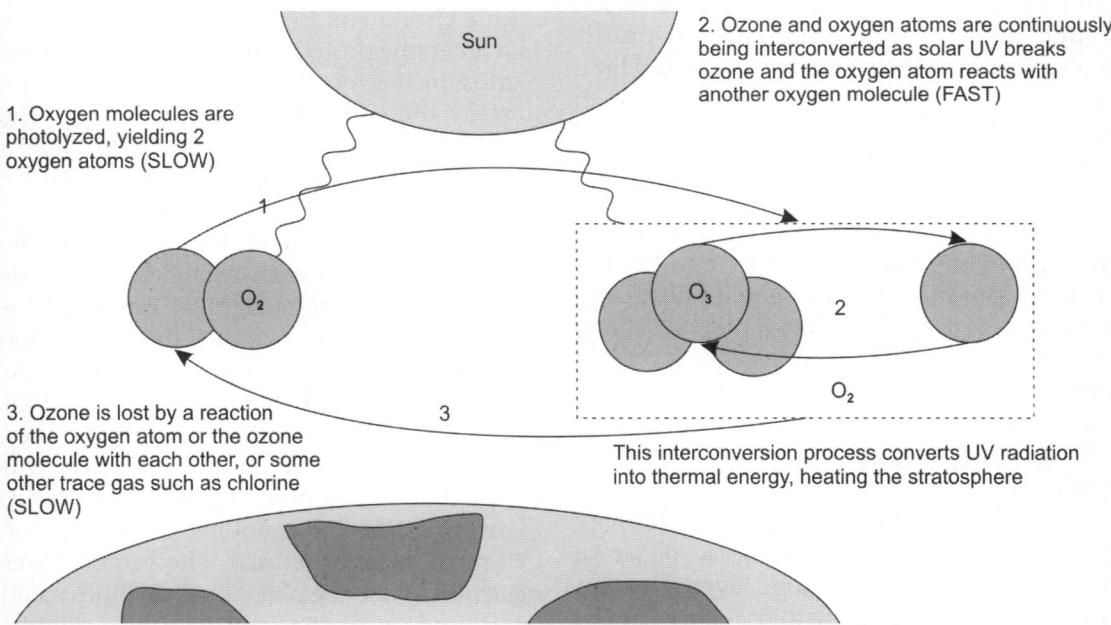

Fig. 10.9: Schematic presentation of ozone layer depletion

major cause of blindness in the world. A sustained 10% thinning of the ozone layer is expected to result in almost two million new cases of cataracts per year, globally (Environment Canada, 1993). Weakening of the human immune system (immunosuppression). Early findings suggest that too much UV radiation can suppress the human immune system, which may play a role in the development of skin cancer (Fig. 10.9).

4. **Animals:** In domestic animals, UV over-exposure may cause eye and skin cancers. Species of marine animals in their developmental stage, (e.g. young fish, shrimp larvae and crab larvae) have been threatened in recent years by the increased UV radiation under the Antarctic ozone hole.

5. **Materials:** Wood, plastic, rubber, fabrics and many construction materials are degraded by UV radiation. The economic impact of replacing and/or protecting materials could be significant.

10.6.4 Control

Encourage growth of plants that produce oxygen and discourage deforestation.

1. **Limit private vehicle driving:** A very easy way to control ozone depletion would be to limit or reduce the amount of driving as vehicular emissions eventually result in smog which is a culprit in the deterioration of the ozone layer. Car pooling, taking public transport, walking, using a bicycle would limit the usage of individual transportation. It would be a great option to switch to cars/vehicles that have a hybrid or electric zero-emission engine.

2. **Use eco-friendly household cleaning products:** Usage of eco-friendly and natural cleaning products for household chores is a great way to prevent ozone depletion. This is because many of these cleaning agents contain toxic chemicals that interfere with the ozone layer. A lot of supermarkets and health stores sell cleaning products that are toxic-free and made out of natural ingredients.

3. **Avoid using pesticides:** Pesticides may be an easy solution for getting rid of weed, but are harmful for the ozone layer. The best solution for this would be to try using natural remedies, rather than heading out for pesticides. You can perhaps try to weed manually or mow your garden consistently so as to avoid weed growth.

4. **Developing stringent regulations for rocket launches:** The world is progressing in scientific discoveries by leaps and bounds. A lot of rocket launches are happening the world over without consideration of the fact that it can damage the ozone layer if it is not regulated soon. A study shows that the harm caused by rocket launches would outpace the harm caused due to CFCs. At present, the global rocket launches do not contribute hugely to ozone layer depletion, but over the course of time, due to the advancement of the space industry, it will become a major contributor to ozone depletion. All types of rocket engines result in combustion by products that are ozone-destroying compounds that are expelled directly in the middle and upper stratosphere layer near the ozone layer.

5. **Banning the use of dangerous nitrous oxide:** Due to the worldwide alarm caused by a study in the late 70s' about the alarming rate at which the ozone was being depleted, nations around the globe got together and formed the Montreal protocol in the year 1989 with a strong aim to stop the usage of CFCs. However, the protocol did not include nitrous oxide which is the most fatal chemical that can destroy the ozone layer and is still in use. Governments across the world should take a strong stand for banning the use of this harmful compound to save the ozone layer.

6. **Montreal protocol:** Following the discovery of the Antarctic ozone hole in the late 1985, governments recognized the need for stronger measures to reduce the production and consumption of a number of CFCs (CFC 11, 12, 113, 114 and 115) and several Halons (1211, 1301, 2402). The Montreal protocol on substances that deplete the ozone layer was adopted on 16th September 1987, at the headquarters of the International Civil Aviation Organization in Montreal. The protocol was designed so that the phase out schedules could be revised on the basis of periodic scientific and technological assessments. Following such assessments, the protocol was adjusted to accelerate the phase out schedules. It has also been amended to introduce other kinds of control measures and to add new controlled substances to the list. The Protocol came into force on 1st January, 1989.

10.7 UNITED NATIONS FRAMEWORK CONVENTION ON CLIMATE CHANGE (UNFCCC)

It is an international environmental treaty negotiated at the united nations conference on environment and development (UNCED), informally known as the earth summit, held in Rio de Janeiro from 3 to 14 June 1992.

The objective of the treaty is to "stabilize greenhouse gas concentrations in the atmosphere at a level that would prevent dangerous anthropogenic interference with the climate system."

The treaty itself set no binding limits on greenhouse gas emissions for individual countries and contains no enforcement mechanisms. In that sense, the treaty is considered legally non-binding. Instead, the treaty provides a framework for negotiating specific international treaties (called "protocols") that may set binding limits on greenhouse gases.

The UNFCCC was opened for signature on 9th May 1992, after an intergovernmental negotiating committee produced the text of the framework convention as a report following its meeting in New York from 30th April to 9th May 1992. It entered into force on 21st March 1994. As of May 2011, UNFCCC has 195 parties.

The parties to the convention have met annually from 1995 in conferences of the parties (COP) to assess progress in dealing with climate change. In 1997, the Kyoto protocol was concluded and established legally binding obligations for developed countries to reduce their greenhouse gas emissions. The 2010 Cancun agreements state that future global warming should be limited to below 2.0 °C (3.6 °F) relative to the pre-industrial level. The 20th COP will take place in Peru in 2014.

One of the first tasks set by the UNFCCC was for signatory nations to establish national greenhouse gas inventories of greenhouse gas (GHG) emissions and removals, which were used to create the 1990 benchmark levels for accession of Annex I countries to the Kyoto protocol and for the commitment of those countries to GHG reductions. Updated inventories must be regularly submitted by Annex I countries.

Treaty: The united nations framework convention on climate change (UNFCCC) was opened for signature at the 1992 united nations conference on environment and development (UNCED) in Rio de Janeiro (known by its popular title, the earth Summit). On 12 June 1992, 154 nations signed the UNFCCC, that upon ratification committed signatories, governments to reduce atmospheric concentrations of greenhouse gases with the goal of "preventing dangerous anthropogenic interference with earth's climate system."

The 2010 Cancún agreements (COP 16) include voluntary pledges made by 76 developed and developing countries to control their emissions of greenhouse gases. At the 2012 Doha climate change talks (COP 18), parties to the UNFCCC agreed to a timetable for a global agreement which will include all countries. The timetable states that a global agreement should be adopted by 2015, and implemented by 2020.

Interpreting Article 2: The ultimate objective of the framework convention is to prevent "dangerous" anthropogenic, (i.e. human) interference of the climate system. As is stated in Article 2 of the convention, this requires that GHG concentrations are stabilized in the atmosphere at a level where ecosystems can adapt naturally to climate change, food production is not threatened, and economic development can proceed in a sustainable fashion.

Human activities have had a number of effects on the climate system. Global GHG emissions due to human activities have grown since pre-industrial times. Warming of the climate system has been observed, as indicated by increases in average air and ocean temperatures, widespread melting of snow and ice cover, and rising global average sea level. As assessed by the intergovernmental panel on climate change (IPCC), "most of the observed increase in global average temperatures since the mid-20th century is very likely due to the observed increase in anthropogenic GHG concentrations". "Very likely" here is defined by the IPCC as having a likelihood of greater than 90%, based on expert judgement.

10.8 THE INTERGOVERNMENTAL PANEL ON CLIMATE CHANGE (IPCC)

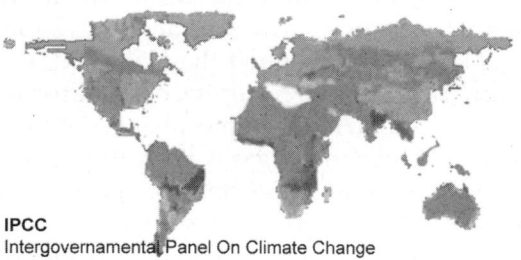

IPCC
Intergovernamental Panel On Climate Change

It is a scientific intergovernmental body, set up at the request of member governments. It was first established in 1988 by two united nations organizations, the world meteorological organization (WMO) and the united nations environment programme (UNEP), and later endorsed by the united nations general assembly through resolution 43/53.

Its mission is to provide comprehensive scientific assessments of current scientific, technical and socio-economic information worldwide about the risk of climate change caused by human activity, its potential environmental and socio-economic consequences, and possible options for adapting to these consequences or mitigating the effects.

Thousands of scientists and other experts contribute (on a voluntary basis, without payment from the IPCC) to writing and reviewing reports, which are reviewed by representatives from all the governments, with a summary for policymakers being subject to line-by-line approval by all participating governments. Typically this involves the governments of more than 120 countries.

The IPCC does not carry out its own original research, nor does it do the work of monitoring climate or related phenomena itself. A main activity of the IPCC is publishing special reports on topics relevant to the implementation of the united nations framework convention on climate change (UNFCCC), an international treaty that acknowledges the possibility of harmful climate change. Implementation of the UNFCCC led eventually to the Kyoto protocol. The IPCC bases its assessment mainly on peer reviewed and published scientific literature. Membership of the IPCC is open to all members of the WMO and UNEP.

Aims: The principles that the IPCC operates under are set out in the relevant WMO executive council and UNEP governing council resolutions and decisions, as well as on actions in support of the UNFCCC process. The aims of the IPCC are to assess scientific information relevant to:

1. Human-induced climate change
2. The impacts of human-induced climate change
3. Options for adaptation and mitigation

The **intergovernmental panel on climate change (IPCC)** is a scientific intergovernmental set up at the request of member governments. It was first established in 1988 by two united nations organizations, the world meteorological organization (WMO) and the united nations environment programme (UNEP), and later endorsed by the United Nations General Assembly through resolution 43/53. Membership of the IPCC is open to all members of the WMO and UNEP. The IPCC is chaired by Rajendra K. Pachauri. The IPCC produces reports that support the united nations framework convention on climate change (UNFCCC), which is the main international treaty on climate change. The ultimate objective of the UNFCCC is to "stabilize greenhouse gas concentrations in the atmosphere at a level that would prevent dangerous anthropogenic, i.e. human-induced interference with the climate system." IPCC reports cover "the scientific, technical and socio-economic information relevant to understanding the scientific basis of risk of human-induced climate change, its potential impacts and options for adaptation and mitigation." The IPCC does not carry out its own original research, nor does it do the work of monitoring climate or related phenomena itself. The IPCC bases its assessment on the published literature, which includes peer-reviewed and non-peer reviewed sources.

The IPCC provides an internationally accepted authority on climate change, producing reports which have the agreement of leading climate scientists and the consensus of participating governments. The 2007 Nobel Peace prize was shared, in two equal parts, between the IPCC and Al Gore.

Operations

The chair of the IPCC is Rajendra K. Pachauri, elected in May 2002. The previous chair was Robert Watson. The chair is assisted by an elected bureau including vice-chairs, working group co-chairs and a secretariat.

The IPCC panel is composed of representatives appointed by governments and organizations. Participation of delegates with appropriate expertise is encouraged. Plenary sessions of the IPCC and IPCC working groups are held at the level of government representatives. Non-governmental and inter-governmental organizations may be allowed to attend as observers. Sessions of the IPCC bureau, workshops, expert and lead authors meetings are by invitation only. Attendance at the 2003 meeting included 350 government officials and climate change experts. After the opening ceremonies, closed plenary sessions were held. The report states there were 322 persons in attendance at sessions with about seven-eighth of participants being from governmental organizations.

Assessment Reports

The IPCC has published four comprehensive assessment reports reviewing the latest climate science, as well as a number of special reports on particular topics. These reports are prepared by teams of relevant researchers selected by the bureau from government nominations. Drafts of these reports are made available for comment in open review processes to which anyone may contribute. The IPCC published its first assessment report in 1990, a supplementary report in 1992, a second assessment report (SAR) in 1995, and a third assessment report (TAR) in 2001. A fourth assessment report (AR4) was released in 2007 and a fifth is due to be issued in 2014.

Each assessment report is in three volumes, corresponding to Working Groups (WG) I, II and III. Unqualified, "the IPCC report" is often used to mean the working group I report, which covers the basic science of climate change.

First Assessment Report

The IPCC first assessment report was completed in 1990, and served as the basis of the UNFCCC. The executive summary of the WG I summary for policymakers report says they are certain that emissions resulting from human activities are substantially increasing the atmospheric concentrations of the greenhouse gases, resulting on average in an additional warming of the Earth's surface. They calculate with confidence that CO_2 has been responsible for over half the enhanced greenhouse effect. They predict that under a "business as usual" (BAU) scenario, global mean temperature will increase by about 0.3 °C per decade during the 21st century. They judge that global mean surface air temperature has increased by 0.3° to 0.6 °C over the last 100 years, broadly consistent with prediction of climate models, but also of the same magnitude as natural climate variability. The unequivocal detection of the enhanced greenhouse effect is not likely for a decade or more.

Second Assessment Report

Climate change 1995, the IPCC Second assessment report (SAR), was finished in 1996. It is split into four parts:

- A synthesis to help interpret UNFCCC Article 2
- *The science of climate change* (WG I)
- *Impacts, adaptations and mitigation of climate change* (WG II)
- *Economic and social dimensions of climate change* (WG III)

Each of the last three parts was completed by a separate working group, and each has a summary for policymakers (SPM) that represents a consensus of national representatives. The SPM of the WG I report contains:

1. Greenhouse gas concentrations have continued to increase
2. Anthropogenic aerosols tend to produce negative radioactive forcings
3. Climate has changed over the past century (air temperature has increased between 0.3° and 0.6 °C, since late 19th century; this estimate has not significantly changed since the 1990 report).
4. The balance of evidence suggests a discernible human influence on global climate (considerable progress since the 1990 report in distinguishing between natural and anthropogenic influences on climate, because of including aerosols; coupled models; pattern-based studies)
5. Climate is expected to continue to change in the future (increasing realism of simulations increases confidence; important uncertainties remain but are taken into account in the range of model projections)
6. There are still many uncertainties (estimates of future emissions and biogeochemical cycling; models; instrument data for model testing, assessment of variability, and detection studies.

Third Assessment Report

The third assessment report (TAR) was completed in 2001 and consists of four reports, three of them from its working groups:

- Working Group I: The scientific basis
- Working Group II: Impacts, adaptation and vulnerability
- Working Group III: Mitigation
- Synthesis report

A number of the TAR's conclusions are given quantitative estimates of how probable it is that they are correct, e.g. greater than 66% probability of being correct. These are "Bayesian" probabilities, which are based on an expert assessment of all the available evidence.

Fourth Assessment Report

The fourth assessment report (AR4) was published in 2007. Like previous assessment reports, it consists of four reports:

- Working Group I: The physical science basis
- Working Group II: Impacts, adaptation and vulnerability
- Working Group III: Mitigation
- Synthesis report

People from over 130 countries contributed to the IPCC Fourth Assessment Report, which took 6 years to produce. Contributors to AR4 included more than 2500 scientific expert reviewers, more than 800 contributing authors, and more than 450 lead authors.

Global warming projections from AR4 are shown below. The projections apply to the end of the 21st century (2090–99), relative to temperatures at the end of the 20th century (1980–99). Add 0.7 °C to projections to make them relative to pre-industrial levels instead of 1980–99. Descriptions of the greenhouse gas emissions scenarios can be found in special report on emissions scenarios.

10.9 KYOTO PROTOCOL TO UNITED NATIONS FRAMEWORK CONVENTION ON CLIMATE CHANGE

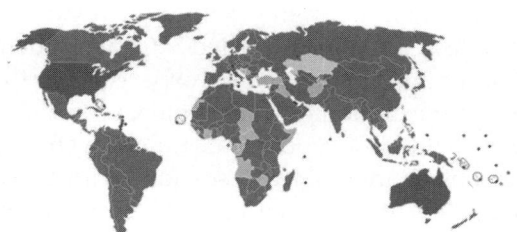

Green = Signed and ratified
Yellow = Signed – ratification pending
Red = Signed – ratification declined
Grey = No problem

Kyoto Protocol, participation (2005)

It is an international treaty that sets binding obligations on industrialised countries to reduce emissions of greenhouse gases.

The UNFCCC is an environmental treaty with the goal of preventing "dangerous" anthropogenic, (i.e. human-induced) interference of the climate system. According to the UNFCCC website, the protocol "recognises that developed countries are principally responsible for the current high levels of GHG emissions in the atmosphere as a result of more than 150 years of industrial activity, and places a heavier burden on developed nations under the principle of "common but differentiated responsibilities." There are 192 parties to the convention, including 191 states (all the UN members, except Andorra, Canada, South Sudan and the United States) and the European Union. The United States signed but did not ratify the protocol and Canada withdrew from it in 2011. The protocol was adopted by Parties to the UNFCCC in 1997, and entered into force in 2005.

As part of the Kyoto protocol, many developed countries have agreed to legally binding limitations/reductions in their emissions of greenhouse gases in two commitments periods. The first commitment period applies to emissions between 2008–2012, and the second commitment period applies to emissions between 2013–2020. The protocol was amended in 2012 to accommodate the second commitment period, but this amendment has (as of January 2013) not entered into legal force.

Emissions Trading

An international emissions trading allows developed countries to trade their commitments under the Kyoto protocol. They can trade emissions quotas among themselves, and can also receive credit for financing emissions reductions in developing countries. Developed countries may use emissions trading until late 2014 or 2015 to meet their first-round targets. Developing countries do not have binding targets under the Kyoto protocol, but are still committed under the treaty to reduce their emissions. Actions taken by developed and developing countries to reduce emissions include support for renewable energy, improving energy efficiency, and reducing deforestation. Under the protocol, emissions of developing countries are allowed to grow in accordance with their development needs.

Green Investment Scheme

A green investment scheme (GIS) refers to a plan for achieving environmental benefits from trading surplus allowances (AAUs) under the Kyoto protocol. The green investment scheme (GIS), a mechanism in the framework of international emissions trading (IET), is designed to achieve greater flexibility in reaching the targets of the Kyoto protocol while preserving environmental integrity of IET. However, using the GIS is not required under the Kyoto protocol, and there is no official definition of the term. Under the GIS a party to the protocol expecting that the development of its economy will not exhaust its Kyoto quota, can sell the excess of its Kyoto quota units (AAUs) to another party. The proceeds from the AAU sales should be "greened", i.e. channeled to the development and implementation of the projects either acquiring the greenhouse gases emission reductions (hard greening) or building up the necessary framework for this process (soft greening).

Amendment and Possible Successors

In the non-binding 'Washington declaration' agreed on 16th February 2007, heads of governments from Canada, France, Germany, Italy, Japan, Russia, the United Kingdom, the United States, Brazil, China, India, Mexico and South Africa agreed in principle on the outline of a successor to the Kyoto protocol. They envisaged a global cap-and-trade system that would apply to both industrialized nations and developing countries, and initially hoped that it would be in place by 2009. The United Nations Climate Change Conference in

Copenhagen in December 2009 was one of the annual series of UN meetings that followed the 1992 Earth Summit in Rio. In 1997 the talks led to the Kyoto protocol, and the conference in Copenhagen was considered to be the opportunity to agree a successor to Kyoto that would bring about meaningful carbon cuts. The 2010 Cancún agreements include voluntary pledges made by 76 developed and developing countries to control their emissions of greenhouse gases. In 2010, these 76 countries were collectively responsible for 85% of annual global emissions. By May 2012, the USA, Japan, Russia, and Canada had indicated they would not sign up to a second Kyoto commitment period. In November 2012, Australia confirmed it would participate in a second commitment period under the Kyoto protocol and New Zealand confirmed that it would not.

10.10 ENVIRONMENTAL POLICY OF THE GOVERNMENT OF INDIA

Environmental policy of the government of India is the policy which drives the government of India and its people on the path of awareness towards the environment and also to take precautionary steps against depletion of the environment.

Functions performed by this policy from the very beginning are:

1. Developed procedures for setting up and protection of reserved forests, protected forests, and village forests.

2. Formalization of national parks, wildlife sanctuaries, *conservation reserves* and *community reserves.* Protection to habitat and wildlife within premises of such protected areas.

3. Development of *National Board for Wildlife and State Boards for Wildlife* for identification of future protected areas.

4. Stating that no state government or other authority shall make any order directing

   i. any reserved forest shall cease to be reserved;

   ii. any forest land or any portion thereof may be used for any non-forest purpose;

   iii. That any forest land or any portion thereof may be assigned by way of lease or otherwise to any private person or to any authority, corporation, agency or any other organization not owned, managed or controlled by government;

   iv. That any forest land or any portion thereof may be cleared of trees which have grown naturally in that land or portion, for the purpose of using it for re-afforestation.

10.11 SOURCES OF CO_2 EMISSION

There are both natural and human sources of carbon dioxide emissions. Natural sources include decomposition, ocean release and respiration. Human sources come from activities like cement production, deforestation as well as the burning of fossil fuels like coal, oil and natural gas. Due to human activities, the atmospheric concentration of carbon dioxide has been rising extensively since the industrial revolution and has now reached dangerous levels not seen in the last 3 million years. Human sources of carbon dioxide emissions are much smaller than natural emissions but they have upset the natural balance that existed for many thousands of years before the influence of humans. This is because natural sinks remove around the same quantity of carbon dioxide from the atmosphere than are produced by natural sources. This had kept carbon dioxide levels balanced and in a safe range. But human sources of emissions have upset the natural balance by adding extra carbon dioxide to the atmosphere without removing any.

10.11.1 Human Sources

Since the industrial revolution, human sources of carbon dioxide emissions have been growing. Human activities such as the burning of oil, coal and gas, as well as deforestation are the primary cause of the increased carbon dioxide concentrations in the atmosphere. 87% of all human-produced carbon dioxide emissions come from the burning of fossil fuels like coal, natural gas and oil. The remainder results from the clearing of forests and other land use changes (9%), as well as some industrial processes such as cement manufacturing (4%) Fig. 10.10.

10.11.2 Fossil Fuel Combustion/Use

The largest human source of carbon dioxide emissions is from the combustion of fossil fuels. This produces 87% of human carbon dioxide emissions. Burning these fuels releases energy which is most commonly turned into heat, electricity or power for transportation. Some examples of where they are used are in power plants, cars, planes and industrial facilities. In 2011, fossil fuel use created 33.2 billion tonnes of carbon dioxide emissions worldwide. The 3 types of fossil fuels that are used the most are coal, natural gas and oil. Coal is responsible for 43% of carbon dioxide emissions from fuel combustion, 36% is produced by oil and 20% from natural gas. Coal is the most carbon intensive fossil fuel. For every tonne of coal burned, approximately 2.5 tonnes of carbon dioxide are produced. Of all the different types of fossil fuels, coal produces the most carbon dioxide. Because of this and it's high rate of use, coal is the largest fossil fuel source of carbon dioxide emissions. Coal represents one-third of fossil fuels' share of world total primary energy supply but is

responsible for 43% of carbon dioxide emissions from fossil fuel use. Anything involving fossil fuels has a carbon dioxide emission ticket attached. So for example, burning these fuels release energy but carbon dioxide also gets produced as a byproduct. This is because almost all the carbon that is stored in fossil fuels gets transformed to carbon dioxide during this process. The three main economic sectors that use fossil fuels are: electricity/heat, transportation and industry. The first two sectors, electricity/heat and transportation, produced nearly two-thirds of global carbon dioxide emissions in 2010.

10.11.3 Electricity/Heat Sector

Electricity and heat generation is the economic sector that produces the largest amount of man-made carbon dioxide emissions. This sector produced 41% of fossil fuel related carbon dioxide emissions in 2010. Around the world, this sector relies heavily on coal, the most carbon-intensive of fossil fuels, explaining this sector giant carbon footprint.

Almost all industrialized nations get the majority of their electricity from the combustion of fossil fuels (around 60%–90%). Only Canada and France are the exception. Depending on the energy mix of your local power company you probably will find that the electricity that you use at home and at work has a considerable impact on greenhouse gas emission. Below (Table 10.1) is a chart for percentage of electrical energy produced by fossil fuel combustion for the G8 nations.

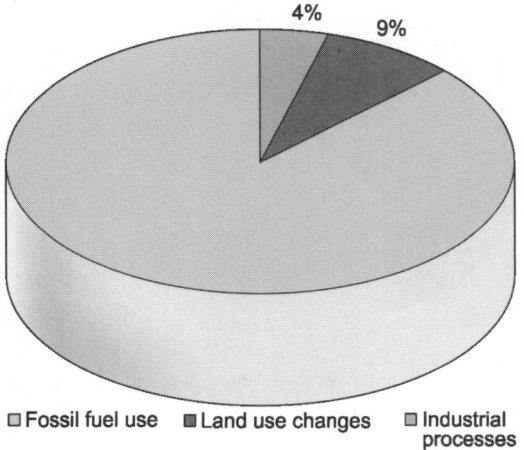

4% 9%

☐ Fossil fuel use ■ Land use changes ☐ Industrial processes

Fig.10.10: Human sources of carbon dioxide

Table 10.1 Electrical energy produced by fossil fuel combustion

G8 nation	Fossil fuel combustion	Total	%
Canada	154.55	569.41	27.1%
France	52.23	535.45	9.8%
Germany	354.78	561.57	63.2%
Italy	223.16	268.18	83.2%
Japan	640.17	982.76	65.1%
Russia	569.72	869.07	65.6%
UK	278.21	373.26	74.5%
United States	2,758.65	3,891.72	70.9%

The industrial, residential and commercial sectors are the main users of electricity covering 92% of usage. Industry is the largest consumer of the three because certain manufacturing processes are very energy intensive. Specifically, the production of chemicals, iron/steel, cement, aluminum as well as pulp and paper account for the great majority of industrial electricity use. The residential and commercial sectors are also heavily reliant on electricity for meeting their energy needs, particularly for lighting, heating, air conditioning and appliances.

10.11.4 Transportation Sector

The transportation sector is the second largest source of anthropogenic carbon dioxide emissions. Transporting goods and people around the world produced 22% of fossil fuel related carbon dioxide emissions in 2010. This sector is very energy intensive and it uses petroleum based fuels (gasoline, diesel, kerosene, etc.) almost exclusively to meet those needs. Since the 1990s, transport related emissions have grown rapidly, increasing by 45% in less than 2 decades. Road transport accounts for 74% of this sector's carbon dioxide emissions. Automobiles, freight and light-duty trucks are the main sources of emissions for the whole transport sector and emissions from these three have steadily grown since 1990. Apart from road vehicles, the other important sources of emissions for this sector are marine shipping and global aviation.

Marine shipping produces 14% of all transport carbon dioxide emissions. While there are a lot less ships than road vehicles used in the transportation sector, ships burn the dirtiest fuel on the market, a fuel that is so unrefined that it can be solid enough to be walked across at room temperature. Because of this, marine shipping is responsible for over 1 billion tonnes of carbon dioxide emissions. This is more than the annual emissions of several industrialized countries (Germany, South Korea, Canada, UK, etc.) and this sector

continues to grow rapidly. Global aviation accounts for 11% of all transport carbon dioxide emissions. International flights create about 62% of these emissions with domestic flights representing the remaining 38%. (Fig. 10.11). Over the last 10 years, aviation has been one of the fastest growing sources of carbon dioxide emissions. Aviation is also the most carbon-intensive form of transportation, so it's growth comes with a heavy impact on climate change.

Figure highlights one of the most alarming trends in today's modern economy. Emissions caused by the transportation of people and goods has grown so rapidly that it has surpassed emissions from the industrial sector, which has had a huge impact on climate change. This trend started in the 1990's and has continued ever since causing an increase in indirect emissions. The emissions caused by the transportation of goods are examples of indirect emissions since the consumer has no direct control of the distance between the factory and the store. The emissions caused by people traveling (by car, plane, train, etc.) are examples of direct emissions since people can chose where they are going and by what method. Since the distance traveled by goods during production is continuing to grow, this

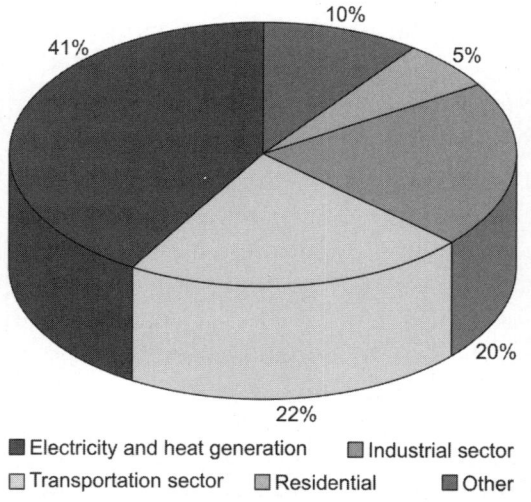

Fig. 10.11: CO_2 emissions from fossil fuel combustion.

is putting more pressure on the transportation industry to bridge the gap and ends up creating more indirect emissions. What's worse is that 99% of the carbon dioxide emissions caused by transportation of people and goods all over the world comes from the combustion of fossil fuels.

10.11.5 Industrial Sector

The industrial sector is the third largest source of man-made carbon dioxide emissions. This sector produced 20% of fossil fuel related carbon dioxide emissions in 2010. The industrial sector consists of manufacturing, construction, mining, and agriculture. Manufacturing is the largest of the 4 and can be broken down into 5 main categories: paper, food, petroleum refineries, chemicals, and metal/mineral products. These categories account for the vast majority of the fossil fuel use and CO_2 emissions by this sector. Manufacturing and industrial processes all combine to produce large amounts of each type of greenhouse gas but specifically large amounts of CO_2. This is because many manufacturing facilities directly use fossil fuels to create heat and steam needed at various stages of production. For example factories in the cement industry, have to heat up limestone to 1450 °C to turn it into cement, which is done by burning fossil fuels to create the required heat.

10.11.6 Land Use Changes

Land use changes are a substantial source of carbon dioxide emissions globally, accounting for 9% of human carbon dioxide emissions and contributed 3.3 billion tonnes of carbon dioxide emissions in 2011. Land use changes are when the natural environment is converted into areas for human use like agricultural land or settlements. From 1850 to 2000, land use and land use change released an estimated 396–690 billion tonnes of carbon dioxide to the atmosphere, or about 28–40% of total anthropogenic carbon dioxide emissions. Deforestation has been responsible for the great majority of these emissions. Deforestation is the permanent removal of standing forests and is the most important type of land use change because its impact on greenhouse gas emissions. Forests in many areas have been cleared for timber or burned for conversion to farms and pastures. When forested land is cleared, large quantities of greenhouse gases are released and this ends up increasing carbon dioxide levels in three different ways. Trees act as a carbon sink. They remove carbon dioxide from the atmosphere via photosynthesis. When forests are cleared to create farms or pastures, trees are cut down and either burnt or left to rot, adds carbon dioxide to the atmosphere. Since deforestation reduces the amount of trees, this also reduces how much carbon dioxide can be removed by the earth's forests. When deforestation is done to create new agricultural land, the crops that replace the trees also act as a carbon sink, but they are not as effective as forests. When trees are cut for lumber the wood is kept which locks the carbon in it but the carbon sink provided by forests is reduced because of the loss of trees.

10.11.7 Industrial Processes

There are many industrial processes that produce significant amounts of carbon dioxide emissions as a by product of chemical reactions needed in their production process. Industrial processes account for 4% of human carbon dioxide emissions and contributed 1.7 billion tonnes of carbon dioxide emissions in 2011. Many industrial processes emit carbon dioxide directly through fossil fuel combustion as well indirectly through the use of electricity that is generated using fossil fuels. But there are four main types of industrial process that are a significant source of carbon dioxide emissions: the production and consumption of mineral products such as cement, the production of metals such as iron and steel, as well as the production of chemicals and petrochemical products.

Cement production produces the most amount of carbon dioxide amongst all industrial processes. To create the main ingredient in cement, calcium oxide, limestone is chemically transformed by heating it to very high temperatures. This process produces large quantities of carbon dioxide as a byproduct of the chemical reaction. So making 1000 kg of cement produces nearly 900 kg of carbon dioxide is evolved. Steel production is another industrial process that is an important source of carbon dioxide emissions. To create steel, iron is melted and refined to lower its carbon content. This process uses oxygen to combine with the carbon in iron which creates carbon dioxide. On average, 1.9 tonnes of CO_2 are emitted for every tonne of steel produced.

Fossil fuels are used to create chemicals and petrochemical products which leads to carbon dioxide emissions. The industrial production of ammonia and hydrogen most often use natural gas or other fossil fuels as a starting base, creating carbon dioxide in the process. Petrochemical products like plastics, solvents, and lubricants are created using petroleum. These products evaporate, dissolve, or wear out over time releasing carbon dioxide.

10.11.8 Natural Sources

Apart from being created by human activities, carbon dioxide is also released into the atmosphere by natural processes. The earth's oceans, soil, plants, animals and volcanoes are all natural sources of carbon dioxide emissions.

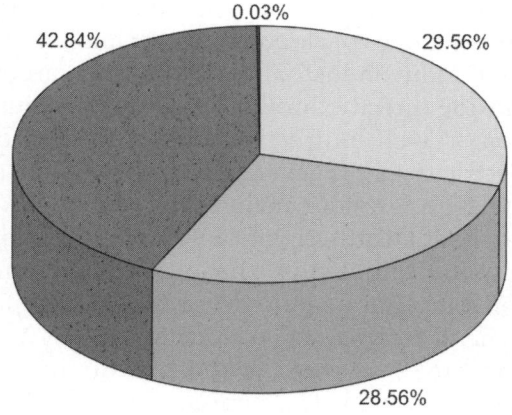

- ■ Ocean atmosphere exchange
- ▨ Plant and animal respiration
- ☐ Soil respiration and decomposition
- ■ Volcanic eruptions

Human sources of carbon dioxide are much smaller than natural emissions but they upset the balance in the carbon cycle that existed before the industrial revolution. The amount of carbon dioxide produced by natural sources is completely offset by natural carbon sinks and has been for thousands of years. Before the influence of humans, carbon dioxide levels were quite steady because of this natural balance.

42.84% of all naturally produced carbon dioxide emissions come from ocean-atmosphere exchange. Other important natural sources include plant and animal respiration (28.56%) as well as soil respiration and decomposition (28.56%). A minor amount is also created by volcanic eruptions (0.03%).

10.11.9 Ocean-Atmosphere Exchange

The largest natural source of carbon dioxide emissions is from ocean-atmosphere exchange. This produces 42.84% of natural carbon dioxide emissions. The oceans contain dissolved carbon dioxide, which is released into the air at the sea surface. Annually this process creates about 330 billion tonnes of carbon dioxide emissions. Many molecules move between the ocean and the atmosphere through the process of diffusion, carbon dioxide is one of them. This movement is in both directions, so the oceans release carbon dioxide but they also absorb it. The effects of this movement can be seen quite easily, when water is left to sit in a glass for long enough, gases will be released and create air bubbles. Carbon dioxide is amongst the gases that are in the air bubbles.

10.11.10 Plant and Animal Respiration

An important natural source of carbon dioxide is plant and animal respiration, which accounts for 28.56% of natural emissions. Carbon dioxide is a byproduct of the chemical reaction that plants and animals use to produce the energy they need. Annually this process creates about 220 billion tonnes of carbon dioxide emissions. Plants and animals use respiration to produce energy, which is used to fuel basic activities like movement and growth. The process uses oxygen to break down nutrients like sugars, proteins and fats. This releases energy that can be used by the organism but also creates water and carbon dioxide as by-products.

10.11.11 Soil Respiration and Decomposition

Another important natural source of carbon dioxide is soil respiration and decomposition, which accounts for 28.56% of natural emissions. Many organisms that live in the earth's soil use respiration to produce energy. Amongst them are decomposers who break down dead organic material. Both of these processes releases carbon dioxide as a byproduct. Annually these soil organisms create about 220 billion tonnes of carbon dioxide emissions. Any respiration that occurs below-ground is considered soil respiration. Plant roots, bacteria, fungi and soil dwelling organisms use respiration to create the energy they need to survive but this also produces carbon dioxide. Decomposers that work underground breaking down organic matter (like dead trees, leaves and animals) are also included in this. Carbon dioxide is regularly released during decomposition.

10.11.12 Volcanic Eruptions

A minor amount carbon dioxide is created by volcanic eruptions, which accounts for 0.03% of natural emissions. Volcanic eruptions release magma, ash, dust and gases from deep below the earth's surface. One of the gases released is carbon dioxide. Annually this process adds about 0.15 to 0.26 billion tonnes of carbon dioxide emissions. The most common volcanic gases are water vapour, carbon dioxide, and sulfur dioxide. Volcanic activity will cause magma to absorb these gases, while passing through the earth's mantle and crust. During eruptions, the gases are then released into the atmosphere.

Index